MANUAL OF SATELLITE COMMUNICATIONS

Emanuel Fthenakis

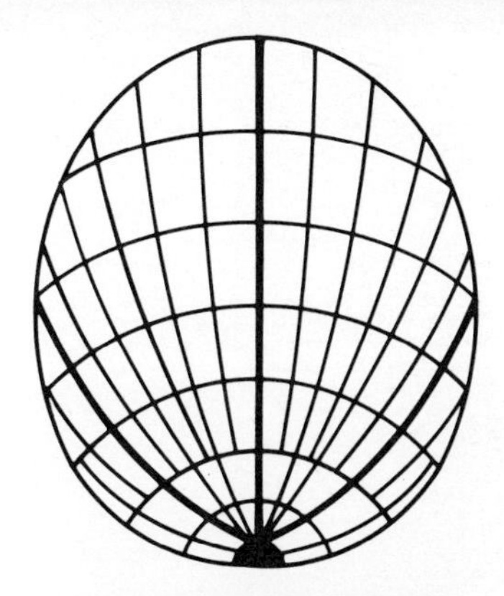

MANUAL OF SATELLITE COMMUNICATIONS

McGraw-Hill Book Company

New York St. Louis San Francisco Auckland Bogotá Hamburg Johannesburg London Madrid Mexico Montreal New Delhi Panama Paris São Paulo Singapore Sydney Tokyo Toronto

Library of Congress Cataloging in Publication Data

Fthenakis, Emanuel.
Manual of satellite communications.

Bibliography: p.
Includes index.
1. Artificial satellites in telecommunication.
I. Title.
TK5104.F8 1984 621.38′0422 83-842
ISBN 0-07-022594-X

1 2 3 4 5 6 7 8 9 0 KGPKGP 8 9 8 7 6 5 4 3

ISBN 0-07-022594-X

The editors for this book were Harry Helms and George Watson, the designer was Elliot Epstein, and the production supervisor was Thomas G. Kowalczyk. It was set in Caledonia by University Graphics, Inc.

Printed and bound by The Kingsport Press.

To the memory of my father

CONTENTS

APPENDIXES

PREFACE

The design of a good, reliable satellite communications system presents a significant challenge to the designer. Low available signal power, limited bandwidth, nonlinear operation, proximity of adjacent channels, high cost, severe weight restrictions, and inaccessibility for maintenance are only some of the technical problems encountered. Restrictions stemming from governmental regulations and nationalistic interests have further complicated the system designer's task. Fortunately, however, innovative minds have coped with these complex factors and have succeeded in bringing the space communications era to the office, computer room, or television set of the user at competitive prices.

Engineers, especially new graduates, in their efforts to adapt their general academic training to this new field of satellite communications find it necessary to take additional specialized courses. The material in this book is based on lecture notes which I used in teaching such a course for a first-year graduate curriculum. The lectures were prepared for engineers who were just starting their careers in the satellite communications industry and who thus had little or no practical experience in the field. In addressing this audience I found it necessary to include material on certain fundamental principles of network and communications theory which are widely used in the design of satellite communications systems. Since there are several excellent references specializing in one or more of the topics on fundamentals covered in this material, I did not attempt to treat these subjects with strict mathematical rigor. In addition, because of the wide variety of topics that had to be covered in a single semester, the depth of presentation had to be limited in some cases.

As a result of my association with several technical organizations specializing in satellite communications, it became obvious that the lecture material could be adapted to the use of a much larger group of practicing professional engineers. I was encouraged by many of my colleagues to publish this material, and as a result this book was written. The following paragraphs present a brief chapter-by-chapter description of the text.

The first chapter summarizes the development of satellite communications systems in the period from 1962 to 1982. The emphasis is on U.S. systems and the evolution of commercial communications via satellite, including some of the regulatory aspects. The second chapter is descriptive in nature and acquaints the reader with the terminology and the function of the key elements of a complete system, including cost elements. Some of the unique characteristics of satellite communications are also examined.

Chapter 3 covers satellite orbits and some typical launch sequences. The purpose of this chapter is to acquaint the reader with terminology and fundamental relationships of the various orbital parameters. The spacecraft, considered as a platform designed to carry a communications repeater station, is covered in Chapter 4. Again the main purpose of this chapter is to acquaint the reader with the terminology and the critical parameters that affect the platform's main function, i.e., to house a long-life communications repeater.

Chapter 5 discusses the communications repeater and defines the terminology used in two-port network theory, which is helpful in the characterization of the low-noise front section of the repeater. Since noise is a critical problem, an elementary treatment of noise sources is also included. Finally, the nonlinear characteristics of the power amplifier are discussed, including an elementary illustration of the intermodulation problem. A similar approach is used in Chapter 6, which discusses communications earth stations.

Chapter 7 develops the space-link equations.

Chapters 8 to 12 examine some of the fundamentals of communications system theory that are directly applicable to satellite communications. Appendixes B, C, and D summarize the mathematical tools utilized in these chapters.

The intention of Chapter 13 is to provide better understanding of the most significant transmission impairments.

Finally, Chapter 14 addresses the characteristics of voice, video, and data channels. It treats the baseband signals and the techniques used for designing satellite communications networks.

The field of communications via satellite is still developing. Both new applications and new technology are strong driving forces in this respect. In addition, alternative advanced communications concepts are emerging every day, creating pressures for continuous improvement of the economic aspects. Consequently, there is a great need for continuous innovation and improvement, and at the same time, excellent opportunities for contribution by the practicing engineer in the field of satellite communications.

Many of my colleagues and associates have assisted me in preparing this book. Their contributions range from assistance with certain theoretical elements to suggestions and recommendations on the basic material, provision of illustrations, and review of drafts.

It would be impractical to acknowledge all the contributors by name, but I wish nonetheless to express here my sincere appreciation for their help. In particular I wish to thank several of my colleagues at Fairchild Space and Electronics Company and American Satellite Corporation who made material contributions toward the creation of this book. I am also indebted to Mary Shaw for her valuable assistance in organizing, producing, and correcting the drafts.

Emanuel Fthenakis

MANUAL OF SATELLITE COMMUNICATIONS

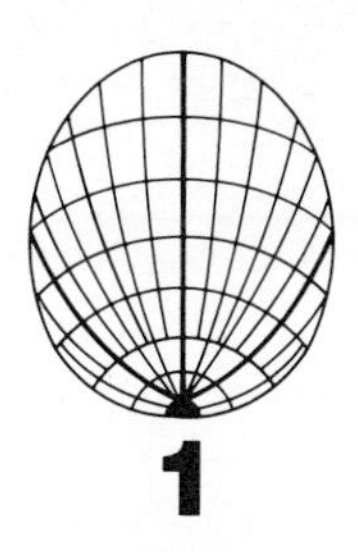

1

THE FIRST 20 YEARS OF COMMUNICATIONS SATELLITE DEVELOPMENT (1962–1982)

During the 20 years from 1960 to 1980, the capability for worldwide satellite communications developed dramatically to become an integral part of the various operating communications networks. At the same time, the regulatory and legal concepts governing the communications industry were revolutionized in order to accommodate the new technological capabilities.

ORIGINAL SYSTEMS

The first serious efforts by the United States for the development of a satellite communications capability date back to the late 1950s. At that time the U.S. Department of Defense initiated a number of projects, such as Project Courier, which finally resulted in the Initial Defense Communications Satellite Program (IDCSP). The IDCSP satellites were launched by the U.S. Air Force and were placed in operation in 1966. The system was used for strategic communications by the Defense Communications Agency (DCA). By 1982 the third generation of a Defense Satellite Communications System (DSCS-III) had been developed and placed in operation.

Simultaneously, the National Aeronautics and Space Administration (NASA) also initiated a number of experimental satellite programs, such as Relay and Syncom, which culminated in the first operational commercial communications satellite, Early Bird, launched in 1965 for the Intelsat consortium. The Early Bird was immediately followed by the launch of the Intelsat II in 1966. Intelsat, a consortium of many countries of the western world, including the United States, the United Kingdom, France, and Germany, utilized this system to provide international communications across the oceans. By 1982 Intelsat had developed and placed in operation its fifth generation of communications satellites.

At about the same time, the Soviet Union placed the first of many communications satellites in operation. The Molniya satellite was launched and placed in operation in 1965.

Intelsat, although providing satellite communications services for commercial use, did not represent a real competitive commercial venture since all the members of the consortium were government-controlled entities. The first U.S. commercial satellite communications system to operate under the forces of the free marketplace and to be financed with risk capital went into operation during 1974. The Federal Communications Commission (FCC) accepted applications for U.S. domestic satellite communications systems in 1971. After the FCC's adoption of a limited "open skies" policy, three domestic carriers initiated operations within months of each other during 1974: American Satellite Corporation, a subsidiary of Fairchild Industries, Inc.; Americom of RCA; and Western Union. Initially all three carriers utilized the 12-transponder, spin-stabilized Westar I and II satellites as their space segment. The previous year Canada had procured the U.S.-built and -launched satellite, Anik, and, through special agreements, transponders in the Anik spacecraft were leased to the U.S. domestic carriers. During 1976, RCA-Americom launched the 24-transponder, three-axis-stabilized satellite Satcom I, the first of a series of similar spacecraft. American Telephone and Telegraph (AT&T) and General Telephone and Electronics (GTE) shared a space segment, Comstar I and II, also launched in 1976. The sixth U.S. domestic carrier, Satellite Business Systems (SBS), a partnership formed by IBM, Comsat, and Aetna Insurance, launched its first spacecraft in 1981 and initiated service shortly thereafter. Eventually American Satellite Corp. acquired a 20 percent ownership in the Westar I, II, III, IV, and V space segment. Table 1-1 tabulates these six original U.S. domestic carriers.

Until the launch of the SBS space segment, every commercial system, with the exception of some experimental systems, operated at C band (4 GHz/6 GHz). SBS initiated operations utilizing the K_u frequency band (11 GHz/14 GHz). The Westar, Comstar, and SBS satellites are spin-stabilized whereas Satcom is a three-axis, body-stabilized spacecraft with deployable solar arrays. With the exception of Comstar, which utilized the Atlas-Centaur as a launch vehicle, all spacecraft were launched with a Thor-Delta booster.

TABLE 1-1 U.S. DOMESTIC SATELLITE CARRIERS

Operational date	Carrier	Space segment	Frequency
1974 1974	American Satellite Western Union	Westar I, II, III	C band
1974	RCA-Americom	Satcom (1976) I, II, IIIR, IV	C band
1976 1976	AT&T GTE	Comstar D-1, D-2, D-3, D-4	C band
1981	SBS	SBS F-1, F-2	K_u band

FIGURE 1-1 Initial Defense Communications Satellite Program (IDCSP): single spacecraft. ***(Photograph courtesy of Ford Aerospace and Communications Corporation.)***

Each of the six original satellite carriers has expended in 1978 dollars somewhere between $200 and $600 million in capital and start-up costs for its original system. This figure includes both the space and ground segment. It is estimated that the total amount spent by the six carriers is in the neighborhood of $2 billion. Figures 1-1 to 1-3 depict some of the above-mentioned satellites.

SECOND-GENERATION U.S. DOMESTIC CAPABILITY

During the 1980–1982 time frame a number of new applications for new systems or for expansion of existing systems were filed with the Federal Communications Commission. Table 1-2 tabulates these proposed systems, indicating the owner and the approximate year of launch. The two new entrants in the list are Hughes Aircraft Company and Southern Pacific Communications Corporation (SPCC). (For several previous years Southern Pacific Communications had been operating as a terrestrial carrier.) With the exception of the SBS sat-

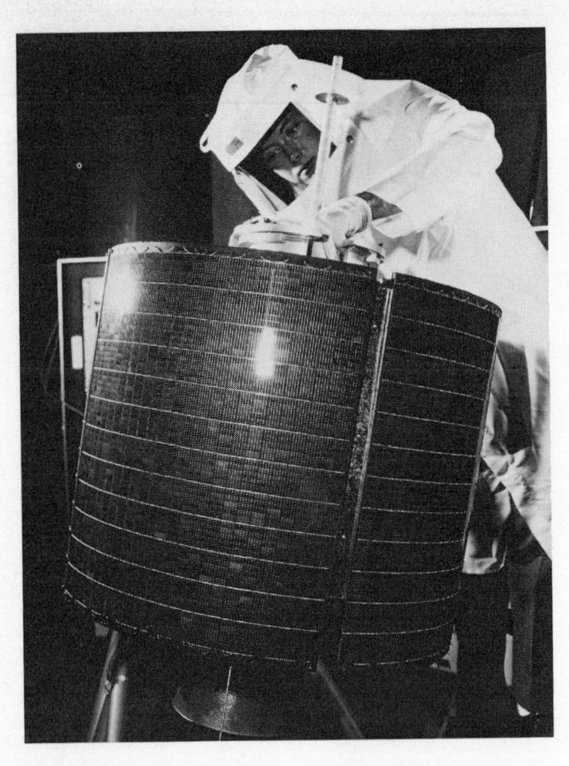

FIGURE 1-2 Early Bird. *(Photograph courtesy of Hughes Aircraft Company.)*

TABLE 1-2 SECOND-GENERATION PROPOSED U.S. DOMESTIC SPACE SEGMENTS

Satellite system	Owner	Starting year	Frequency
Galaxy H-1, H-2, H-3	Hughes (new)	1982	C band
Spacenet I, II, III	SPCC (new)	1983	C band, K_u band
Advanced Westar F-1, F-2, F-3, F-4 (TDRSS)	Spacecom (new), American Satellite, Western Union	1983	C band, K_u band
G-Star I, II	GTE (new)	1984	K_u band
Telstar 3-A, 3-B, 3-C	AT&T (new)	1983	C band
American Satellite I, II	American Satellite (new)	1985	C band, K_u band
Westar IV, V, VI	Western Union (new), American Satellite	1982	C band
SBS III, IV	SBS	1982	K_u band

FIGURE 1-3 Westar II and electronics and propulsion sections of Westar III. ***(Photograph courtesy of Hughes Aircraft Company.)***

ellites, the space segments represent a new design of spacecraft configuration employing at least 24 transponders, some utilizing both the C and K_u bands simultaneously.

If one excludes the Advance Westar of the Tracking and Data Relay Satellite System (TDRSS) space segment, the aggregate cost in 1981 dollars of all the proposed additional U.S. domestic space segments in Table 1-2 will be about $1.5 billion, which does not include the establishment of additional ground segments to serve the new entrants. In its original configuration, the TDRSS/Advance Westar space segment consisted of four spacecraft in orbit and two on the ground and was to be dedicated to both the U.S. government and domestic commercial service on approximately a 3:1 ratio. The total cost of the space segment was estimated to be about $1.5 billion.

Each of the initial five domestic carriers before 1976 developed a backbone ground segment of major earth stations. Each of these ground segments involved somewhere between four and ten major earth stations. Subsequently,

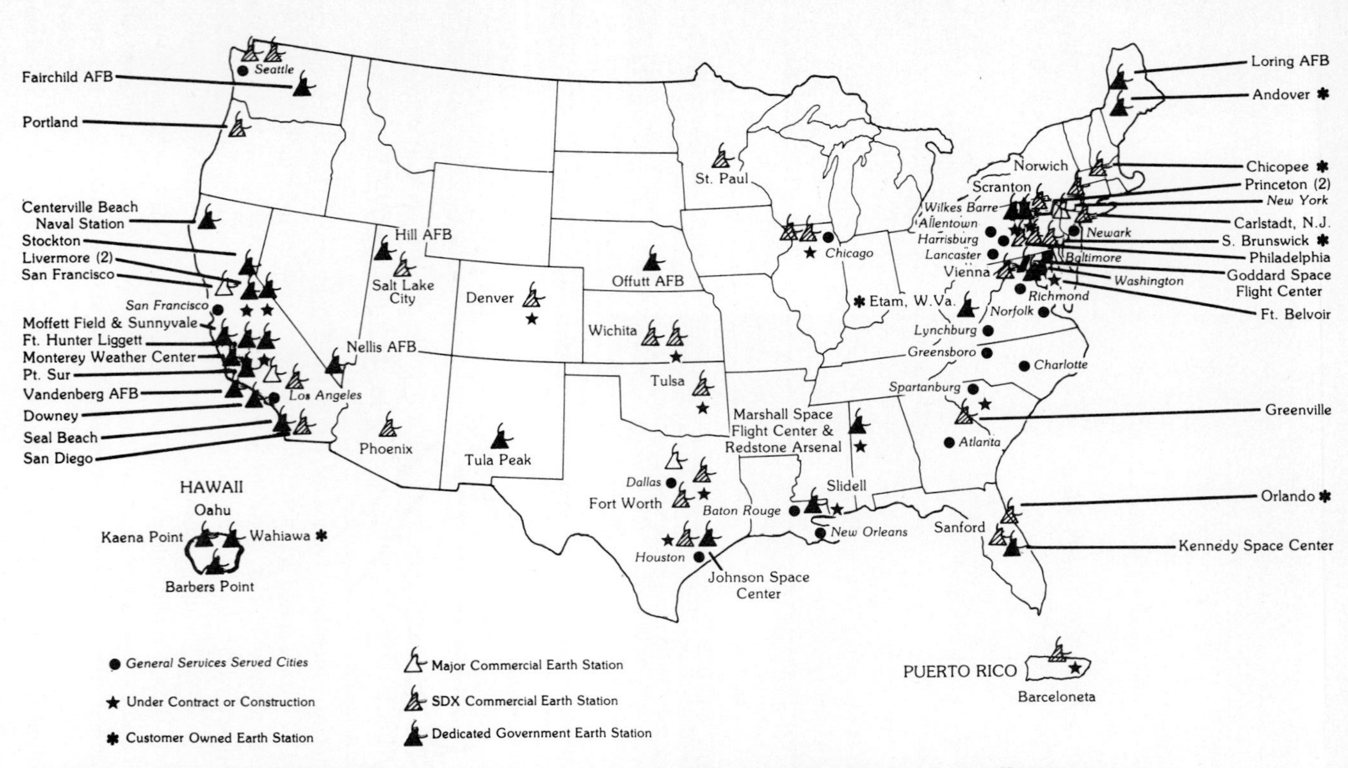

FIGURE 1-4 American Satellite Company communications network, 1981. *(Courtesy of American Satellite Company.)*

a substantial amount of additional ground segment was established for specific business users, government networks, receive-only stations for television operations, etc. Figure 1-4 depicts the 1981 network configuration utilized by American Satellite Corp. It consisted of over 100 earth stations of which only four belonged to the original backbone network. A typical business-dedicated network (for First Interstate Bancorp) is represented in Fig. 1-5. By 1982 literally thousands of stations had been deployed all over the United States with multichannel receive-only capabilities for the service of the television entertainment industry.

Some of the terrestrial networks were modified to be fully digital by utilizing time-division, multiple-access (TDMA) techniques and digital voice transmission. For example, by 1980 the American Satellite backbone network used a 63-Mb/s-per-transponder TDMA system. The voice was compressed from the usual 64-kb/s pulse-code modulation (PCM) to 32 kb/s per channel by near-instantaneous companding (a modified PCM technique). SBS utilized a 48-Mb/s TDMA system.

Table 1-3 summarizes the major services provided by the U.S. domestic satellite carriers.

It is interesting to note that the problem of satellite delay for high-speed interactive data was solved as early as 1978 and that high throughput efficiencies of about 98 percent have been achieved. In addition several dedicated networks—such as banking networks—utilize special encryption techniques for

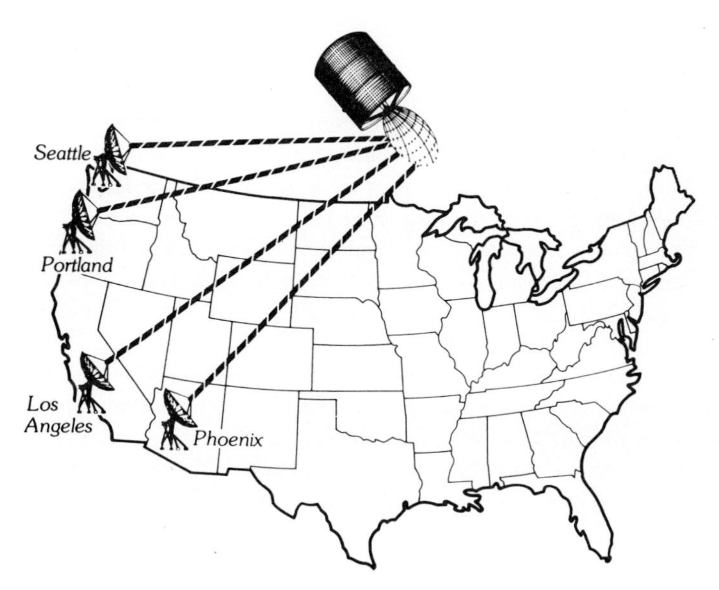

FIGURE 1-5 First Interstate Bancorp Data Processing Company network map.

TABLE 1-3 SERVICES PROVIDED BY U.S. DOMESTIC SATELLITES

Services	Technology
Telephony (voice, low-speed data)	Voice analog Digital voice Digital compression, statistical interpolation
High-speed data	Interactive data Delay compensation
Dedicated networks	Frequency-division multiple access (FDMA) Single channel per carrier (SCPC) Time-division multiple access (TDMA)
Video conference	Digital voice Analog video
TV broadcast	Analog Digital
Radio broadcast	Analog Digital

TABLE 1-4 TRENDS FOR U.S. DOMESTIC SATELLITE SYSTEMS

Business-specialized dedicated networks
Long-haul bulk carriers
Entertainment industry
Direct broadcast

data transmission to ensure privacy. Video conferencing could become cost-effective by utilizing digital techniques with data compression. For full-motion video, rates of about 1.5 to 6 Mb/s are utilized, and, for lesser performance, rates in the range of kilobits per second are used.

In 1978 satellite carriers' revenues were about $88 million, and by 1986 their revenues are expected to increase to over $800 million. This figure does not include the revenues for cable TV served by satellites, which has been one of the fastest-growing segments of the satellite market.

Table 1-4 depicts the trends foreseen for the U.S. domestic satellite systems in the period 1982 to 1990. It is interesting to note that a substantial amount of what is perceived as terrestrial long-haul traffic for switched message service will use satellite transmission; therefore, the long-haul bulk carrier segment will grow substantially. In addition, aggressive plans are being made for the establishment of direct broadcast systems to reach millions of households throughout the United States. A typical direct broadcast system will require an investment of approximately $600 million. With the development of the personal computer and the automation of the retailing industry, private data links between households and consumer outlets will probably become a reality in the 1980s. In addition, electronic mail and videotext will grow substantially during this decade. It is expected that the volume of communications services will grow by almost $50 billion per year by 1990. A substantial portion of this volume will be served by domestic satellites.

THE EARLY INTERNATIONAL, REGIONAL, AND FOREIGN SYSTEMS

International communications via satellite were initiated by Intelsat I (Early Bird) in 1965. Intelsat II was launched in 1966, and it was followed by Intelsat series III, IV, and IVA. Each series consisted of several spacecraft, all spin-stabilized. The first three-axis-stabilized Intelsat spacecraft with deployable solar array was launched in 1981 with the initiation of the Intelsat V series. In addition, Intelsat V (see Fig. 1-6) with its dual frequencies, i.e., C band and K_u band, first introduced K_u capability to the Intelsat system. Some of the key characteristics of the Intelsat satellites, among others, are tabulated in Appendix E. All of the Intelsat systems were managed and operated by the Intelsat consortium, which, by 1982, included over 100 nations.

FIGURE 1-6 Intelsat V. ***(Photograph courtesy of Ford Aerospace and Communications Corporation.)***

Canada initiated the Anik series for domestic use in 1973. Telesat of Canada, an organization partially owned by the Canadian government, was the manager and operator of the Canadian systems. By 1980 Canada had developed substantial domestic technology for producing key subsystems in the communications satellite field.

Europe also initiated an aggressive program in the field of rocketry and satellite systems. This cooperative program among the European nations included a communications satellite capability. Initially it involved experimental systems, such as the Symphonie satellite launched in 1974 with the U.S. launch vehicle Thor-Delta. The European launch vehicle, Ariane, provided a serious alternative for launching communications satellites in the post-1980 time frame.

A number of regional systems were studied and some were actually deployed in the late 1970s and early 1980s. A good example of such a regional system is the Palapa system which was launched in 1976 by the U.S. launch vehicle Thor-Delta. This system, deployed on behalf of Indonesia, was developed with U.S. technology.

The Soviet Union developed and deployed a number of satellite systems that followed the initial Molniya launched in 1965. The Loutch series represents a very versatile capability for telephony, television, and audio broadcasting. It is capable of covering a wide area including links to cities such as Havana, Belgrade, and Moscow. The Statsionar Raduga and Statsionar Ekran are two configurations of the Statsionar series first launched in 1975. This U.S.S.R. series is also very versatile, covering a wide range of services.

THE EVOLUTION OF LEGAL AND REGULATORY ASPECTS IN THE UNITED STATES

The Communications Act of 1934 was the initial legislative act in the United States to provide the basis for the regulation of the communications service industry. The Federal Communications Commission has been the agency responsible for administering the regulatory and legal aspects of communications services. Until 1970, when deregulation was initiated, a de facto monopoly existed in this industry, with the FCC regulating the overall national aspects and the local regulatory authorities in each state overseeing the local aspects.

The Bell System, consisting of the American Telephone and Telegraph Company, the Bell Telephone Laboratories (research), the Western Electric Company (manufacturing), and several local telephone companies, dominated the telecommunications service market. A substantial number (close to 1500) of independent telephone companies such as GTE and United also enjoyed a de facto monopoly in their territories; however, the Bell System revenues accounted for more than 80 percent of the total U.S. revenues. Table 1-5 depicts the various original communications common carriers before a number of new

TABLE 1-5 MAIN U.S. COMMON CARRIERS BEFORE DEREGULATION

AT&T and the Bell System companies
Independent telephone companies
Western Union
International record carriers:
Global Communications (Globcom)—RCA
World Communications (Worldcom)—ITT
Western Union International (WUI)
Communications Satellite Corporation (Comsat)

competitive carriers were established as a result of the deregulation process initiated in the 1970s.

The Communications Satellite Act of 1962 addressed international communications via satellite. As a result of this act the Communications Satellite Corporation (Comsat) was created as the only authorized agent to represent the United States' interests in the international consortium, Intelsat. Comsat, in addition to being the U.S. signatory in Intelsat, was also the original manager of the Intelsat organization as well as the main technical and operating arm. Since almost all of the other nations considered communications services to be an activity of the state, they were represented in Intelsat by government-controlled entities. By 1981 the Intelsat membership included approximately 100 nations.

After the Carterfone decision, the deregulation process in the United States was initiated, and the FCC licensed a number of other common carriers (OCCs) to operate and offer certain specialized services. These carriers are often known as the "specialized common carriers."

With respect to satellite communications, early in 1970 the FCC announced its intention of accepting applications for the establishment and operation of several U.S. domestic satellite systems.

In 1972 the Second Report and Order was issued adopting a "limited open skies" policy under which all legally, technically, and financially qualified applicants would be authorized to provide certain services, such as private-line services (i.e., dedicated point-to-point service). Docket no. 16495, 35 FCC 2d 844, covers these proceedings. One of the most important aspects of this decision was the interconnection policy between the new satellite carriers and the existing monopoly terrestrial carriers. In effect the FCC ruled that the existing terrestrial carriers should provide terrestrial facilities for the local distribution of the long-haul satellite service to the end users.

A typical example of a satellite circuit is depicted in Fig. 1-7. The part of

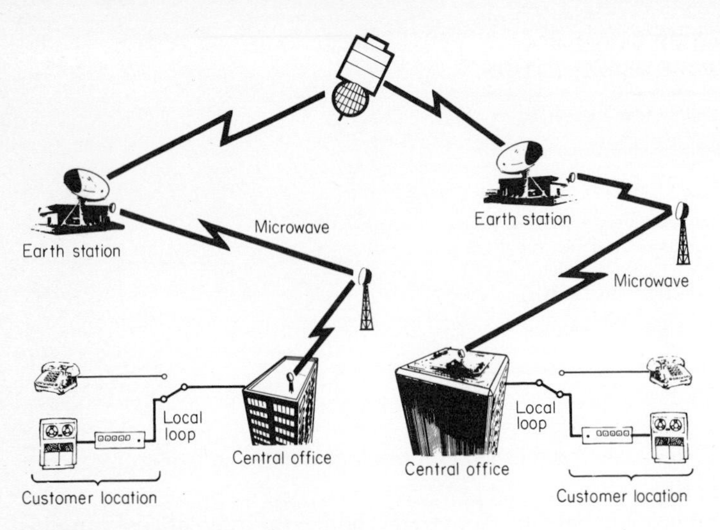

FIGURE 1-7 Typical satellite circuit.

the circuit utilized for distribution of the long-haul service to the end user from a central distribution office is commonly known as a "local loop." A substantial controversy was associated with the interpretation of the interconnection policy until 1975. In 1975 the FCC, in the Memorandum Opinion and Order in Docket no. 20099, 52 FCC 2d 727, approved an agreement between AT&T and the other common carriers and resolved most of the disputes generated by different interpretations of this policy.

THE ADMINISTRATION OF SATELLITE SERVICE AUTHORIZATIONS

In general, an FCC authorization is required for establishment and operation of channels of communication. If the entity owning and operating the satellite or ground facilities is a carrier, the carrier must obtain an initial blanket authorization pursuant to §214 of the Communications Act. Unless such a blanket carrier authorization has been obtained, an independent operator must obtain an authorization for each circuit or transponder to be operated.

The Federal Communications Commission assigns orbital positions for all geostationary communications or communications-related spacecraft. In so doing it also regulates the spacing from spacecraft to spacecraft in order to avoid interference between systems operating at the same frequency. The spacing of the first-generation satellites was established at 4 or 5° of the orbital arc. The dish size of the communicating earth stations is also a critical parameter in intersystem interference for each operating frequency. For example, at C band, the smallest ground-station dish authorized by the FCC for simultaneous receive-transmit (RT) operation is 5 m in diameter.

Establishment and operation of satellite service requires the following:

1. Space segment

   a. Construction permit—allows the manufacturer of the spacecraft to initiate and complete construction (§319a of the Communications Act). Such a permit can be waived temporarily by the FCC (pursuant to §319d of the Communications Act) so that the prospective operator can commit funds toward construction before receiving the construction permit.

   b. License—required to own and operate the spacecraft (§301 and §307 of the act).

   c. Launch authority and assignment of orbital location.

2. Earth stations, including telemetry, tracking, and command (TTC) stations.

   a. Construction permit—conditions and waiver possibilities similar to those in the satellite case.

   b. License—required in order to own and operate earth stations.

In the case of the earth stations, the prospective operator, to obtain a construction permit, must coordinate with all other operators in the vicinity of the desired location in order to avoid frequency interference. The results of this coordination must be submitted to the FCC. In special situations involving receive-only earth stations, a construction permit is not required. Since the station is not transmitting, it cannot interfere with other operators. However, the operator has no protection against interference from other operators. If such a protection is required, the prospective operator must file with the FCC an abbreviated filing omitting all the financial and character information.

Since the regulatory environment is continuously changing, the requirements for obtaining authorization as well as the types of authorizations needed are constantly changing. Prospective operators must carefully review the existing rules in each case.

REFERENCES

Babcock, J. H.: "Architecture and Management of DOD Satellite Communications Systems," *Signal*, vol. 30, no. 6, March 1976.

Brown, Martin P., Jr. (ed.): *Compendium of Communication and Broadcast Satellites 1958 to 1980*, IEEE Press, New York, 1981.

Deerkoski, L. F.: "Tracking and Data Relay Satellite System (TDRSS) Telecommunications Services," *EASCON '75 Convention Record*, September 1975.

Edelson, B. I.: "Global Satellite Communications," *Scientific American*, vol. 236, no. 2, February 1977.

Feldman, N. E., and C. M. Kelly (eds.): *Communication Satellites for the 70's Systems*, The MIT Press, Cambridge, Mass., 1971.

Kadar, Ivan (ed.): *AIAA Selected Reprint Series*, vol. XVIII: *Satellite Communication Systems*, American Institute of Aeronautics and Astronautics, New York, 1976.

Pfieffer, B., and P. Viellard: "The Experimental Telecommunication Satellite Project Symphonie," *Journal of the British Interplanetary Society*, vol. 26, no. 2, February 1973.

Special Issue on INTELSAT IV, *COMSAT Technical Review*, vol. 2, no. 2, Fall 1972.

Van Trees, Harry L. (ed.): *Satellite Communications*, IEEE Press, New York, 1979.

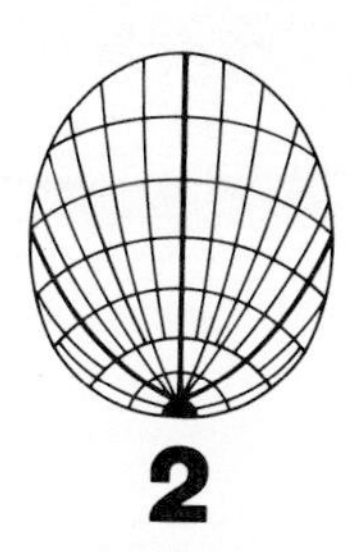

2
SYSTEM DESCRIPTION, DEPLOYMENT, AND COST

A typical satellite communications system includes the following elements:

Space segment: One or several spacecraft with in-orbit spare capability. The spare capability could be transponders on the same spacecraft or it could be a whole spare spacecraft. Usually there is at least one spare spacecraft on the ground ready to be launched.

Earth stations for communications: A system may include a great variety of earth stations. These stations may vary in size of dish, transmitting power, receiving sensitivity, capacity, access mode (FDMA, TDMA, CDMA), etc. Usually a certain number of these stations may constitute a subnetwork dedicated to a specific service. It is most likely that the earth stations of a subnetwork will be similar, and all must operate in the same access mode.

Telemetry, tracking, and command (TTC) stations: Usually two of these stations are utilized per space segment. These stations track the position of the satellite, send commands for station keeping and attitude control, and receive telemetry indicating the status of the spacecraft.

Terrestrial distribution including central offices and local loops: The earth station may or may not be located at the premises of the end user. If it is not, a terrestrial distribution system is required to transmit the signal to the location of the end user. This terrestrial transmission system may be a microwave system, a fiber-optics system, a single pair of copper wires, etc. If there are several end users at a certain location, bulk terrestrial transmission may be used to transmit to a central office location where, after signal conditioning, local loops are used to arrive at each end-user terminal.

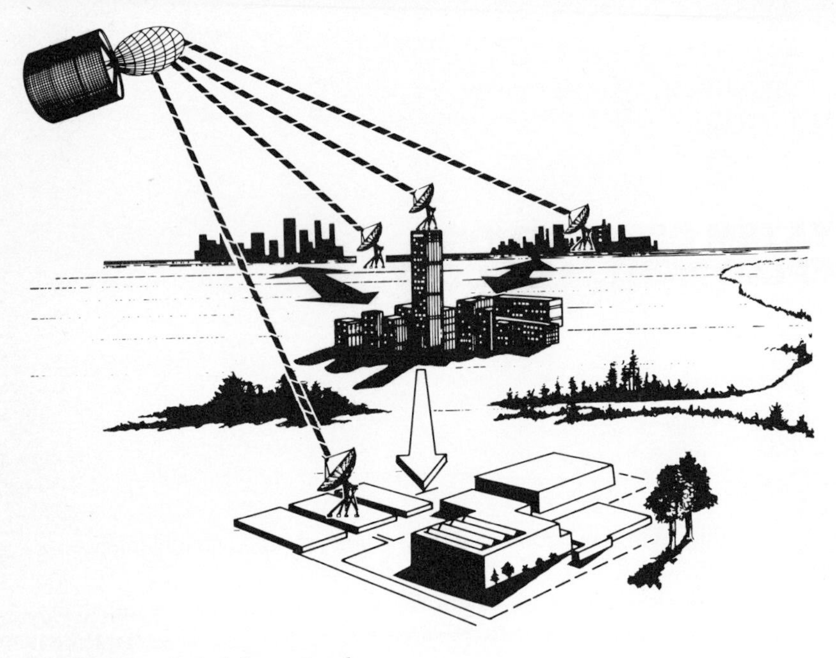

FIGURE 2-1 Typical data network.

Network control center: This facility is not mandatory but is highly desirable. This center monitors the operational status of the system. Depending on operational requirements and maintenance philosophy, the center may have various degrees of complexity, sophistication, and capability to monitor in detail even minute elements of the system. In case of failure it usually has the capability to switch redundant elements by remote control or to switch out of the system a failed element that may interfere with the rest of the operation. For a network with a small number of earth stations, one of the earth stations may be expanded to include the network control center. It may provide visual or audible alarms in case of emergency and, in general, assists in the overall everyday operations.

Figure 2-1 depicts a typical system configuration.

SPACE SEGMENT AND LAUNCH CONSIDERATIONS

Most of the present-day communications satellite systems utilize spacecraft placed in geostationary orbits. This means that the spacecraft is in a circular orbit on a plane identical with the earth's equatorial plane and at such altitude that the orbital period is identical with the earth's rotational period. This altitude is approximately 35,860 km (22,300 mi) over the surface of the earth, which corresponds to a distance of 42,230 km (26,260 mi) from the center of the earth, since the earth's radius is 6370 km (3960 mi).

It is true that not all satellite communications systems utilize spacecraft in geostationary orbits. For example, the original Molniya system deployed by the Soviet Union consisted of three spacecraft, each placed in a highly elliptical orbit with 65° inclination. This system is depicted in Fig. 2-2.

The United States' first-generation military communications satellite system DSCS-I (IDCSP) consisted of a large number of satellites in an equatorial subsynchronous orbit, each spacecraft drifting from west to east at an approximate rate of 30° per day.

The geostationary configuration is the one most often used because of a number of significant inherent advantages. The fact that it does not require tracking earth stations simplifies the earth-station design and maintenance and results in a substantially lower cost of the ground segment. In addition it simplifies the operational aspects by avoiding the hand-over problem from satellite to satellite. From now on in this text, only geostationary systems will be considered since they are used universally, with very few exceptions.

Most of the initial domestic systems utilized three spacecraft in their first deployment. Two of these spacecraft were placed in orbit within a short time of each other, and the third was used as a ground spare, ready to be launched in case of a serious failure of the orbiting spacecraft.

A satellite in geostationary orbit looks at the earth with a cone angle of 17.3°, corresponding to an arc of 18,080 km along the equator. This is adequate for most domestic or regional systems. Figure 2-3 depicts the coverage on the earth's surface by a spacecraft placed in 90°W longitude geostationary orbit over the South American continent. The depicted area will be illuminated by a beam with a 17.3° cone angle, usually known as a "global beam," with the cone axis along the local vertical.

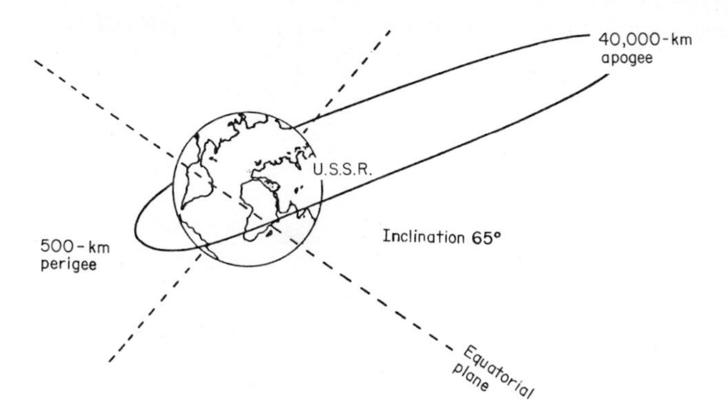

FIGURE 2-2 U.S.S.R. Molniya satellite.

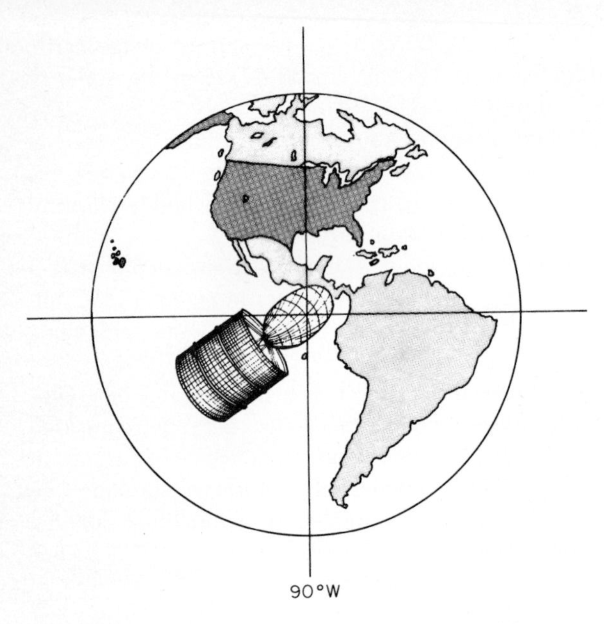

FIGURE 2-3 Satellite field of view at 90°W longitude.

The area illuminated by a global beam of a satellite placed in 130°W longitude is depicted in Fig. 2-4. As it can be seen from this figure, the edge of the beam also illuminates Puerto Rico and the northern tip of the North American continent. If the satellite were moved farther west, the global beam would miss part of the eastern territory. Similarly, the edge of the global beam of the satellite positioned at 90°W illuminates the tip of Alaska. Again, if the satellite were moved farther east, the beam would miss Alaska. Satellites placed within an arc extending from 89°W to 139°W can provide communications for all 50 states. This assumes a minimum 5° elevation for the earth stations (elevation = angle of antenna axis to horizontal). This arc is the primary arc for the U.S. domestic systems. The arc from 53°W to 139°W will cover 48 states (excluding Alaska and Hawaii) and is said to provide Conus coverage (Conus = contiguous United States).

At least three operational spacecraft are required for global coverage. Figure 2-5 depicts this geometry. As a result, international systems like Intelsat require an initial complement of three "hot" spacecraft plus the necessary spares.

If three spacecraft are positioned symmetrically 120° apart, as in Fig. 2-5, the global beams will overlap on the earth's surface. The length of the overlap measured as an arc will be 4745 km, corresponding to an angle of 42.7°. An earth station placed in the middle of this arc will see the adjacent spacecraft with an elevation angle of 22°.

The launch vehicle and the launch operations up to this date have been provided by a government or government-controlled agency. In the United States, the military launches its own satellites. Nongovernment carriers contract with the National Aeronautics and Space Administration (NASA) for these services. Other nations with launch capability have similar arrangements.

Until the U.S. space transportation system (STS), known as the shuttle, reached an operational status, expendable launch vehicles such as the Thor-Delta, Atlas-Centaur, and Titan were used exclusively for launching communications satellites. As a result the total weight and volume of each spacecraft were limited by the capability of the launch vehicles. Of course, since the launch is one of the most costly elements of the overall expense, each spacecraft designer attempted to take advantage of the maximum capability of the corresponding launch vehicle. As a result, only a few discrete sizes of satellites were developed, corresponding to the available launch vehicle capability.

The shuttle, with a multiple payload capability, provides a choice of continuous growth of the satellite in size until the maximum capability is reached. Since the shuttle can place its payload in a low earth orbit only, the maximum capability is also determined by the size of the second expendable propulsion stage used to boost the spacecraft into a synchronous orbit. This second operation again by necessity dictates discrete sizes of payloads, such as communications satellites, even when the shuttle is used.

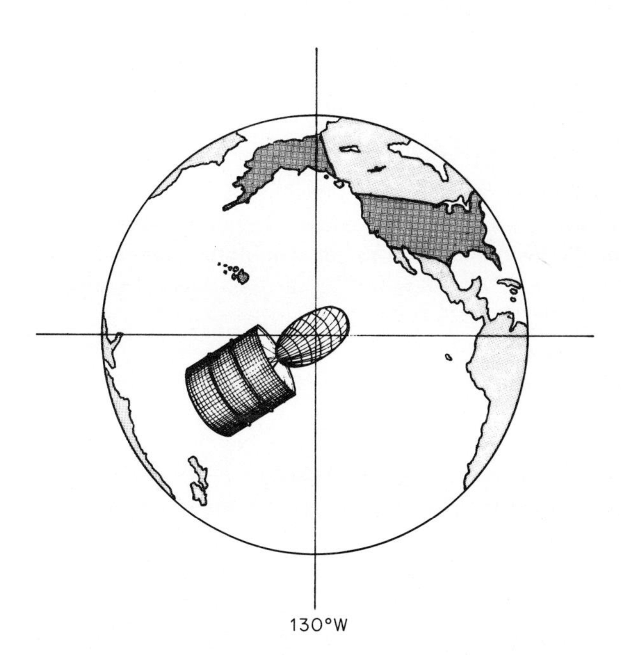

FIGURE 2-4 Satellite field of view at 130°W longitude.

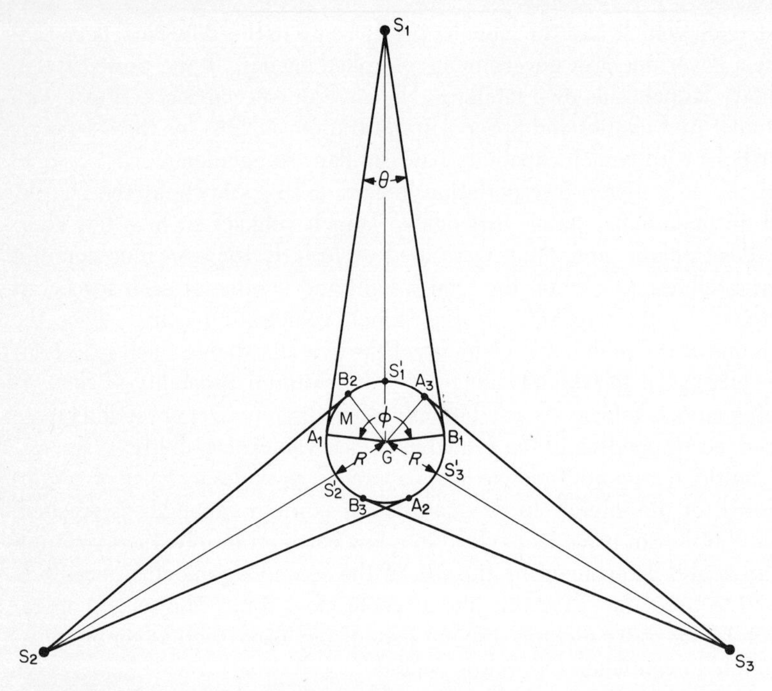

FIGURE 2-5 Geostationary system: geometric distances and angles on the equatorial plane.

Earth's radius: $R = GA_1 = 6370$ km

Satellite to earth surface: $S_1S_1' = S_2S_2' = S_3S_3' = 35{,}860$ km

Satellite to earth center: $S_1G = S_2G = S_3G = 42{,}230$ km

Satellite maximum distance to earth: $S_1A_1 = S_1B_1 = S_2A_2 = S_2B_2 = S_3A_3 = S_3B_3 = 41{,}747$ km

Look angle from satellite (global-beam angle): $\theta = 17.3°$

Angle illuminated on earth: $\angle A_1GB_1 = \phi = 162.7°$ ($= 180° - 17.3°$)

Arc length illuminated on earth: $A_1S_1'B_1 = 18{,}080$ km

Earth's circumference = 40,004 km

Overlap angle on earth: $\angle A_1GB_2 = 42.7°$

Of course, the number of economically sized spacecraft from the launch vehicle standpoint could be increased by launching more than one spacecraft simultaneously.

The life of a spacecraft in orbit, barring component failures, is limited by the amount of fuel required for "station keeping," i.e., maintaining the allocated orbital spot, and by the degradation of the solar-cell power-generating system due to cosmic radiation. In other words, when the spacecraft runs out of fuel or electric power, its life ends.

GROUND SEGMENT

The ground segment of a satellite communications system consists of the earth stations, the terrestrial distribution, and the network control facility. Earth stations in their more common configuration utilize an antenna system with a solid dish, a microwave feed, a pedestal, and equipment for adjusting the dish orientation in azimuth and elevation. Depending on the size of the dish, the condition of the soil, and the requirements for shielding, a substantial amount of civil works may be required in order to provide an adequate mount for the antenna system. Concrete slabs in many thicknesses, together with a variety of anchoring devices, have been used extensively.

Most of the terrestrial facilities, including the earth stations, require a backup power system in case of commercial power failure. The term "uninterruptable power supply" is used for special designs that switch automatically to the backup power-generating system in case of commercial power failure; storage batteries absorb the load during the switching period. Such backup power systems can be very expensive, depending on the size and duration of the load that they have to carry and the sophistication of the design. A typical earth station is depicted in Fig. 2-6.

Local loops are usually provided by the local telephone companies, utilizing the public telephone network. In some cases, however, when adequate facilities do not exist, special construction may be required. This special interconnection

FIGURE 2-6 Commercial 10-m earth station in San Francisco, California. ***(Photograph courtesy of American Satellite Company.)***

FIGURE 2-7 Rooftop 5-m earth station, First Interstate Bank, Portland, Oregon. *(Photograph courtesy of American Satellite Company.)*

facility may or may not be a part of the public network. A dedicated earth station is depicted in Fig. 2-7.

Even in cases in which the earth station is located at the premises of the end user, a short interconnecting facility may be required to reach the computer room, the private automatic branch exchange (PABX) room, or the service equipment that requires the satellite transmission service. Networks of earth stations with the corresponding terrestrial distribution facilities may be dedicated to a particular user and they may also be simultaneously interconnected to a general-purpose network including public telephone facilities.

CHARACTERISTIC ADVANTAGES AND PROBLEMS

Satellite communications systems are characterized by certain unique advantages and also by certain unique problems. Some of these unique characteristics are discussed below.

Broadcast Capability

A geostationary satellite with a global beam looks at the earth with a 17.3° angle. The geometry is depicted in Fig. 2-5. This global beam will cover an arc of 18,080 km along the equator (corresponding to a longitude angle of 162.7°)

and it will radiate up to 81.35° of latitude on either side of the equator. Any earth station within these limits could communicate with any other earth station within the same limits utilizing this one global beam. This is a substantial advantage over terrestrial microwave radio communications requiring line-of-sight capability with repeaters spaced every 20 to 40 mi from each other. For point-to-multipoint transmissions over long distances, satellite networks are definitely recommended. In fact, the television broadcast industry has made extensive use of satellite transmission.

Signal Quality (Low Bit Error Rate)

Most terrestrial systems require repeaters placed at relatively short distances from each other. Each repeater is a source of noise, introducing some signal degradation, and the total transmission is degraded by the product of all the individual degradations. The satellite system with only one repeater in space can provide signal quality which is better by orders of magnitude than that provided, for example, by microwave systems. A typical satellite bit error rate (BER) in the case of digital transmission is 10^{-8} versus 10^{-5} for a microwave system.

Broadband Bulk Transmission

Satellite communications systems can be designed to transmit, in a very cost-effective manner, long-haul bulk transmission over a wide bandwidth or, in the case of digital transmission, at a very high bit rate with a relatively low bit error rate.

Flexible Networking

The point-to-multipoint characteristic coupled with the broadband capability provides the necessary characteristics for design of switched networks with on-demand instantaneous network reconfiguration capability. A typical node configuration is depicted in Fig. 2-8.

Echo

Since a geosynchronous satellite is placed 35,860 km above the surface of the earth, a microwave signal must travel more than twice this distance from earth station to earth station. Because of the finite speed of propagation (the velocity of light, 2.99×10^8 m/s), the signal will be delayed for over 0.24 s and over twice this amount, that is, 0.48 s, for a round trip. Since some degree of mismatch will exist in each terminal, an echo will be generated. Echo delay of 0.48 s is perceptible by the human ear and is very disturbing in a two-way

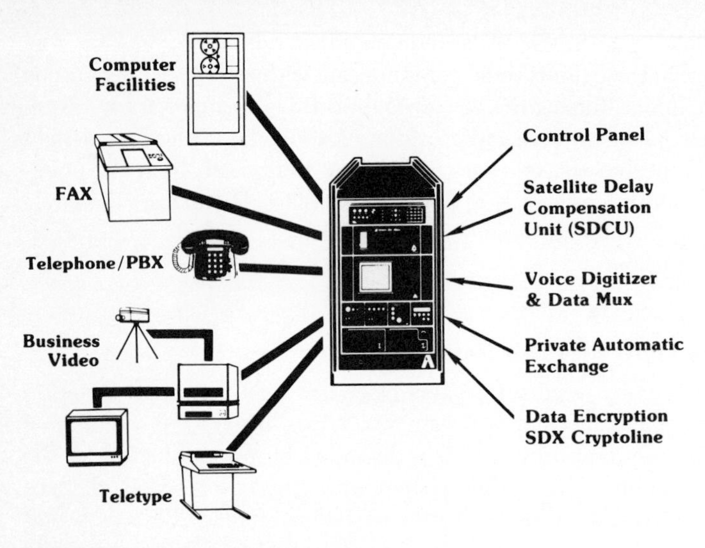

FIGURE 2-8 Communications node.

telephone conversation. As a result, sophisticated echo-canceling techniques must be used in order to obtain satisfactory two-way conversation. Four-wire systems with different transmitting and receiving paths do not suffer from this delayed echo problem.

Systems with high return loss minimize the echo problem. Echo suppressors, by voice activation, switch off the return path, allowing only transmission from the talker to the listener. However, since talkers are not disciplined to speak in sequence, severe clipping may take place. Echo cancelers, although more sophisticated and expensive, are more effective in actual operation. Echo cancelers work by comparing the waveforms in both directions and eliminating (canceling) a returning waveform that resembles one just transmitted.

Data Throughput Efficiency

In the case of interactive data transmission with relatively short block length, the transmission delays will reduce the total throughput capability of the system substantially. Special delay-compensating devices are required for such systems. Most modern data systems are built with protocols capable of minimizing this effect without delay-compensating devices.

American Satellite Corporation has developed a unique delay-compensation device named the wideband satellite delay compensation unit (WSDCU). Figure 2-9 depicts this unit's performance characteristics over a specific data link. This device, located at the transmitting end of the link, has enough memory to store each transmitted block until an acknowledgment has been received from the receiving end. If an error is acknowledged, the device will retransmit the appropriate block. In addition, by providing the necessary acknowledgment, even if it has not yet been received, the WSDCU causes the transmitting end to transmit continuously.

SYSTEM AND EQUIPMENT ECONOMIC CONSIDERATIONS

In the face of rapidly advancing space and ground communications-equipment technology, including launch vehicle capability, the economic considerations of satellite communications systems are significantly and continuously changing. Moreover, a number of additional factors, such as subsidies due to competitive pressures, changing regulations of the telecommunications industry, and an inflationary economy, introduce significant complexities in any attempt to perform a comprehensive economic analysis.

The second-generation systems of the 1980s utilize more powerful launch vehicles than those employed by the first generation in the 1970s. As a result, and because of orbital spot scarcity, in the 1980s satellites for domestic or regional commercial systems are being configured with a capacity of at least 24 equivalent transponders vs. the 12 transponders of the 1970s.

Table 2-1 compares the cost of a typical space segment launched in 1984 with the cost of a space segment launched in 1974. Since the quoted costs ignore inflationary factors, the appropriate adjustments must be made for a legitimate

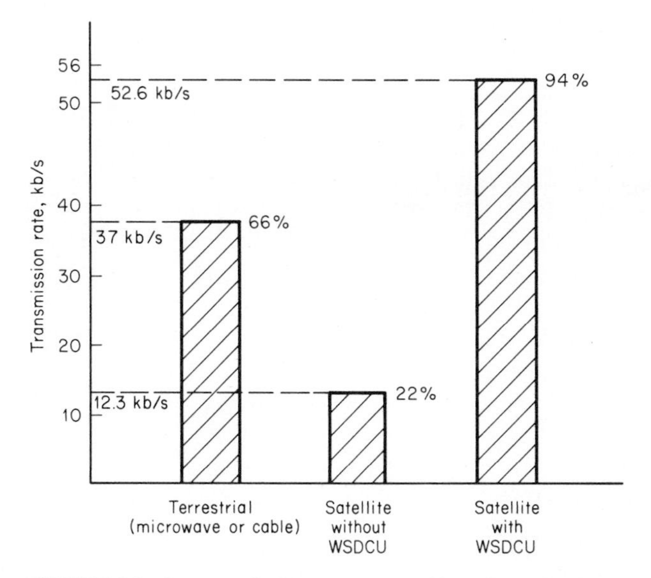

FIGURE 2-9 Impact of WSDCU on a 56-kb/s channel. Block size, 1000 bytes; BER, 10^{-6}.

TABLE 2-1 SPACE SEGMENT COSTS (Domestic System)

Element	First-generation space segment,* cost in millions of 1973 dollars†	Second-generation space segment,‡ cost in millions of 1982 dollars†
Three spacecraft (including test equipment and services)	35	130
Two launch vehicles	20¶	70¶
Two TTC stations	5	12
Other costs (insurance, launch, operations, etc.)	15	30
Total initial deployment costs	75	242

*Twelve transponders. Date of launch, 1974.

†Inflation factor between 1973 and 1982, approximately 2:1.

‡Twenty-four transponders. Date of launch, 1984.

¶Both of the Thor-Delta type.

comparison. In both cases an initial system consists of two spacecraft in orbit and one on the ground. Table 2-1 indicates that the launch vehicle cost is a substantial portion of the overall space-segment cost. In the 1982 time frame the shuttle also became available for general use with a launch price significantly below the Thor-Delta or Atlas-Centaur costs. In addition, at this same time lower prices were advertised for the European Ariane launcher. However, since in both cases these lower prices represent introductory efforts for the 1982–1985 time frame, they most likely do not represent realistic long-term costs.

Various publications by Intelsat experts indicate that the Intelsat IV spacecraft with a 4000-circuit capacity (launched in 1971) had a space-segment capital cost per circuit-year of $1200, while the Intelsat V spacecraft with a 12,000-circuit capacity (launched in 1981) had a corresponding cost per circuit-year of $850. Both spacecraft have a design life of 7 years. If one wishes to take into consideration the inflationary factor from 1971 to 1981, the 1981 figure should be multiplied by approximately 0.5.

The space-segment cost is, of course, only one element of the overall cost for end-to-end service. Another major cost consideration is that of the earth stations, which generally consist of the following elements per station:

Antenna subsystem
High-power amplifier (HPA)
Low-noise receiver (LNR)
Up- and down-converter

Modem
Controller
Civil works
Electric power

Tables 2-2 through 2-6 tabulate 1982 costs of the first five elements. Since the configurations of earth stations can vary widely, depending upon the capa-

TABLE 2-2 ANTENNA SYSTEMS COSTS (Two-Port Transmit/Receive with Anchor Bolts)

	Average cost, thousands of 1982 dollars			
Dish size, m	Main structure	Deice	Installation and erection	Typical civil works
5	9	11	7.5	7.5
7	25	15	9	10
10	45	26	11	15

TABLE 2-3 HIGH-POWER AMPLIFIER COSTS

	Average cost, thousands of 1982 dollars		
Power rating, W	Single unit	Redundant unit with switch	Remarks
5	8		Solid state
75	16	37	TWT amplifier
125	21.5	49	TWT amplifier
400	29	64	TWT amplifier
1500	37	85	Klystron
3300	39	88	Klystron

TABLE 2-4 LOW-NOISE AMPLIFIER COSTS

	Average cost, thousands of 1982 dollars	
Noise temperature, K	Single unit	Redundant unit with switch
120	1	4
80	3	11
45*	20	42
33*	30	56

*Both units are parametric amplifiers.

TABLE 2-5 UP- AND DOWN-CONVERTERS
(Two to Three Channels)

Type	Average cost, thousands of 1982 dollars	
	Single unit	Redundant unit with switch
Up-converted	15	32
Down-converted	9	30

TABLE 2-6 MODEM COSTS

Data rate	Average costs, thousands of 1982 dollars	
	Hard-decision decoding (single/redundant)	Soft-decision decoding (single/redundant)
50 kb/s	10/40	14/42
56 kb/s	10/40	14/42
1.544 Mb/s	10/40	14/42

bilities required, their costs, too, can differ significantly. The earth-station controller could also have a variety of missions, such as controlling time or controlling power. For example, time-division multiple-access (TDMA) controllers are used for controlling the timing and duration of the transmission from each earth station in a system. Since a specially designed controller is required for each application, generalized cost figures for controllers are not meaningful. Civil works and electric power costs, as previously discussed, can vary substantially from case to case, sometimes representing a significant percentage of the overall earth station cost.

Small receive-only (RO) earth stations utilized in the television industry can be obtained for under $5000 (1982 prices), including installation.

The remainder of the ground segment, i.e., terrestrial distribution, central offices, etc., also contributes substantially to the overall cost of the end-to-end service.

In most cases the total ground-segment cost per circuit exceeds the corresponding space-segment cost. As a result, it is of primary importance to develop a cost-effective design of the ground segment. An iterative design process involving all the elements of the system is a necessary step in the establishment of the architecture of each network. Since the space segment cannot be modified once launched, the optimization of each network mainly involves the design of the ground segment and the method by which the ground segment accesses the fixed but channelized space segment. The space segment is designed to be more or less transparent to the various modes of communication.

REFERENCES

Helder, G. K: "Digital Echo Suppressors and the CCITT," International Conference on Communications, 1976.

Members of the Technical Staff of Bell Telephone Laboratories: *Transmission Systems for Communications*, rev. 4th ed., Bell Telephone Laboratories, Winston Salem, N.C., 1971.

Onufry, M.: "Echo Control in Telephone Communications," *Proc. IEEE National Telecommunications Conference*, vol. 1, 1976.

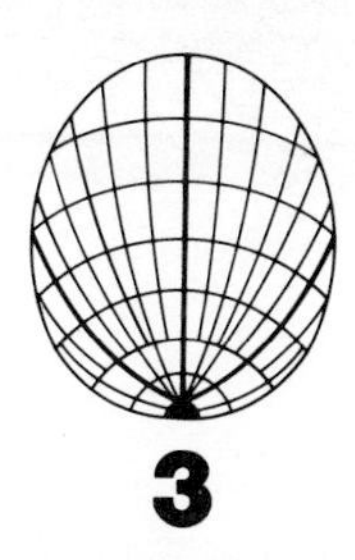

3

ORBITS, LAUNCH SEQUENCE, AND POSITIONING ON STATION

ORBITS

This section will consider only earth orbits, i.e., the orbits of satellites around the earth. Certain simplifying assumptions will be made, the most important ones being the following:

- There are only two bodies, namely, the earth and the satellite, without any internal or external forces except the gravitational forces.
- Both earth and satellite are considered as point masses with the mass of each located at the respective center of mass.

With the first assumption the influence of all the other celestial bodies such as the sun and the moon is omitted. With the second assumption various other small effects are omitted, one example of which is the effects due to the oblateness of the earth.

The Two-Body Problem

Under the assumptions outlined above, the gravitational force on each of the two bodies is given by Newton's law,

$$\mathbf{F} = -\mathbf{r}_o G \frac{Mm}{r^2} \tag{3-1}$$

where

$\mathbf{r}_o$ = the unit vector along the line joining the two masses

G = the universal gravitational constant

$= 6.67 \times 10^{-8}\ \mathrm{dyn \cdot cm^2/g^2} = 6.67 \times 10^{-11}\ \mathrm{N \cdot m^2/kg^2}$

M = the mass of the earth
= 5.98×10^{24} kg
m = the mass of the satellite
r = the distance between the two centers of mass

Figure 3-1 depicts a coordinate system with origin O (the center of the earth) and polar coordinates r and ϕ, where r is the radius vector to the satellite center of mass and the angle ϕ is measured from an arbitrary axis x. The velocity vector $\mathbf{v}$ of the satellite is tangential to the satellite path. The velocity $\mathbf{v}$ could also be expressed as $\mathbf{v} = d\mathbf{r}/dt$, and because $\mathbf{r} = \mathbf{r}_o r$ and $d\mathbf{r}_o = \mathbf{n}_o\, d\phi$, where $\mathbf{n}_o$ is the unit vector normal to $\mathbf{r}_o$, we have

$$\mathbf{v} = \mathbf{r}_o \frac{dr}{dt} + \mathbf{n}_o r \frac{d\phi}{dt} \tag{3-2}$$

The acceleration $\mathbf{a} = d\mathbf{v}/dt = d^2\mathbf{r}/dt^2$ could be similarly expressed in terms of the two components a_r and a_ϕ since $d\mathbf{n}_o = -\mathbf{r}_o\, d\phi$. Therefore,

$$\mathbf{a} = \mathbf{r}_o a_r + \mathbf{n}_o a_\phi$$

where

$$a_r = \frac{d^2r}{dt^2} - r\left(\frac{d\phi}{dt}\right)^2$$
$$a_\phi = \frac{1}{r}\frac{d}{dt}\left(r^2 \frac{d\phi}{dt}\right) \tag{3-3}$$

Because of Newton's law, as expressed in Eq. (3-1), the equation of motion of the satellite with respect to the center of the earth could be formulated as

$$m \frac{d^2\mathbf{r}}{dt^2} = -\mathbf{r}_o G \frac{Mm}{r^2} = 0$$

or

$$\frac{d^2\mathbf{r}}{dt^2} + \mathbf{r}_o \frac{\mu}{r^2} = 0 \tag{3-4}$$

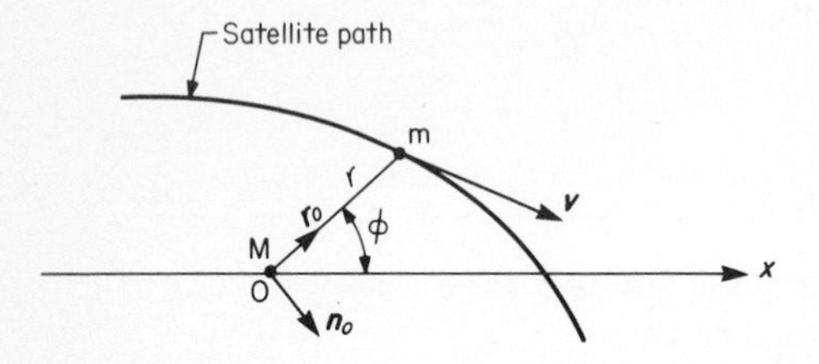

FIGURE 3-1 Two-body system in polar coordinates.

where

$$\mu = GM = \text{gravitational parameter}$$
$$= 3.99 \times 10^{14}\ \text{m}^3/\text{s}^2 = 1.407 \times 10^{16}\ \text{ft}^3/\text{s}^2$$

The motion expressed by Eq. (3-4) is known as "central force" motion.

Angular Momentum

The angular momentum **H** is defined as

$$\mathbf{H} = \mathbf{r} \times (m\mathbf{v}) \tag{3-5}$$

If we define **h** as the specific angular momentum, i.e., angular momentum per unit mass, we will have

$$\mathbf{h} = \mathbf{r} \times \mathbf{v} \tag{3-6}$$

Substituting the value of **v** from Eq. (3-2), we find

$$\mathbf{h} = \mathbf{h}_o r^2 \frac{d\phi}{dt} \tag{3-7}$$

where $\mathbf{h}_o = \mathbf{r}_o \times \mathbf{n}_o$ is the unit vector perpendicular to the plane of motion, because $\mathbf{r}_o \times \mathbf{r} = 0$.

But from the equation of motion, Eq. (3-4), and Eqs. (3-3) we also obtain

$$\frac{d^2\mathbf{r}}{dt^2} + \mathbf{r}_o \frac{\mu}{r^2} = \mathbf{r}_o a_r + \mathbf{n}_o a_\phi + \mathbf{r}_o \frac{\mu}{r^2} = 0$$

and consequently $a_\phi = 0$, or

$$\frac{d}{dt}\left(r^2 \frac{d\phi}{dt}\right) = 0 \qquad \text{and} \qquad r^2 \frac{d\phi}{dt} = \text{constant}$$

Comparing this conclusion with Eq. (3-7), we also obtain

$$\mathbf{h} = \text{constant} \tag{3-8}$$

Since the area covered in time dt by the radius r is $\frac{1}{2}r^2\, d\phi$, the conclusion is that the radius r covers equal areas at equal times on the plane of the orbit. This is in fact the celebrated Kepler second law of motion of celestial bodies.

Energy

The total energy of the mass m can be expressed as the sum of the potential and kinetic energies. The change in potential energy E_p will be $dE_p = -\mathbf{F}\, d\mathbf{r}$,

and

$$E_p = \int_{-\infty}^{r} G\frac{Mm}{r^2}\,dr = -\frac{GMm}{r} = -m\frac{\mu}{r}$$

But since the kinetic energy is defined as $E_K = \frac{1}{2}mv^2$, the total energy E_T along the trajectory will be

$$E_T = E_p + E_K = m\left(\frac{1}{2}v^2 - \frac{\mu}{r}\right)$$

and by calling E = specific mechanical energy, i.e., total energy per unit mass, we have

$$E = \frac{1}{2}v^2 - \frac{\mu}{r} \tag{3-9}$$

The specific mechanical energy E remains constant along the trajectory since there is neither generation nor loss of energy.

Trajectory

We can eliminate the variable t from the equation of motion, Eq. (3-4), by utilizing Eqs. (3-3) and (3-7). The result is

$$\frac{h^2}{\mu}\left(\frac{d^2u}{d\phi^2} + u\right) = 1 \tag{3-10}$$

where $u = 1/r$.

Equation (3-10) is the differential equation of the trajectory of mass m in polar coordinates r, ϕ with center at the center of the earth O. By using the total energy relationship for determining the constants of integration, we can integrate this equation, finding

$$r = \frac{p}{1 + e\cos\phi} \tag{3-11}$$

where

$$p = \frac{h^2}{\mu} \tag{3-12}$$

$$e = \sqrt{1 + \frac{2h^2E}{\mu^2}} = \sqrt{1 + \frac{2pE}{\mu}}$$

Equation (3-11) represents a conic section by a plane. The value of e determines the type of conic section; specifically,

$e = 0$ results in a circle
$0 < e < 1$ results in an ellipse
$e = 1$ results in a parabola
$e > 1$ results in a hyperbola

Elliptical Orbits Figure 3-2 depicts an elliptical orbit ($0 < e < 1$) expressed by Eq. (3-11). This relationship can be rewritten as

$$r = \frac{a(1 - e^2)}{1 + e \cos \phi} \tag{3-13}$$

where a is the ellipse major semiaxis. The following relationships apply to the various parameters of the ellipse:

$$\begin{aligned}
&\text{Major semiaxis} = a = -\frac{\mu}{2E} \\
&\text{Minor semiaxis} = b \\
&\text{Eccentricity} = e = \sqrt{\frac{a^2 - b^2}{a}} = \frac{c}{a} = \sqrt{1 + \frac{2pE}{\mu}} \\
&\text{Semi-latus rectum} = p = a\,(1 - e^2) = \frac{h^2}{\mu}
\end{aligned} \tag{3-14}$$

The origin of the polar system considered in Fig. 3-1 will be located on F, one of the two foci F and F′. There are two extreme points A and P called the "apses," with P, nearest to the prime focus F, called the "periapsis" and A called the "apoapsis." For earth orbits with the earth located at F, the periapsis is called the "perigee" and the apoapsis is called the "apogee." Similarly, for heliocentric trajectories, like the earth's trajectory around the sun (termed the "ecliptic"), P is called the "perihelion" and A the "apohelion."

The angle of the orbital plane with respect to the earth's equatorial plane is called the "inclination" of the orbit.

The plane of the earth's orbit around the sun has an inclination of 23.5° with respect to the earth's equatorial plane. The line of the intersection of the ecliptic

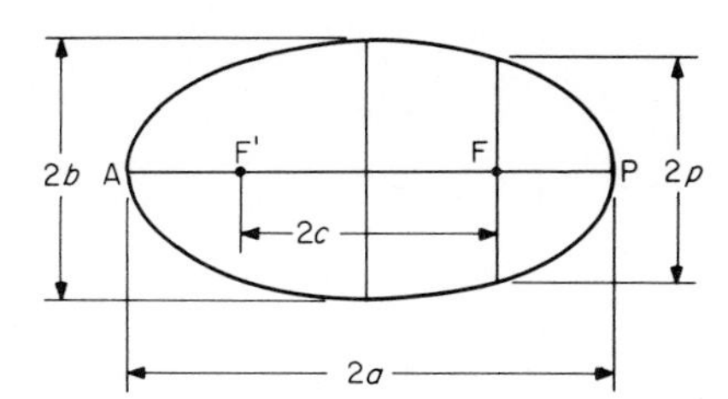

FIGURE 3-2 Elliptical orbit.

with the earth's equatorial plane defines the x axis, one end of which points toward the vernal equinox (i.e., the point on the ecliptic where the first day of spring begins) and the other end of which points toward the autumnal equinox (i.e., the point on the ecliptic where the first day of autumn begins).

Figure 3-3 depicts a geocentric system of coordinates in which the x axis is on the equatorial plane and points toward the vernal equinox and the z axis points toward the north pole.

The extension of the earth's equatorial plane to infinity, intersecting a fictitious sphere with infinite radius, defines a plane called the "celestial equator," while the fictitious sphere is called the "celestial sphere."

The line determined by the intersection of the orbital plane and the celestial equator is called the "line of the nodes," which pierces the celestial sphere at two points, one in each direction, called the "nodes." The ascending node is the node where the satellite moves from south to north, and the opposite, or descending node, is the node where the satellite moves from north to south. Because of the oblateness of the earth, the line of the nodes regresses at a rate depending on the orbit inclination and altitude. The orbit is called "direct" if the satellite in it moves easterly, i.e., in the same direction as the earth's rotation. When the satellite moves in the direction opposite to the earth's rotation, the orbit is called "retrograde."

The position of the satellite on the orbit at a given time t_o is called the "epoch." The angle Ω on the equatorial plane between the x axis and the line of the nodes is called the "longitude of the ascending node." The angle ω in the plane of the orbit between the line of the nodes (in the direction of the ascending node) and the periapsis is called the "argument of the periapsis"; for earth orbits it is called the "argument of the perigee."

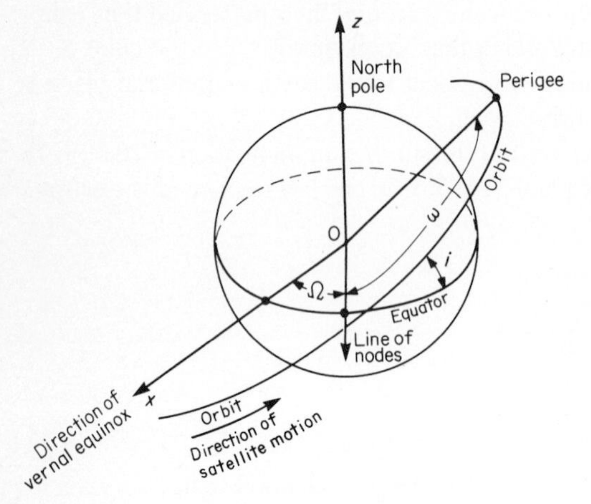

FIGURE 3-3 Geocentric equatorial coordinate system; i = inclination.

The velocity of the satellite along its orbit is given by the relationship

$$v = \left[2\mu \left(\frac{1}{r} - \frac{1}{2a} \right) \right]^{1/2} \tag{3-15}$$

This is easily seen by inspecting Eqs. (3-9) and (3-14). The period T of the orbit is

$$T = \frac{2\pi a^{3/2}}{\sqrt{\mu}} = \frac{2\pi a^{3/2}}{\sqrt{GM}} \tag{3-16}$$

A special case is a circular orbit when $a = b$ and $e = 0$. If R is the radius of the orbit, then $a = R$. The velocity of the satellite is constant along the orbit and is given as

$$V_{\text{circ}} = \left(\frac{\mu}{R} \right)^{1/2} \tag{3-17}$$

and the period is

$$T_{\text{circ}} = \frac{2\pi R^{3/2}}{\sqrt{\mu}} \tag{3-18}$$

The angle on the plane of the satellite's orbit, between the perigee and the position of the satellite at a particular time t_o, is called the "true anomaly at epoch."

From the above relationships it is easy to verify Kepler's laws: namely,

1. The orbit of each plane is an ellipse with the sun as a focus.
2. The line joining the planet with the sun sweeps equal areas in equal times.
3. The square of the period of a planet is proportional to the cube of its mean distance from the sun.

LAUNCH SEQUENCE

When a communications spacecraft is placed in a geosynchronous orbit, a number of steps occur:

1. The main launch vehicle is ignited, and after lift-off, during the main boost phase, the craft is powered through the earth's atmosphere. In order to avoid the effect of severe aerodynamic forces during the powered flight, only limited and absolutely necessary maneuvering is performed, and the launch vehicle is guided with a small angle of attack.

2. As soon as the upper stages of the launch vehicle exit the atmosphere, the fairing (shroud) is jettisoned. The fairing, which covers the spacecraft on top of the launch vehicle to protect it from aerodynamic forces during powered flight, is no longer needed.

3. The final stage of the launch vehicle ignites; after burnout, the spacecraft is injected into the transfer orbit near the perigee. This orbit is a highly elliptical one with an apogee at synchronous altitude, that is, 22,282 mi (19,300 nautical miles, or 35,860 km). The plane of the orbit is usually inclined. The inclination angle depends on the launch site and the launch vehicle capability and on how the flight trajectory is optimized. A typical inclination angle for launches from Cape Kennedy with U.S. launch vehicles is 28.5° with respect to the earth's equatorial plane.

4. After the payload has been tracked from the ground and a precise determination made of the orbital parameters—a task that takes several orbital periods—the apogee kick motor (AKM) is ignited at the apogee in order to remove the inclination of the orbital plane (orbital plane correction) and, at the same time, to circularize the orbit. Before ignition, the thrust vector is oriented appropriately by commanding, from the ground, the attitude orientation system of the spacecraft. For launch vehicles (such as the Titan IIIC) which have a restartable upper stage, attitude control capability, and adequate total impulse, the upper stage is used instead of a spacecraft solid propulsion unit such as the AKM. At the end of this launch phase, the spacecraft has been placed in a near-synchronous equatorial orbit and is drifting slowly, since its

TABLE 3-1 PAYLOAD DELIVERY CAPABILITY

Launch vehicle	Geostationary transfer orbit, lb	Inclination, degrees	Geostationary orbit,* lb
Delta 2914	1550	28.5	820
Delta 3914	2000	28.5	1060
Delta/PAM-D	2750	28.5	1475
Atlas-Centaur	4400	28.5	2350
Titan IIIC	—	—	3100
SSUS-D	2750	28.5	1475
SSUS-A	4400	28.5	2020
Ariane			
AR-2	4400	9.6	
AR-3	5280	9.6	3300

*Computed by assuming AKM specific impulse = 295 s.

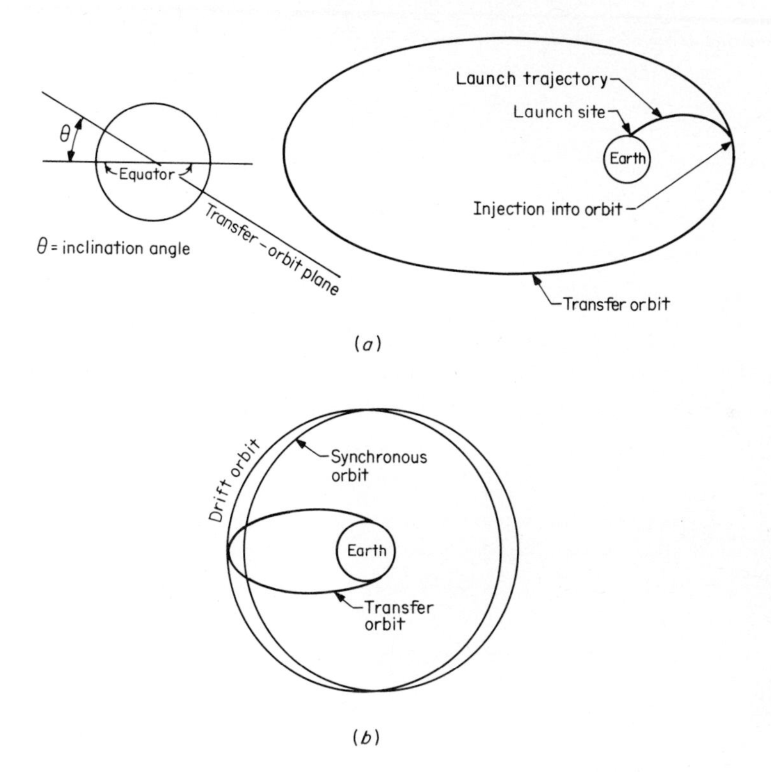

FIGURE 3-4 Launch sequence: (*a*) transfer orbit, (*b*) transition from transfer orbit to synchronous orbit.

orbital velocity is not precisely the one required for absolute synchronous performance. For this reason this orbit is often called a "drift orbit."

5. The spacecraft is attitude-stabilized and is set to operate on its own electric power. Over a period of several days the spacecraft is tracked from the ground and—through its own secondary propulsion system activated by ground commands—is positioned on station (i.e., in the preassigned orbital spot) in order to commence operations.

Table 3-1 tabulates the payload delivery capability of the various launch vehicles. Figure 3-4 gives a graphic representation of the launch sequence described above. Figure 3-5 depicts the IDCSP multiple payload on a dispenser structure, the jettison of the Titan IIIC fairing, and the "seeding" of the satellites in orbit.

(*a*)

FIGURE 3-5 (*a*) The Initial Defense Satellite Program payload on a dispenser. (*b*) Jettison of Titan III-C fairing and the IDCSP payload. (*c*) Titan IIIC transstage "seeding" the IDCSP satellites in orbit. ***(Courtesy of Ford Aerospace and Communications Corporation.)***

STATION KEEPING: ORBIT GEOMETRY

The orbital velocity in a circular orbit is given by the relationship

$$V_c = \left(\frac{GM}{R}\right)^{1/2}$$

where R = radius of orbit, and the orbital period is

$$T = \frac{2\pi R^{3/2}}{(GM)^{1/2}}$$

Since

$$\begin{aligned} M &= \text{the earth's mass} \\ &= 5.98 \times 10^{24}\ \text{kg} \end{aligned}$$

(*b*)

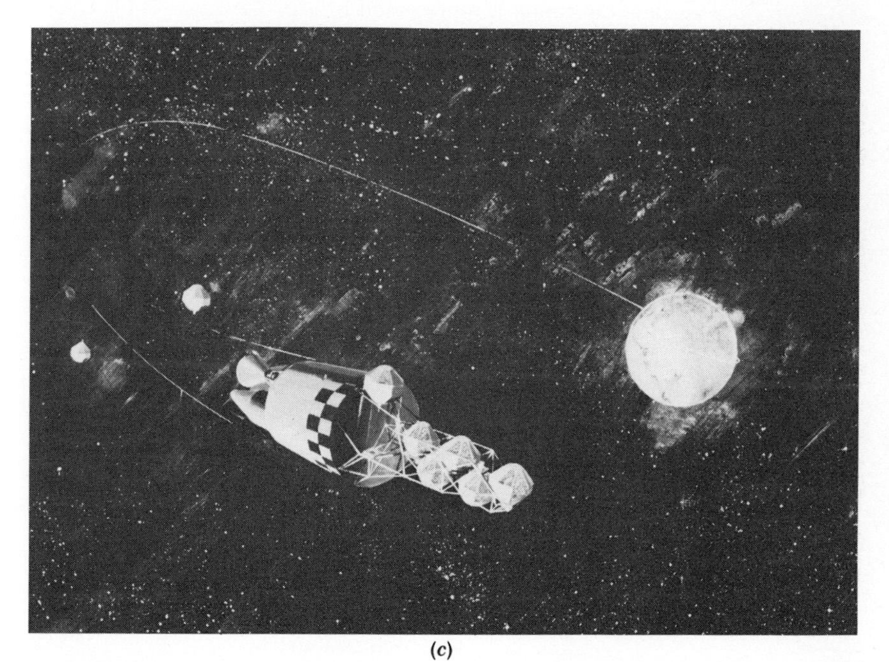

(*c*)

G = Newton's gravitational constant
$= 6.67 \times 10^{-11}\ \text{N}\cdot\text{m}^2/\text{kg}^2$
$\mu = GM$ = gravitational parameter
$= 3.99 \times 10^{14}\ \text{m}^3/\text{s}^2 = 1.407 \times 10^{16}\ \text{ft}^3/\text{s}^2$

for a synchronous orbit ($T = 24$ h), $R = 42{,}230$ km and $V_{syn} = 11{,}070$ km/h $= 3074$ m/s.

A number of external forces will move the spacecraft from its station unless the spacecraft periodically makes the necessary corrections. East-west corrections will maintain the longitude and north-south corrections will maintain the orbit inclination. Ground commands initiate these corrections.

Drift along the orbit, resulting in a longitudinal error, is caused by the nonuniformity of the earth's gravitational field. The orbital inclination is changed as a result of solar and lunar gravitational forces. A variation of 0.86° per year in the inclination plane was caused by these forces for the years close to 1980. A velocity increment of approximately 155 ft/s per year is required to correct this anomaly. In addition, magnetic field forces as well as limited accuracy in measuring satellite position and execution of commands cause a certain drift from the desired station.

Since all corrections have a degree of imperfection, the subsatellite point (projection of the satellite on the earth's surface along the local vertical) will describe a figure 8 around the desired station. See Fig. 3-6 for a representation of the motion over a 24-h period. The geometry of the various distances from the earth for a satellite on station is represented in Fig. 3-7 (see also Fig. 2-5).

Since the velocity of light $c = 186{,}000$ mi/s, or 2.99776×10^8 m/s, the minimum travel time (delay) of a radio signal from the ground to the spacecraft will be $2SS'/c = 0.2396$ s and the maximum delay will be $2SA/c = 0.2790$ s. For a given position on the earth, this delay can be derived by knowing the elevation angle or the angle to the local horizontal for an observer looking

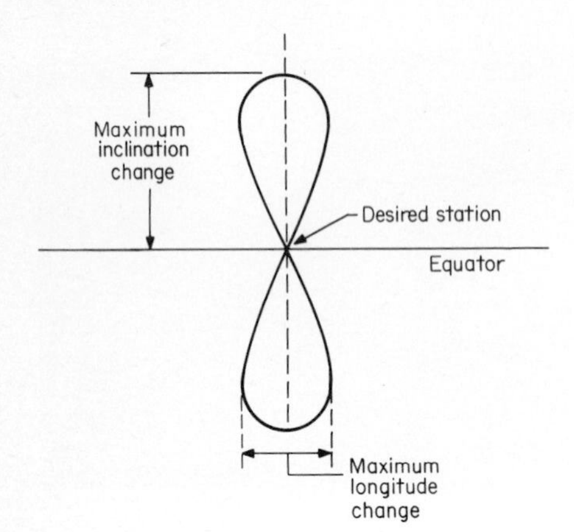

FIGURE 3-6 Satellite footprint due to drift from station.

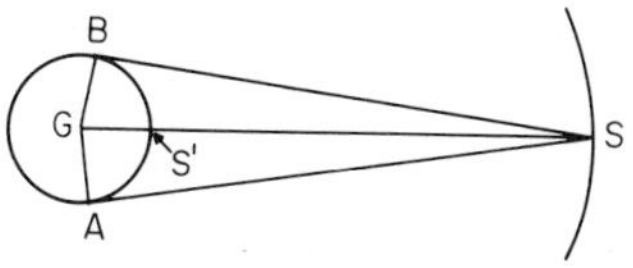

FIGURE 3-7 Geometrical distances for geostationary satellites.

∠ ASB = 17.3°

SS′ = 35,860 km (S′ = subsatellite point)

GS′ = 6370 km (earth's radius, *R*)

SG = 42,230 km

SB = SA = 41,760 km (maximum distance to earth, tangential)

straight at the satellite. From a north or south latitude of 81.25° the elevation angle is 0°, and from a latitude of 77.25° it is 5°. An elevation angle of 5° or more must be maintained by the earth station antenna if reasonable communications capability is expected.

EXAMPLE 3-1

Assume a transfer orbit with apogee at synchronous altitude, perigee at 240 km, and zero inclination. What is the required impulse at the apogee in order to achieve a circular synchronous orbit?

The synchronous altitude is 35,860 km, and because the earth's radius is 6370 km, the apogee of the transfer orbit will be at a distance $r_a = 35{,}860 + 6370 = 42{,}230$ km from the center of the earth and the perigee will be at a distance $r_p = 240 + 6370 = 6610$ km. Consequently, the major semiaxis of the elliptical orbit will be

$$a = \frac{r_a + r_p}{2} = \frac{42{,}230 + 6610}{2} = 24{,}420 \text{ km}$$

The velocity at apogee will be

$$V_a = \left[2\mu\left(\frac{1}{r_a} - \frac{1}{2a}\right)\right]^{1/2}$$

$$= \left\{2(3.99 \times 10^{14})\left[\frac{1}{42{,}230 \times 10^3} - \frac{1}{2(24{,}420 \times 10^3)}\right]\right\}^{1/2} \text{ m/s}$$

$$= 1599 \text{ m/s}$$

Since the synchronous velocity is $V_s = (\mu/r_a)^{1/2}$ or

$$V_s = \left(\frac{3.99 \times 10^{14}}{42{,}230 \times 10^3}\right)^{1/2} \text{ m/s} = 3074 \text{ m/s}$$

The required impulse will be

$$\Delta V = V_s - V_a = 1475 \text{ m/s}$$

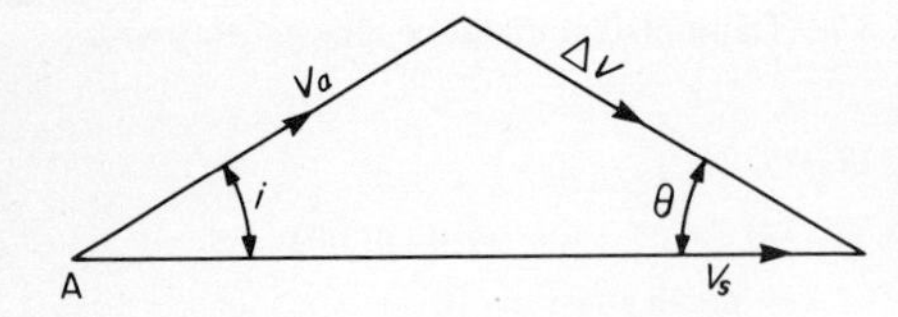

FIGURE 3-8 Velocity diagram (Example 3-2). V_a = 1599 m/s, V_s = 3074 m/s, i = 28.5°.

EXAMPLE 3-2

Assume the apogee of the transfer orbit considered in Example 3-1 to be at the equatorial plane but the inclination of the orbit to be 28.5°. What is the required impulse at the apogee to achieve an equatorial circular synchronous orbit?

The velocity diagram on a plane normal to the radius vector from the center of the earth to the apogee is depicted in Fig. 3-8.

We have

$$(\Delta V)^2 = V_a^2 + V_s^2 - 2V_aV_s \cos i$$
$$= (1599)^2 + (3074)^2 - 2(1599)(3074) \cos 28.5$$

or

$$\Delta V = 1835 \text{ m/s}$$

This impulse must be applied in the direction indicated by the velocity diagram in Fig. 3-8, which is at an angle θ in relation to the equatorial plane, such that

$$\sin \theta = (\sin i) \frac{V_a}{\Delta V} \qquad \text{or} \qquad \sin \theta = 0.416$$

and

$$\theta = 24.6°$$

REFERENCES

Bate, R. R., D. D. Mueller, and J. E. White: *Fundamentals of Astrodynamics*, Dover Publications, New York, 1971.

Finlay-Freundlich, E.: *Celestial Mechanics*, Pergamon Press, New York, 1958.

Wintner, Aurel: *The Analytical Foundations of Celestial Mechanics*, Princeton University Press. Princeton, N.J., 1947.

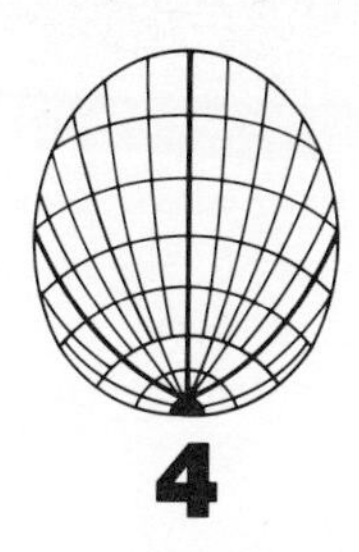

4
THE COMMUNICATIONS SPACECRAFT

A communications spacecraft is usually stabilized in attitude so that the communications antenna points toward the earth. In addition, the solar-energy-collecting system points toward the sun for efficient collection of solar energy.

A spacecraft bus of this kind usually requires a structure housing the following subsystems:

Stabilization (or attitude control)

Electric power (with solar array and storage battery)

Secondary propulsion (for station keeping and attitude control)

Telemetry, tracking, and command (for the satellite housekeeping functions)

Thermal control (to avoid extreme temperatures)

The payload of this bus consists of the communications repeaters and associated communications antennas having area beams and/or spot beams, steerable or fixed.

STABILIZATION

Two distinct stabilization concepts have been used in communications satellite design, namely, spin stabilization and three-axis or body stabilization.

Spin Stabilization

Usually a cylindrical spacecraft is spinning around its main axis of symmetry like a spinning wheel. The spacecraft is balanced so that this axis (the y axis) is also one of the principal axes of inertia; the other two are the x and z axes (Fig. 4-1). While the spacecraft is in a synchronous equatorial orbit, the y axis of rotation is parallel with the earth's geographic polar axis (Fig. 4-2).

For a stable spinning condition, the ratio of the moments of inertia of the spacecraft must satisfy the relationships

$$\frac{I_y}{I_x} > 1 \qquad \text{and} \qquad \frac{I_y}{I_z} > 1$$

As long as no external forces exist, the momentum vector along the y axis will maintain its direction in space and the spacecraft will be spin-stabilized.

Typical inertia ratios I_y/I_x range from 1.1 to 1.3, and a typical spin speed is 100 r/min.

A number of external forces such as solar forces, magnetic-field forces, and forces due to moving elements within the spacecraft (the friction of motor bearings, etc.) will generate disturbing torques. Solar heating could create thermal distortions which give rise to disturbances. Disturbances will cause both precession and nutation of the momentum vector, i.e., the spin axis of the spacecraft.

Solar and magnetic torques will create precession depending on the actual characteristics of the spacecraft. The attitude error buildup could be significant. For example, an attitude error of 0.1° per 10-day period could easily be accumulated. Such an error is corrected by thrusting the jets of the secondary propulsion system of the spacecraft every 10 days in order to generate a correction torque. Attitude correction is either automatic or commanded from the ground through the TTC subsystem. Sun sensors could be used to determine the direction of the spin-vector direction.

Nutation dampers are used to eliminate the effects of nutation-causing disturbances, such as mismatch and misalignment of thrusters and motor-bearing friction. A nutation damper is an energy-absorbing device that dissipates nutation energy to the environment. Several such devices have been developed, the operations of which depend on magnetic friction, fluid friction, etc.

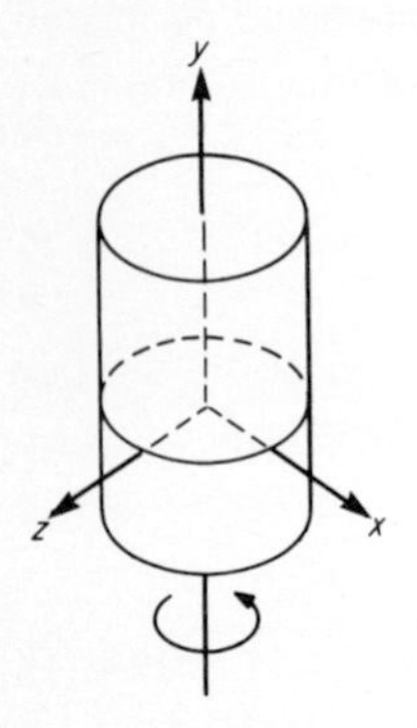

FIGURE 4-1 Spinning body: principal axes.

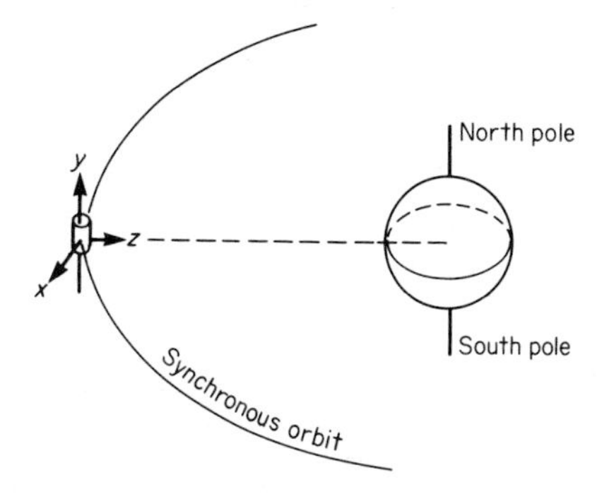

FIGURE 4-2 Inertial axis orientations of a spin-stabilized spacecraft; x is the roll axis (local horizontal in the orbit plane), y is the pitch axis (normal to the orbit plane), and z is the yaw axis (local vertical).

Spin-rate maintenance can be achieved through metered impulses by thrusting the secondary propulsion jets. Spinning satellites can achieve attitude accuracy of the order of a few tenths of a degree.

While the spacecraft is spinning, the communications antenna must be pointed toward the earth. Usually a parabolic dish or a horn or a similar element is used as a communications antenna.

The typical case is one in which a parabolic dish for an area beam must be oriented so that the main-lobe axis forms a preset small angle with the local vertical; i.e., it is directed almost toward the center of the earth. In order to achieve this, one must, first, detect the local vertical and, second, decouple the antenna mount from the spinning main body and despin the mount. This despin function can be achieved either electronically by utilizing phased-array techniques or mechanically by despinning the antenna mount. Mechanically despinning the antenna mount results in a "dual-spin" spacecraft design and requires a bearing which must operate under the environmental conditions of space (a hard vacuum, cosmic radiation, etc.). Lubricating a bearing under hard-vacuum conditions presents a significant problem since normal lubricants would quickly evaporate and the bearing would freeze. Present-day technology has successfully solved this problem with special bearings having a life in excess of 10 years. This solution has prevailed over electronic despinning, since electronic despinning results in a nonuniform radiation pattern that, in effect, interferes with the communications signal.

Detecting the local vertical and using that information in the control loop to orient and stabilize the antenna along the local vertical is a general problem in spacecraft design. Similar requirements exist for many categories of spacecraft other than communications satellites with stabilization systems of varying degrees of complexity. This problem has been successfully solved by detecting with satellite instruments the contrast between the relatively warm earth and the surrounding cold space or by utilizing beacon signals from the ground and operating on the strengths of the signals.

With the conventional representation of **i**, **j**, **k** unit vectors along the x, y, z axes and the angular velocity ω given as

$$\omega = \mathbf{i}\omega_x + \mathbf{j}\omega_y + \mathbf{k}\omega_z$$

the fundamental equations governing the motion of a spinning body are the angular momentum and kinetic energy relationships, namely,

$$\mathbf{H} = \mathbf{i}I_x\omega_x + \mathbf{j}I_y\omega_y + \mathbf{k}I_z\omega_z$$

and

$$E = \tfrac{1}{2}I_x\omega_x^2 + \tfrac{1}{2}I_y\omega_y^2 + \tfrac{1}{2}I_z\omega_z^2$$

where **H** and E are respectively the angular momentum and kinetic energy of the spacecraft. Moreover, if **T** is the total torque applied on the spacecraft,

$$\mathbf{T} = \dot{\mathbf{H}}$$

With these relationships analytical solutions could be obtained in specific cases by using simplifying assumptions.

Figure 1-3 depicts the Westar spin-stabilized satellite placed in orbit in 1974 for domestic communications.

Three-Axis Stabilization

Communications spacecraft in geostationary orbits are often stabilized in three axes with respect to a geocentric system that rotates with the earth. In Fig. 4-3 the three principal inertial axes through the center of gravity of the spacecraft (O) coincide respectively with the local vertical (z axis, or yaw axis), the orbital velocity direction tangential to the orbit (x axis, or roll axis), and the axis normal to the orbital plane (y axis, or pitch axis).

For a body-stabilized spacecraft, these three axes appear to be stationary to an observer located at the earth's surface (point P). With respect to the earth's center (G), the observer at P rotates in the equatorial plane with an angular

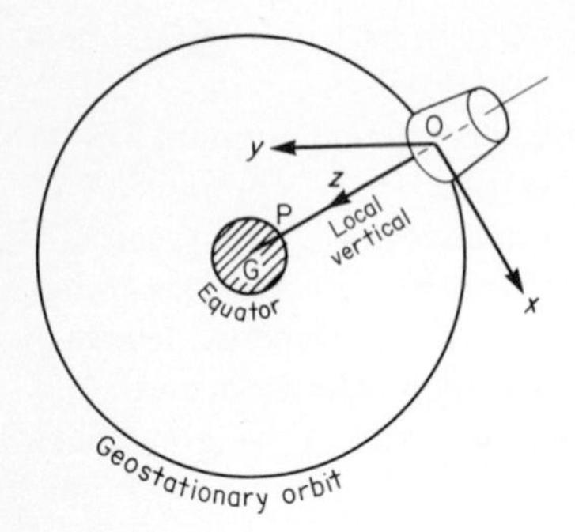

FIGURE 4-3 Three-axis-stabilized satellite (axes of reference, equatorial plane section); x is the roll axis (local horizontal in the orbit plane); y is the pitch axis (normal to the orbit plane), and z is the yaw axis (local vertical).

velocity of 360° per day, while the spacecraft center of gravity O also orbits on the same plane with the same angular velocity with respect to the center of the earth G.

Consequently, a three-axis-stabilized spacecraft, although appearing to an observer at P to be attitude-stabilized in all three axes, has in effect an angular motion with respect to inertial space. More specifically, in order to maintain its yaw axis z along the local vertical GO, the spacecraft must possess an angular velocity Ω around the pitch axis y. This angular velocity is in general called "pitch rate" or "inertial rate" and for a synchronous orbit is equal to the earth's angular velocity; i.e.,

$$\Omega = \frac{2\pi}{24 \times 3600} \text{ rad/s} = 7.27 \times 10^{-5} \text{ rad/s}$$

The sensory system utilized to detect the local vertical for attitude control purposes is the same discussed for the spin-stabilized spacecraft. It uses horizon sensors to detect the earth's horizon by discriminating cold space from the heat-radiating earth. One horizon sensor is used for sensing excursions due to angular motion around the x axis and a second one for sensing excursions due to angular motion around the y axis. Reaction torques are applied either through jet firings or through acceleration of momentum wheels aligned along these axes. A momentum package consists of a motor coupled to a flywheel. By accelerating or decelerating the flywheel, a torque is applied on the spacecraft. When the flywheels reach their maximum speed, they must be unloaded by an impulse torque applied through the secondary propulsion jets. The secondary propulsion jets are fed either from compressed gas such as nitrogen stored under pressure or from a combustion chamber burning a propellant fuel. Propellant fuel systems exclusively are used for long-life spacecraft.

Angular motion around the yaw axis z could be sensed through sun-tracking sensors properly programmed or through a rate sensor such as a gyroscope mounted in such a way as to sense any change in the inertial rate Ω resulting from the rotation of the pitch axis y around the yaw axis z. The same information could be derived by sensing the spacecraft's x-axis component of the spacecraft's "inertial rate." In fact, the coupling of the yaw and roll axes has extensively been used in "momentum-bias" systems.

Usually a body-stabilized spacecraft will deploy a solar array mounted on solar booms extended from the main spacecraft body. Such a spacecraft is shown in Fig. 4-4. The solar paddles very often rotate around the spacecraft in order to track the sun for optimum collection of solar energy. Sun sensors are used to detect the direction of the sun, generating the appropriate signals to drive the motors used to rotate the solar-array booms.

The overall block diagram of a three-axis-stabilized spacecraft is depicted in Fig. 4-5. Momentum-bias systems do not require a separate yaw sensor, and

FIGURE 4-4 ATS-6 satellite, three-axis-stabilized. *(Courtesy of Fairchild Industries, Inc.)*

yaw control is provided in response to roll errors, taking advantage of the roll-yaw coupling. An example of a momentum-bias system is a system utilizing a gimbaled flywheel, gimbaled around the x axis and spinning around the z axis. Yaw error signal is not used, but by limiting the gimbal motion due to precession, the appropriate control of yaw is achieved. Momentum-bias systems do not provide tight control around the yaw axis but limit the maximum amplitude of the yaw error.

Sensing the Attitude of the Spacecraft

Sensors on board the spacecraft have been extensively used for sensing the spacecraft attitude. Most of these sensors could be placed in one of two categories, i.e., sensors that use the radiation emanating from celestial bodies (earth, sun, stars) or sensors based on the inertial characteristics of gyroscopic devices.

Horizon, or Local-Vertical Sensors One common requirement of communications spacecraft is the pointing of the communications antenna toward the earth. Therefore the ability to determine the "local vertical" is very useful for either spinning or three-axis-stabilized satellites.

Detecting the earth's horizon (on the assumption of a spherical earth) and determining the center of the horizon circle provide the direction of the local

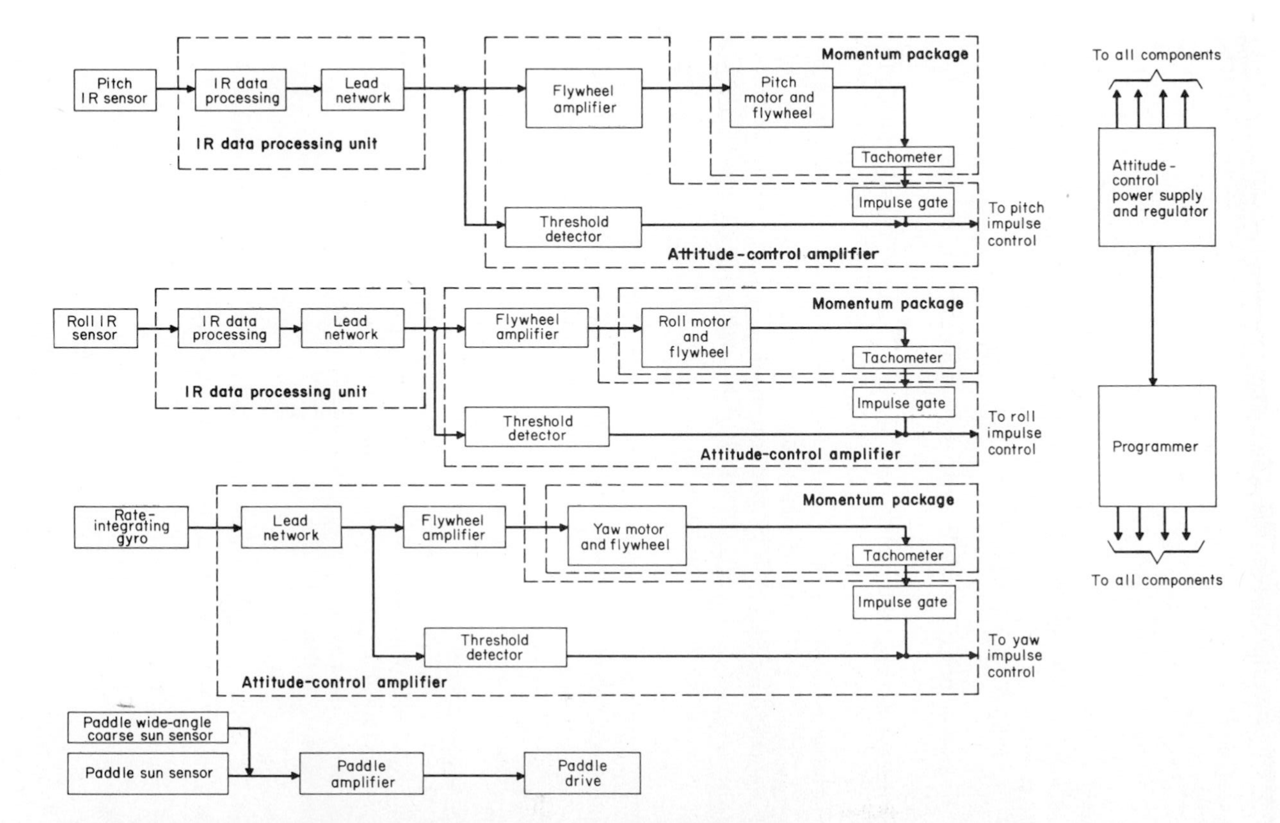

FIGURE 4-5 Attitude control subsystem (functional block diagram).

vertical through the satellite. The assumption of a spherical earth results in an angle error:

$$\Delta\phi = \frac{\alpha}{2\sqrt{h^2 + 2hR}} \quad \text{rad} \tag{4-1}$$

where

h = satellite altitude over the surface of the earth
R = 6370 km (earth's radius)
α = local anomaly, km

For a geostationary satellite we have h = 35,860 km, and therefore $\Delta\phi = 1.2\alpha \times 10^{-5}$ rad.

From synchronous altitude the earth looks like a disk. The view angle from the satellite to the ends of the diameter of that disk is approximately 17.3°. The demarcation of the edge of the earth with cold space is not sharp, because of the atmosphere and clouds at various altitudes. However, the infrared radiation of the earth in contrast to cold space can be detected accurately enough with sensitive infrared detectors (bolometers). The sensors that have been developed to provide the horizon information are called "horizon sensors"; usually two are required per spacecraft.

The horizon can be observed optically by visible or infrared radiation. Infrared is much more attractive because it can be used both day and night. A first approximation of the earth's thermal radiation is easily computed from heat-balance relationships. By considering the earth as a blackbody (with mean temperature T in kelvins), the heat H_R radiated from the earth's surface is

$$H_R = \sigma(4\pi R_o^2)T^4 \tag{4-2a}$$

where σ is the Stefan-Boltzmann constant, 5.6686×10^{-8} W/m$^2\cdot$K^4, and R_o is the earth's radius, 6370 km. If one assumes that the only source of energy is the solar energy, the heat radiated by the earth must be

$$H_R = E_o(1 - \alpha)(\pi R_o^2) \tag{4-2b}$$

where E_o is the solar energy constant, 1.395 kW/m^2, and α is the earth's planetary albedo, 0.34. From Eqs. (4-2a) and (4-2b) we have

$$H_R = \sigma(4\pi R_o^2)T^4 = E_o(1 - \alpha)(\pi R_o^2)$$

$$T = \left[\frac{E_o(1 - \alpha)}{4\sigma}\right]^{1/4} = 254 \text{ K}$$

The radiation resulting from a mean blackbody temperature of 254 K peaks at approximately 9.5 μm. The radiation from specific layers of the atmosphere, for example the carbon dioxide layer, has also been used to obtain the horizon demarcation. Actual measurements confirm the theoretical results within the measurement accuracy of ±10 percent.

Thermistor cells (bolometers) have been used very effectively as detectors in most horizon-sensor designs. The lenses of the optics in front of the bolometer are usually made out of germanium and are properly coated to provide the necessary filtering. Figure 4-6 depicts the positioning of a particular horizon sensor relative to the main axis of a three-axis-stabilized spacecraft. This particular configuration utilizes a rotating prism which allows the optical system to scan the earth-sky horizon. The cone angle of scan is wide enough (for example, 90°) and is referenced at the point where the scan circles intersect the pitch and yaw planes. The prism lens field of view is of the order of 1° × 1°. This 1° × 1° "window" is rotated around the 90° cone to provide high-voltage output for earth signals and low-voltage output for sky signals. The output is amplified by a very sensitive low-noise amplifier. Time differences between the resulting earth and space pulses and the "null" defining the pitch and roll axes are used to derive the control signal.

Sensing the Direction of the Orbital Plane (Yaw Signal) By sensing the orbital plane direction, information is generated for control of the third axis of the vehicle, the other two axes being controlled from horizon information. An inertial component, such as a rate-integrating gyro, is used for this purpose. The

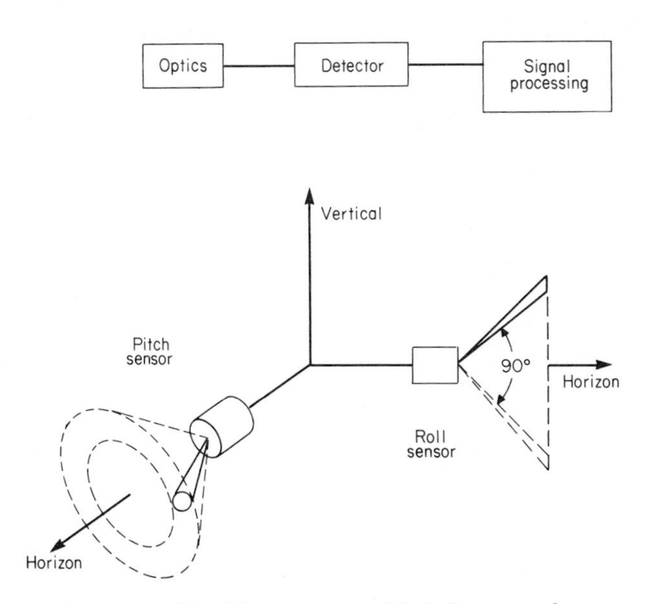

FIGURE 4-6 Infrared horizon sensors (block diagram and mounting for a three-axis-stabilized spacecraft).

preferred system, of course, is the one in which drift errors with time are not cumulative (so that accuracy is not degraded with time). An installation schematic of an azimuth gyro for yaw control is depicted in Fig. 4-7. This gyro obtains its primary reference information from the orbital velocity vector.

As the vehicle orbits around the earth, its attitude is aligned with respect to the local vertical by means of horizon sensors and associated control equipment. In order to maintain the alignment with the local vertical, the vehicle must rotate about its center of mass at a rate equal to the rate of rotation of the vehicle about the center of the earth. The axis of rotation is perpendicular to the plane of the orbit. For a circular orbit, the magnitude of the rate of rotation (which we shall call "inertial rate") may be expressed as

$$\Omega = \sqrt{\frac{GM}{R^3}}$$

where G is the gravitational constant, M is the mass of the earth, and R is the orbit radius. For a 35,860-km altitude orbit the inertial rate is 360° per day.

If the roll axis of the vehicle is in the orbit plane as desired, the inertial rate is about the vehicle pitch axis only. If the vehicle has a yaw error, the roll axis is not in the orbit plane, and a component of the orbital inertial rate appears about the vehicle roll axis. This fact can be utilized to synthesize a simple and

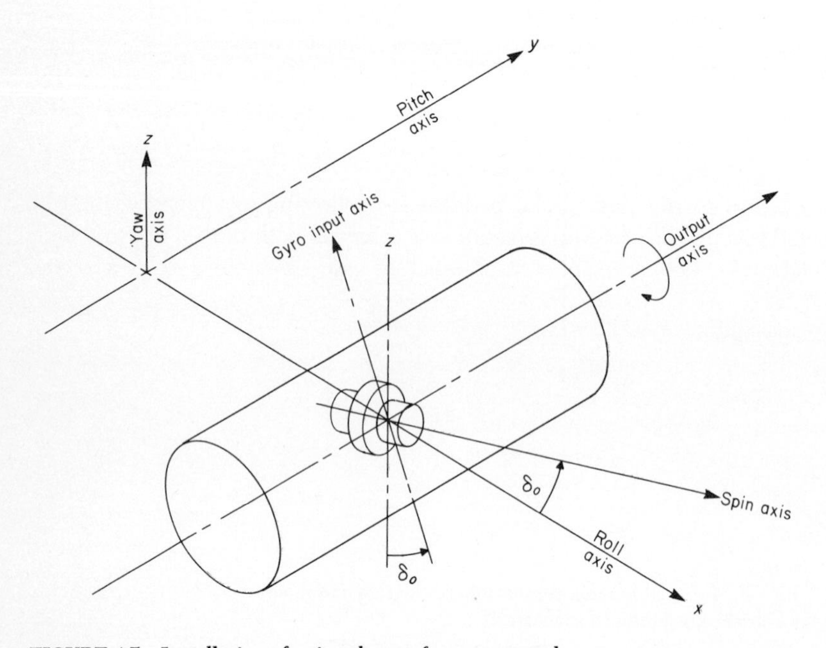

FIGURE 4-7 Installation of azimuth gyro for yaw control.

yet effective yaw-control system. The single-axis integrating gyro is used as the sensor for yaw motions about the local vertical. The gyro is mounted in the vehicle as shown in Fig. 4-7. Here the output or gimbal axis is along the vehicle pitch axis. The nominal position of the gimbal at which the pick-off signal is null is such that the rotor spin axis is at an angle δ_o from the vehicle roll axis. Hence, the gyro input axis is nominally at an angle δ_o from the vehicle yaw axis in the roll-yaw plane.

The gyro pick-off signal is proportional to the integral of angular rates about the gyro input axis. If δ_o is small, the gyro pick-off signal is predominantly proportional to the integral of yaw rates and thus provides a signal with good frequency response for control of yaw motions. At the same time the gyro pick-off signal is proportional to a component of the integral of roll rate. If there is a yaw error due to initial misalignment with orbit plane or due to gyro drift, the roll-rate signal will contain a continuous component of the inertial-rate signal. The gyro pick-off signal resulting from the integration of the continuous roll rate will activate yaw-control torques to correct the yaw error.

Initial misalignment with the orbit plane will be corrected with a time constant τ (the predominant time constant of the transfer function):

$$\tau = \frac{1}{\Omega \tan \delta_o} \tag{4-3}$$

Gyro drift will cause the vehicle to yaw until the component of orbital pitch rate seen by the gyro is equal to the gyro drift rate. Thus gyro drift rate does not produce a continuously increasing error. In momentum-bias systems with gimbaled flywheels, the rate-integrating gyro is in effect replaced by the flywheel gimbal, which provides control torques; an explicit measurement of yaw error is not made.

Orientation to the Sun Solar paddles, for collecting solar energy, must be oriented toward the sun. Sun orientation is achieved with the aid of sun sensors which provide the necessary control signal. In some cases, third-axis (yaw) control is achieved not by sensing the orbital plane, but by obtaining the necessary control signal through sun sensors and computing the necessary corrections.

In general, the resulting geometry is complicated since the earth follows the ecliptic around the sun, which is inclined approximately 23.5° from the equatorial plane, and the satellite rotates around the earth. Appendix A examines the apparent motion of the sun with respect to an earth-oriented satellite with the third axis looking toward the sun.

Control-Torque Generation

The methods devised for providing control torques fall in two categories: inertial and expulsive. Inertial includes gimbaled flywheels, contrarotating flywheels, and simple flywheels. Expulsive methods usually differ from each other only in the working fluid or material expelled. One major advantage of inertially driven torques is that system weight does not increase as a function of time in orbit. Such a system is limited only by the lifetime of the moving components and the adequacy of the electric power.

The simplest inertial device is the variable-speed flywheel, the spin axis of which is fixed with respect to the vehicle. A set of three such flywheels, with spin axes mutually perpendicular, will provide reaction torques for three-axis control. A gimbaled flywheel in roll could provide control torques in yaw because of the roll-yaw coupling and thus would eliminate the need for an extra flywheel.

The major disadvantage of an all-inertial system of control is that the spacecraft is a conservative system; that is, any initial angular momentum is not removed from the system but is merely transferred from the spacecraft proper to the flywheels. If large initial spacecraft rates are anticipated, the flywheels must have a large moment of inertia or be capable of high speed, or both.

One reliable mass-expulsive system expels high-pressure gas through nozzles to obtain reaction force. The gas may be generated by burning fuel in a combustion chamber or by depleting a storage tank containing cold gas. This system is not conservative and can remove the initial spacecraft angular momentum. It cannot do this precisely, however, because of practical limitations on rate-measuring instruments.

In addition, expulsive systems are usually operated in an on-off control mode because of the unavailability of proportional control valves that have sufficiently low leakage rates.

Dead-band oscillations occur about the desired control value. Oscillation rates must be kept extremely low to maintain efficient use of propellant. This requires highly accurate rate information or the capability of metering small, precise impulses. Additionally, the cyclic changes in spacecraft momentum required for solar-collector orientation substantially increase gas consumption. Transfer from high spacecraft rates to the very low dead-band rates during initial stabilization and when the sun is reacquired also consumes a large quantity of fuel. Figure 4-8 is a block diagram of a stored-gas system.

The desirable features of the expulsive and inertial system can be combined. The flywheels are used to provide the cyclic changes of vehicle momentum required, while the gas system provides initial stabilization and removes cumulative momentum that may result from disturbances such as solar pressure, radiation pressure, magnetic-field interactions, and gravity-gradient effects. Cumulative momentum is then removed on the basis of measured flywheel speed; a metered impulse is released when this speed reaches a preselected value. The magnitude of the metered impulse is designed to reduce the flywheel speed to zero or cause a reversal of direction. A stepping motor or linear torquers rotate the solar paddles with respect to the vehicle.

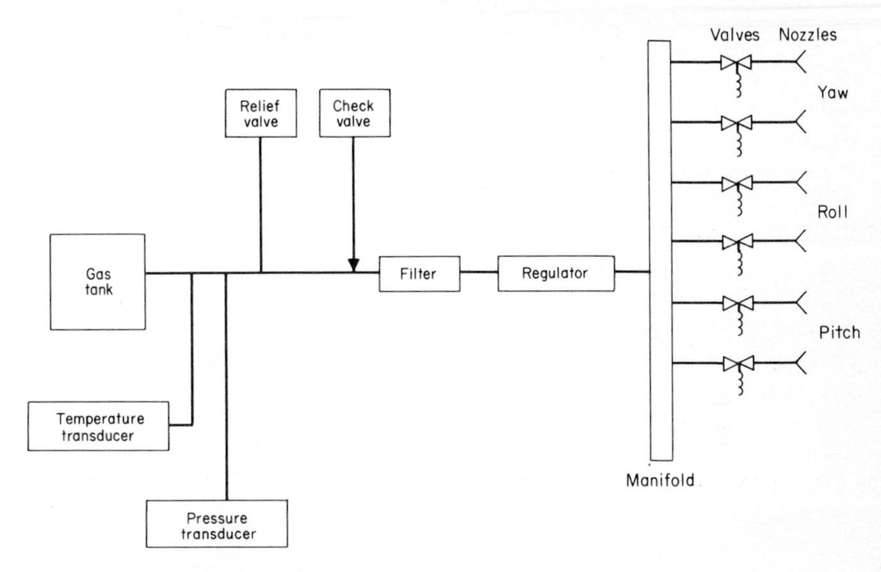

FIGURE 4-8 Stored-gas system (block diagram).

Transfer Characteristics of the Control Loops

Figure 4-5 is a block diagram of a control system utilizing the techniques that have been described thus far. The major parts of the system are flywheel motors, infrared (IR) sensors, a gyro, sun sensors, and a fuel-jet thrusting system.

The pitch- and roll-control channels are identical and basically consist of a combined expulsive and flywheel system. During initial orientation the expendable fuel system supplies control torques to orient the spacecraft to the earth and to reduce the initial separation angular rates. These torques will bring the spacecraft within a predetermined limit cycle with corresponding rate on the order of 0.1° per second. This is accomplished by actuating the thrust control through the threshold detector. When the input to the threshold detector drops below a predetermined level, the firing is cut off.

After initial stabilization, the system switches to the inertial mode, in which the control torques are derived from the motor-driven flywheel. The flywheel is accelerated in one direction by an error signal from the IR horizon sensor through the lead network, and thus the spacecraft accelerates in the other direction. The result is a replacement of the spacecraft momentum by flywheel momentum. The pitch and roll flywheels, therefore, speed up to a value determined by the spacecraft angular rate. The vector sum of the angular momenta of pitch and roll would remain constant if it were not for the effects of disturbances such as solar pressures, radiation pressures, magnetic disturbances, gravity gradients, intermittent operation of payload rotating components, and meteoroids.

This will result in an increased flywheel speed to absorb the additional momentum. To prevent the flywheel speed from becoming excessive or saturating, the metered firing of the expulsive system provides an impulse which is triggered by the flywheel speed sensor. Figure 4-5 shows a tachometer whose voltage is proportional to the flywheel speed feeding the impulse gate circuit. When the flywheel reaches a certain speed, the impulse gate circuit activates the firing of the secondary propulsion system with a gate of predetermined duration. This metered impulse is calculated to cancel the momentum stored in the flywheel. AC induction motors were selected for driving the flywheels because they are more reliable than dc motors which require brushes, although brushless dc motors have been developed. Figure 4-9 is a block diagram of a flywheel attitude control system for pitch and roll.

The system response to an error signal E as a function of the complex frequency s is

$$\frac{\Theta_v}{E} = \frac{K_l K_m (1 + \tau s)}{I_v s(1 + \alpha\tau s)(1 + T_m s)} \tag{4-4}$$

If $K = K_l K_m / I_v$, the closed-loop response is

$$\frac{\Theta_v}{R} = \frac{K(1 + \tau s)}{\alpha\tau T_m s^3 + (T_m + \alpha\tau)s^2 + (\tau K + 1)s + K} \tag{4-5}$$

The response to a disturbance impulse T_o is

$$\frac{\Theta_v}{T_o} = \frac{(1 + \alpha\tau s)(1 + T_m s)}{s[(I_v T_m \alpha\tau)s^3 + (I_v \alpha\tau + I_v T_m)s^2 + (I_v + K_l K_m \tau)s + K_l K_m]} \tag{4-6}$$

This relationship is derived by placing $R = 0$ and $E = -\Theta_v$.

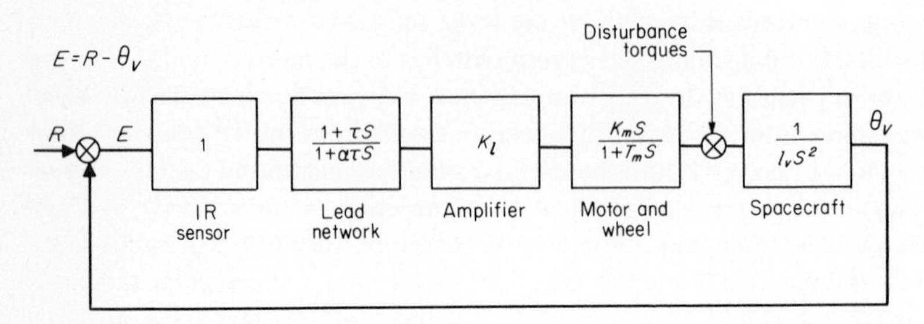

FIGURE 4-9 Flywheel attitude-control system for a three-axis-stabilized spacecraft.

The disturbances expected are such that they produce a torque on the spacecraft. A disturbing torque may be a continuous torque, such as the solar-pressure torque, the gravity-gradient torque, or the torque due to the earth's magnetic field. It may also be an impulsive torque, such as that caused by meteorites.

For T_o, an impulse, the steady-state error is

$$\Theta_{v,ss} = \frac{T_o}{K_l K_m} \quad \text{rad} \tag{4-7}$$

where

K_l = sensor and amplifier gain, V/rad

K_m = motor gain constant, lb·ft·s/V

$$= \frac{(I \text{ flywheel})(\text{torque constant})}{\text{damping factor}}$$

$$K = \frac{K_l K_m}{I_v}$$

$T_o = \Theta_v I_v$ = disturbance impulse, lb·ft·s

$\dot{\Theta}_v$ = spacecraft rate, rad/s

For a unit-step torque the steady-state error is infinite and obviously will lead to saturation of the wheel speed, since

$$T_o = \text{torque} \times t = \dot{\Theta}_v I_v = \omega_{FW} I_{FW}$$

$$\omega_{FW} = \frac{T_o}{I_{FW}}$$

The attitude error of the spacecraft (see "Three-Axis Stabilization," this chapter) builds up as a function of the wheel speed required to absorb the initial conditions and the cumulative disturbances.

The size of the flywheel and the top speed will determine the nominal time between saturations of the flywheel. This will be mainly a consideration of disturbances, and most of these are not affected by the attitude error.

The magnitude of the allowable attitude error in each of the control channels will be determined by the system requirements.

The transfer functions of the third-axis gyro system can be defined by calling ψ the yaw angle around the local vertical, θ the angle about the horizontal axis perpendicular to the orbital plane, and ϕ the angle about the horizontal axis in the orbit plane. The basic moment relationship about the gyro gimbal axis, for an inertial spacecraft rate Ω and small attitude errors, is given by the relationship

$$\Sigma M = I(\ddot{\theta} + \ddot{\delta}) + H(\Omega\phi \cos \delta_o + \dot{\psi} \cos \delta + \Omega\psi \sin \delta_o - \sin \delta_o) + D\dot{\delta} = 0 \tag{4-8}$$

Then the transfer functions of the gyro with respect to the angles ψ and ϕ will be

$$\frac{\delta}{\psi} = \frac{(H/D)\cos\delta_o(s + \Omega\tan\delta_o)}{s[1 + (I/D)s]} \tag{4-9}$$

$$\frac{\delta}{\phi} = \frac{(H/D)(-s\sin\delta_o + \Omega\cos\delta_o)}{s[1 + (I/D)s]} \tag{4-10}$$

The gyro signal caused by ϕ produces a yaw correction. At frequencies below the crossover frequency,

$$\frac{\psi}{\phi} = \frac{(\cot\delta_o)[1 - (s/\Omega)\cot\delta_o]}{1 + (s/\Omega)\tan\delta_o}\,\frac{\psi}{\psi_o}$$

where ψ/ψ_o is the closed-loop transfer function of the yaw loop. The steady-state yaw error will be

$$\frac{\psi}{\delta_{\text{drift}}} = \frac{1}{(H/D)\Omega\sin\delta_o} \tag{4-11}$$

and the predominant time constant will be

$$\tau = \frac{1}{\Omega\tan\delta_o} \tag{4-12}$$

where

- I = moment of inertia of gyro about its gimbal axis
- H = angular momentum of gyro rotor
- D = damping of gyro about gimbal axis
- δ = gimbal deflection from nominal position
- δ_o = nominal angle between gyro spin axis and spacecraft roll axis

If ω_x, ω_y, and ω_z are the spacecraft angular rates about the roll, pitch, and yaw axes, respectively, these rates will be given in terms of the Euler angles ϕ, θ, and ψ as follows:

$$\omega_x = -\Omega\psi + \dot{\phi} \qquad \omega_y = \Omega\phi + \dot{\psi} \qquad \omega_z = -\Omega + \dot{\theta}$$

for small angular deviations.

Two-Sensor Control System with Momentum Bias

A three-axis-stabilized spacecraft in a synchronous orbit will have the pitch axis (i.e., the axis normal to the orbital plane) maintain at all times the same direction with respect to inertial space. This is obvious since the pitch axis must remain normal to the orbital plane and the orbital plane is the earth's equatorial plane, which has, for all practical purposes, a fixed direction in inertial space. It is therefore constructive to operate the pitch flywheel nominally not at zero speed, i.e., zero momentum, but at some preselected, relatively high momentum value called the momentum bias. As a result, the spacecraft will act in a manner similar to the spin-stabilized spacecraft depicted in Fig. 4-2, the only difference being that the angular momentum vector is defined by the spinning flywheel instead of the spinning spacecraft itself.

The pitch-axis attitude signal is derived by a horizon sensor. By operating the flywheel along the pitch axis with a momentum bias at all times, the spacecraft could acquire the necessary stiffness to control the roll and yaw angles. In addition, as a result of gyroscope properties, the spacecraft is subject to yaw-roll cross-coupling. However, as in the case of the spin-stabilized spacecraft, it will also be subject to nutation. By obtaining the differential signal—through a lead network—from a roll-axis horizon sensor, the roll or yaw errors could be controlled by firing the secondary yaw and roll propulsion thrusters. As a result, in roll and yaw the spacecraft will be on a controlled limit cycle with dead bands designed to be within accepted limits, i.e., within a small fraction of a degree.

The control system described above requires only two sensors, namely, the pitch and roll horizon sensors. The nominal momentum bias of the pitch flywheel provides pitch control and yaw-axis attitude restraint.

There are several variations of this control concept, which is called the "momentum-bias attitude-control system." These variations of the momentum-bias system have been studied extensively and have actually been implemented, including systems with one, two, and three flywheels. Certain configurations also include flywheels gimbaled with one or two gimbals. In addition, coils interacting with the earth's magnetic field have been studied and used to provide desaturation torques in place of torques provided by the secondary propulsion thrusters. Spacecraft with momentum-bias attitude-control systems sometimes employ sun sensors for tight yaw control during the operation of the stationkeeping thrusters.

ELECTRIC POWER

Solar energy converted to electricity through photovoltaic cells (solar cells) is the primary source of power in communications satellites. Storage batteries take up the load during the periods when the spacecraft is in the earth's shadow.

Solar Arrays

The main element of the solar array is the photovoltaic cell, a semiconductor device. The cells absorb photons when exposed to sunlight (solar radiation) and produce a voltage potential across the junction of the semiconductor.

The most commonly used solar cell is the silicon cell with an *np* junction.

The open-circuit voltage generated across the junction is very small, about 0.4 to 0.8 V, and the short-circuit current is about 140 mA. As a result, a substantial number of cells must be connected in series to produce a practically useful voltage and current supply. The efficiency of converting solar energy to electricity is small, but substantial improvements have been made in recent years, increasing this efficiency from 5 percent to the 12 percent level. New, more advanced cells, such as gallium arsenide, promise much higher efficiency.

This efficiency depends on several factors, such as the temperature of the cell, the time interval during which the cell has been exposed to space radiation (particle radiation), and the contact resistance.

A solar array usually consists of a lightweight structure on which a substantial number of solar cells are bonded in a combination of series and parallel connections in order to produce the desired voltage and current. A protective glass covering over the cells, transparent to solar radiation, serves as a partial protective shield against the effects of space radiation, which degrades the efficiency of the array. Since total protection is neither feasible nor economical, the total power output of the array decreases with the time of exposure to a radiation environment such as the one that exists in space.

The solar constant E_o at the earth's vicinity is approximately 1.39 kW/m^2. (Some recent measurements indicate that it may be 1.35 kW/m^2.) Consequently a solar array with an area $A_s = 2$ m^2 and total efficiency $e = 5$ percent will produce at the beginning of life (BOL)

$$W = eE_oA_s = 0.05 \times 1.39 \times 2 \text{ kW} = 139 \text{ W}$$

However, if the radiation is not normal to the surface of the array, one has to multiply by the cosine of the appropriate angle. With a 12-mil-thick glass cover, this power may degrade as much as 20 percent in a period of 2 years in the space environment because of exposure to space radiation.

In 1980, state-of-the-art solar-cell manufacturers have reached efficiencies for silicon as high as 15 percent, normal production at the 12 to 13 percent level. In 1980 gallium arsenide efficiencies reached as high as 18 percent in limited production. (These efficiencies have been measured at 28°C.) It is expected that during the eighties, silicon cells will reach 17 percent efficiency and gallium arsenide cells will exceed 20 percent.

The graph in Fig. 4-10 depicts the effect on cell output power of the integrated flux of electrons with energy of 1 MeV. From measured data in space, one could compute the total integrated flux over the life of the spacecraft and derive the end-of-life (EOL) power of an array. For example, in a geostationary orbit, the 1-MeV electron flux over a period of 7 years is expected to be approximately 10^{15} electrons per square centimeter. This assumes good back protection from a rigid substrate. If this protection did not exist, as in semirigid arrays, the total flux would be twice as much.

The main problem in the design of a solar array are weight, thermal cycling, and radiation protection. The orbital loads are usually small so that they do not

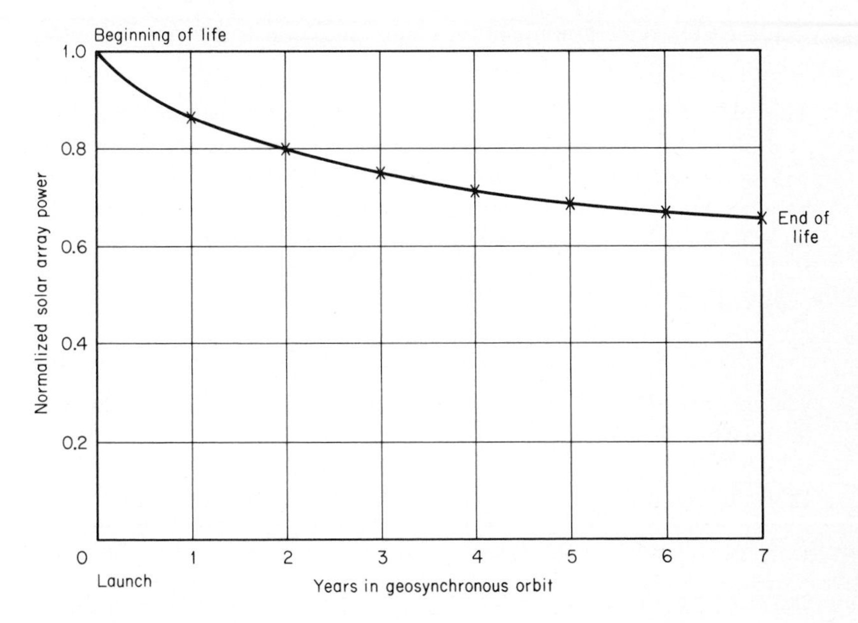

FIGURE 4-10 Normalized solar-array degradation vs. years from launch.

influence the design. Of course, the structure must be capable of withstanding the forces produced by the attitude and orbit control systems. It must also survive the launch environment when stowed within the launch vehicle. This includes acceleration, vibration, accoustic noise, etc. After injection, for deployable arrays, the array must deploy without damaging itself.

Thermal cycling may cause substantial structural problems, especially when dissimilar materials are being used and the bonding materials are sensitive to such thermal stresses.

The array orientation system is usually a system with one degree of freedom, rotating the array once a day about the north-south axis so that it continuously faces the sun at the optimum angle. Because of the 23.5° inclination of the earth's ecliptic, the angle of incidence of the sunlight on the array varies from 0 to 23.5° between equinox and solstice. This results in approximately 8 percent loss of power. The transfer of power from the rotating array to the spacecraft requires the use of slip rings. The proper operation of slip rings in a hard vacuum always presents a problem.

EXAMPLE 4-1

A cylindrical spin-stabilized spacecraft placed in geostationary orbit has a diameter $D = 2$ m. The solar-cell efficiency of the cylindrical solar array is 10 percent, and at the EOL, i.e., after 7 years, it degrades by 35 percent.

If at the EOL the requirement for minimum power is 500 W, what is the minimum height of the cylinder?

The solar energy constant at the vicinity of the earth is approximately $E = 1.39\ \text{W/m}^2$ for normal incidence of the solar rays on the array. This will be the case during the equinox periods. However, the worst case will be when the earth is at one of the solstices, when the equatorial plane of the satellite orbit is at an angle of 23.5° with the arriving solar rays. At that time, if h is the height of the solar array, the total effective power collected will be

$$P_s' = hDE \cos 23.5° \tag{4-13}$$

The product hD represents the effective area of the cylinder presented to the solar-ray flux at any given time. Taking into account the 0.1 efficiency in converting solar energy to electric energy and the 0.35 degradation at EOL, the required minimum power of 500 W will be obtained if

$$P_s = (0.1)(0.65)P_s' = 0.500\ \text{kW}$$

and from Eq. (4-13),

$$h = \frac{0.5}{0.1 \times 0.65 \times 2 \times 1.39 \times 0.92} = 3\ \text{m}$$

Therefore a cylinder height of $h = 3$ m will be required at least.

The EOL power during equinox will be

$$P_E = \frac{P_s}{\cos 23.5°} = \frac{500\ \text{W}}{0.92} = 543.5\ \text{W}$$

i.e., from equinox to solstice there is approximately 8 percent difference.

The BOL power during equinox will be the maximum power at the array, that is, 543.5/0.65 or 836 W, while the BOL power during solstice will be 8 percent less, or 769 W.

It is interesting to note that while an area equal to $\pi Dh = 18.84\ \text{m}^2$ must be covered with solar cells, the effective area of the solar array is only $Dh = 6\ \text{m}^2$. This is a result of the cylindrical shape. Flat solar paddles on three-axis-stabilized spacecraft require less area by a factor of $\pi = 3.14$ for the same amount of output power.

Sun Eclipse

A spacecraft in synchronous equatorial orbit will occasionally fall within the earth's shadow. While the earth shadows the sun from the solar array, storage batteries provide the power to various loads. These batteries will discharge dur-

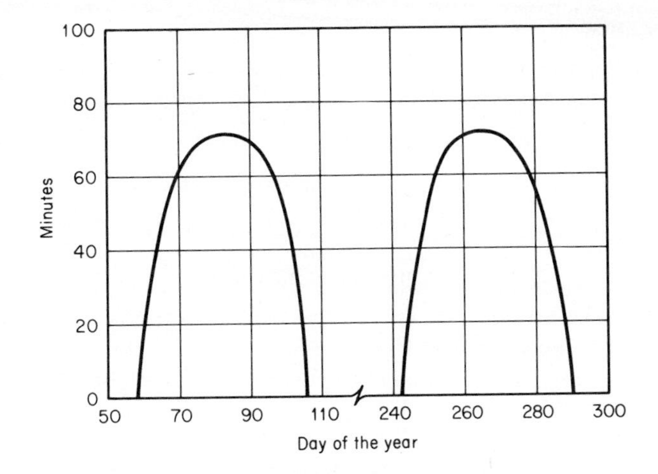

FIGURE 4-11 **Satellite eclipse time.**

ing the eclipse period and charge again from the solar array during the illumination periods. The eclipses take place during the autumnal and vernal equinoxes and last anywhere from 10 to 72 min. The eclipse begins approximately 23 days prior to equinox and ends 23 days after the equinox. Figure 4-11 depicts the eclipse time in minutes vs. the day of the year.

Storage Batteries

Storage batteries provide electric power during the time that the solar array cannot be illuminated by the sun. The depth of discharge during that period is an important design parameter and influences the selection of the type of batteries and the design of the overall power subsystem. In present systems, the weight of the batteries is approximately half the weight of the total power subsystem.

Nickel-cadmium batteries have been extensively used in satellite applications. With 1982 state-of-the-art Ni-Cd batteries, one can obtain 6 to 8 W·h/lb. A significant improvement is expected in the coming years with the use of nickel-hydrogen batteries, which are rated at 10 to 15 W·h/lb. The life of the battery is limited by both the depth of discharge and the number of discharge cycles. Depths of discharge of 60 percent are normally used with Ni-Cd batteries for a life expectancy of 7 to 10 years in geostationary orbit.

Power Subsystem Electronics

A power subsystem requires a number of functions performed by electronic means. Voltage regulators, current limiters, overload protection, switching relays, etc., are some of the components employed by a power subsystem.

SECONDARY PROPULSION

The main functions of the secondary propulsion system in most of the communications spacecraft that operate from a geostationary orbit are the following:

1. *Final injection into the synchronous equatorial orbit.* The propulsion module for this function is often known as the apogee kick motor (AKM), and it is used to correct any residual orbit inclination so that the orbit becomes equatorial, as well as any orbital velocity deficiency so that the orbit becomes circular. These corrections are required after the burnout of the upper stage of the main launch vehicle. In many instances this propulsion unit, an integral part of the spacecraft, utilizes a solid-propellant motor that is fired once to take care of both corrections. In some instances, when the main launch vehicle has a restartable upper stage capable of orbital maneuvers, the AKM is not required. For example, the Titan IIIC transstage can perform these functions. But for most other cases the AKM is required, and it is approximately one-half the total weight (before burn) of the spacecraft.
2. *Station keeping and positioning in station.* This propulsion system is required to last over the life of the satellite. Its main function is to place the spacecraft in position and to keep it in the preassigned orbital spot. There are several disturbing forces due to such factors as the oblateness of the earth and the gravitational attraction of other celestial bodies. For example, the orbital inclination plane will drift approximately 0.86° per year (unless corrected) because of lunar and solar gravitational forces. A liquid propulsion system utilizing a bipropellant or monopropellant fuel is utilized for this propulsion function. Monopropellant hydrazine (N_2H_4) is quite often used—stored in dual tanks. The amount of station-keeping fuel usually determines the useful life of the spacecraft. Of course the solar-cell degradation due to the radiation environment is the other critical factor in determining the spacecraft's useful life.
3. *Attitude control.* Propulsion is required for correcting the spin-axis precession in spinning spacecraft as well as for maintaining the spin rate. Three-axis-stabilized spacecraft require propulsion for unloading the momentum package when it saturates. Both types require secondary propulsion for attitude maneuvers, such as sun acquisition.

Very often the station keeping and the attitude secondary propulsion utilize the same unit with properly located jets. Stored cold nitrogen gas systems are occasionally used for specific functions.

Table 4-1 gives a typical weight summary for a second-generation commu-

TABLE 4-1 TYPICAL SPACECRAFT SECONDARY PROPULSION WEIGHT BREAKDOWN

Spacecraft and propulsion elements	Weight, lb
Spacecraft lift-off weight	2800
AKM propellant	1330
N_2H_4 trim burns	10
N_2H_4 station-keeping/attitude control	300
Spacecraft EOL weight	1160

nications spacecraft (1982 time period). From this table it is obvious that more than half of the lift-off weight at the ground consists of secondary propulsion equipment.

SPACECRAFT ANTENNAS

The most commonly used spacecraft antenna is the parabolic dish for area coverage. From synchronous altitude the maximum beam width that one can use is approximately 17.3° for a "global beam" (see Fig. 2-5).

For spacecraft positioned to communicate over the North American continent, an area beam of 17.3° will cover the contiguous 48 states of the United States as well as Hawaii, Alaska, and Puerto Rico when the spacecraft is positioned between 139 and 50°W longitude. Figure 4-12 shows the orbital arc

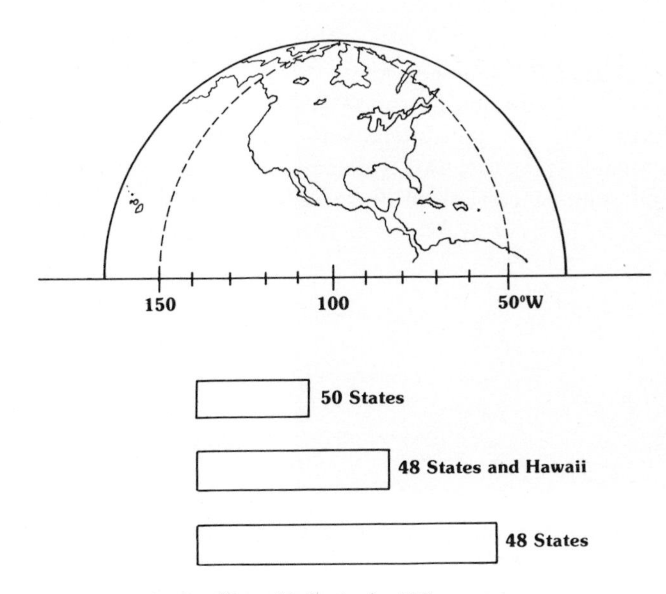

FIGURE 4-12 Satellite orbit limits for U.S. coverage.

requirements for illuminating the major U.S. areas for domestic communications with a 17.3° global beam. Figure 4-13 depicts Westar IV illuminating the United States with a global beam. When it is desirable to concentrate the radiation in a smaller area, regional beams with beam width smaller than 17.3° are used. Finally, spot beams with beam widths of about 1 to 2° are used for covering a specified location or city.

Telemetry, tracking, and command subsystems often make use of omnidirectional antennas so that the spacecraft can be commanded from the ground before the attitude stabilization phase is completed or in the event that stabilization is lost. An antenna may be fixed or steerable. Spin-stabilized spacecraft utilize mechanically or electronically despun antennas oriented toward the earth. Spot-beam antennas are usually steerable through commands from the ground. Electronic techniques for focusing spot beams have been perfected, and a number of today's spacecraft employ electronically steerable beams.

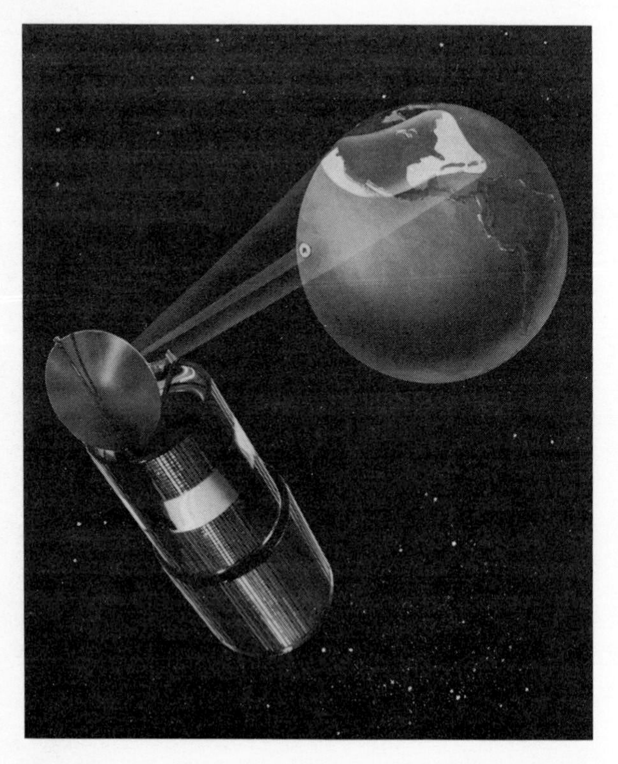

FIGURE 4-13 Westar IV illuminating the United States (artist's concept). ***(Photograph courtesy of Hughes Aircraft Company.)***

Usually the performance of an antenna is characterized from the ratio G/T, i.e., the gain G over the system noise temperature T. This ratio is used to describe the receiving sensitivity. A typical value for a satellite area beam coverage antenna is $G/T = -6$ dB/K. The antenna gain G describes the ability of an antenna to direct power toward a specific direction, in contrast to an omnidirectional antenna.

TELEMETRY, TRACKING, AND COMMAND

Telemetry

In general, the telemetry subsystem consists of various sensors and transducers, the telemetry encoder, the transmitter, and the telemetry antenna.

Various sensors, such as horizon sensors and sun sensors, are used to obtain information about the attitude of the spacecraft in relation to selected celestial bodies. Other sensors register such environmental information as magnetic field, radiation, and meteor impacts. Special transducers record various spacecraft measurements such as temperature, power-supply voltages, and stored-fuel pressure.

All this information is properly formatted at the required signal levels, usually in digital form, and is encoded for transmission. The telemetry encoder multiplexes the various analog and digital channels carrying the information from the various sensors and transducers. In addition, it provides the appropriate signal conditioning. The output of the encoder modulates a subcarrier or carrier of the transmitter for transmission through the telemetry antenna. The telemetry transmitter is usually a phase-modulating source operating at frequencies designated specifically for spacecraft telemetry operations by international treaties. The transmission usually can take place either through an omnidirectional antenna or through a directional antenna. The omnidirectional antenna is used when the spacecraft has not yet been fully stabilized or during emergencies in which stabilization has been lost. A bicone antenna can be used for such a purpose.

Command

The command subsystem consists of the command receiving antennas, the command receiver, the command decoder, and the various driver amplifiers which distribute the appropriate signal for the execution of the command. A high degree of redundancy is used to ensure long-life operation of this subsystem. Omnidirectional antennas are used in nonstabilized spacecraft. The command receiver receives and demodulates the commands from the ground. Again, the frequencies used for this function are specifically allocated by international treaties. In many cases, for reasons of security, the command signal is encrypted.

Various sophisticated techniques are used to eliminate the possibility of unauthorized commands.

The command decoder converts the demodulated signals to the appropriate pulses which operate the various drivers for the execution of the command.

Tracking

Satellite positioning beacons provide tracking signals either through the telemetry subsystem or through the main communication subsystem (by inserting pilot frequencies), or through special tracking antennas.

SPACECRAFT SYSTEMS: GENERAL CONSIDERATIONS

The total electric power requirement is a very important factor in determining the stabilization scheme to be used, i.e., spin or three-axis. For the same solar-power-collecting capability, the spinning spacecraft requires a little over 3 times the solar-array area required by a three-axis-stabilized spacecraft. In order to increase the cylinder surface (i.e., solar-array area) and still maintain a relatively low volume during the launch phase, spinning spacecraft have been designed with "telescoping" extensions of the cylinder which are deployed after the spacecraft has been injected into orbit. However, the increasing cylinder height affects the moment of inertia in such a manner that spinning around the

TABLE 4-2 WEIGHT COMPARISON OF EQUIVALENT 24-TRANSPONDER SPACECRAFT

Element	Weight, lb	
	Three-axis-stabilized	Spin-stabilized
Structure: mechanical assemblies and harness	215*	280
Thermal control	35	55
Propulsion (fuel not included)	60	40
Electric power	280	220
Attitude control	75	55
Antennas	100	120
Communications repeaters	250	220
Telemetry, tracking, and command	50	65
Apogee kick motor	70	70
Other	30	10
Total	1165	1135

*Does not include the weight of solar paddle structure. This weight is included in the electric power entry.

FIGURE 4-14 Westar IV. *(Photograph courtesy of Hughes Aircraft Company.)*

main cylinder axis becomes less stable. For electric power demand of over 1000 W, the spinning design is questionable. Table 4-2 compares the weight distribution of a typical spin-stabilized spacecraft with that of a typical three-axis-stabilized spacecraft at this low range of power generation.

A traveling-wave-tube (TWT) amplifier transponder has an electric power efficiency of approximately 0.3. As a result, a 10-W TWT amplifier will require approximately 35 W of electric power while a 5-W TWT amplifier will require approximately 17 W. For the communications subsystem alone, a 24-transponder spacecraft will require 840 W of electric power to support 10-W TWT amplifiers and 420 W to support 5-W TWT amplifiers. This is an EOL requirement. Since additional power is required for the operation of the rest of the subsystems, a spin-stabilized spacecraft will reach the limit of its capability if a design life of 10 years is desired. If adequate margins are to be maintained, a spinning satellite should be limited to 24 transponders with 7.5-W TWT amplifiers and a life expectancy of 10 years.

Figures 4-14 and 4-15 depict the Westar IV "dual-spin" spacecraft and the internal arrangement of the equipment in the spacecraft.

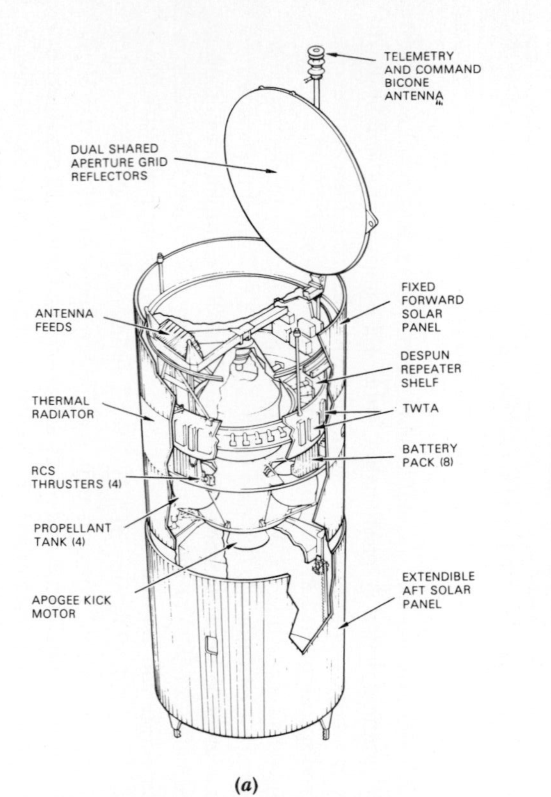

(*a*)

(*b*)

FIGURE 4-15 (*a*) Location of Westar IV components. (*b*) Westar IV as compared with smaller Westar I. *(Photograph courtesy of Hughes Aircraft Company.)*

REFERENCES

Dougherty, J., K. L. Lebsock, and J. J. Rodden: "Attitude Stabilization of Synchronous Communication Satellites Employing Narrow Beam Antennas," *Journal of Spacecraft and Rockets*, vol. 8, no. 8, August 1971, pp. 834–841.

Frieder, M. A.: "Angular Momentum and Nutation Damping," Annual Rocky Mountain Guidance and Control Conference, AAS 82-003, 1982.

Keigler, J. W., W. J. Lindorfer, and L. Muhlfelder: "Momentum Wheel Three-Axis Attitude Control for Synchronous Communications Satellites," *AIAA Selected Reprint Series*, vol. XVIII: *Satellite Communications System*, Ivan Kadar (ed.), 1976.

Lebsock, K. L.: "High Pointing Accuracy with a Momentum Bias Attitude Control System," AIAA Paper 78-569, American Institute of Aeronautics and Astronautics, New York, 1978.

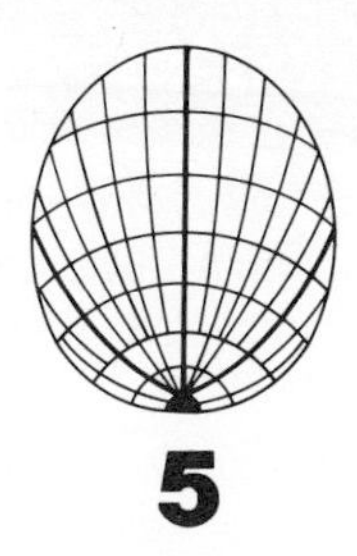

5

SPACECRAFT COMMUNICATIONS SYSTEM

The communications system on board a spacecraft is usually characterized by certain key parameters: namely,

- Frequency (receiving and transmitting)
- Bandwidth utilized by the total number of channels
- Number of discrete bulk communications channels (called "transponders"), each one of a certain bandwidth and power

A repeater with frequency translation and traveling-wave-tube (TWT) amplifiers is the most commonly used configuration on board a spacecraft. The repeater is usually transparent to the signal transmitted from the earth stations, with the uplink frequency being higher than the downlink frequency. The frequency difference minimizes interference between the transmitted and received signals. Occasionally the various transponders of the repeater are configured in a way to optimize the communications links for the expected mode of access by earth stations. Each transponder may be accessed simultaneously by many carriers, with each carrier operating at a different frequency in what is known as frequency-division multiple access (FDMA). Alternatively, each carrier may access the whole transponder for a short period of time specifically allocated to that carrier, in what is known as time-division multiple access (TDMA).

In contrast to the transparent or "passive" repeaters, repeaters with signal-processing capability could be used. The on-board processing of the received signals could take place at radio-frequency (RF) level, intermediate-frequency (IF) level, or at baseband.

On-board switching usually is done at RF level by an RF switch matrix that interconnects each received signal to the desired output port. An example of such a system is the satellite-switched TDMA (SS-TDMA) concept discussed in Chap. 14.

Signal processing at baseband requires complete demodulation, detection, and then reconstruction of the downlink carrier.

The objective of on-board processing is to achieve higher capacity, better connectivity and flexibility, and reduced interchannel interference. Other advantages may also be realized, depending on specific network configurations.

In this chapter we will examine subsystems utilizing transparent repeaters with frequency translation.

COMMUNICATIONS FREQUENCIES AND BANDWIDTH CONSIDERATIONS

Frequencies for satellite communications are allocated internationally at World Administrative Radio Conferences (WARC) under the auspices of the International Telecommunications Union (ITU). Agreements reached in the international conferences after ratification by each participating country have the force of a treaty. In the United States, the Federal Communications Commission has the responsibility of managing the frequency spectrum in accordance with these agreements. Table 5-1 lists the frequency allocations for satellite communications.

TABLE 5-1 FREQUENCY ALLOCATIONS (Values in MHz)

Use	Downlinks	Uplinks
COMMUNICATIONS SATELLITES		
Commercial (C band)	3,700–4,200	5,925–6,425
Military (X band)	7,250–7,750	7,900–8,400
Commercial (K band):		
Domestic	11,700–12,200	14,000–14,500
International	10,950–11,200 11,450–11,700 17,700–21,200	27,500–31,000
OTHER ALLOCATIONS		
Broadcast	2,500–2,535	2,655–2,690
Maritime	1,535–1,542.5	1,635–1,644
Aeronautical	1,543.5–1,558.5	1,645–1,660
TELEMETRY, TRACKING, AND COMMAND		
137.0–138.0, 401.0–402.0, 1,525–1,540		

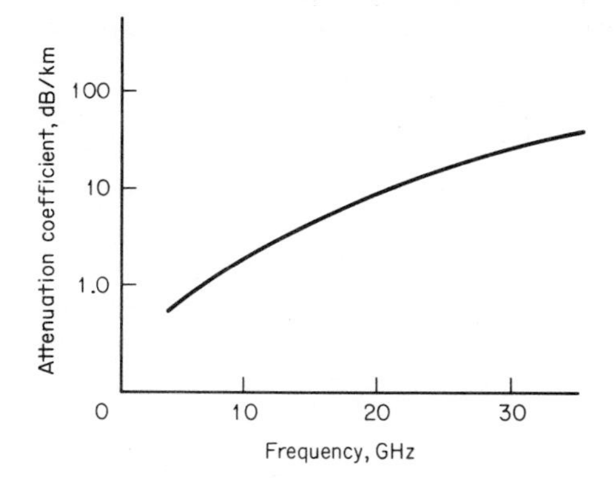

FIGURE 5-1 Rain attenuation (4 in/h, T = 10°C).

Characteristically, frequencies at the lower part of the spectrum are less susceptible to attenuation by rain and require less expensive equipment than frequencies at the higher end of the spectrum. Figure 5-1 gives the attenuation coefficient vs. frequency for rain of 4 in/h. However, as the spectrum at the lower frequencies becomes saturated by an excessive number of users, higher frequencies become the only alternative.

Many systems utilize frequency reuse techniques in order to expand the available bandwidth. This is accomplished by employing orthogonal polarizations for channels operating in the same frequency band. Isolation of over 30 dB can be achieved by this method. Directional beams (spot beams) can also occupy the same frequency band by making use of spatial isolation (see "Spacecraft Antennas" in Chap. 4).

POLARIZATION AND FREQUENCY REUSE

A propagating wave consists of two vectors, the electric-field vector **E** and the magnetic-field vector **H**. In general each of these vectors has three components, each with respect to an arbitrary orthogonal system of reference x, y, z. In the case where the vectors **E** and **H** have no components along the direction of the wave propagation, we have a plane wave. If x is the direction of propagation of a plane wave, then **E** and **H** are located on a plane perpendicular to the direction of propagation x having components only along the y and z directions. In effect, these **E** and **H** vectors are located in the yz plane but, in general, vary with time (for example, obeying a sinusoidal law) and satisfy the relationship

$$E_x = H_x = 0$$

For simplicity let us consider only the electric-field vector **E** on this plane. If both of the components E_y and E_z are present and in phase with each other, the vector **E**, although changing magnitude with time, does not change direction. This is obvious since the ratio E_z/E_y remains constant because of the in-phase assumption. The angle that the vector **E** forms with the y axis is constant, and it can be expressed as $\tan^{-1} E_z/E_y$. This wave is said to be linearly polarized along this constant direction of the vector **E**. If **E** is linearly polarized along a horizontal axis, the wave is called horizontally polarized (HP). If **E** is linearly polarized along the vertical axis, the wave is called vertically polarized (VP). These two polarizations are orthogonal. If two waves, one horizontally polarized and one vertically polarized, were of the same sinusoidal frequency and were also propagating simultaneously, they would not interfere with each other. This again is obvious, since no component of the horizontal wave will project in the vertical direction and vice versa. This principle is employed extensively for reusing the frequency spectrum allowed for satellite transmission. Of course, if in the process one of the waves loses polarization, then it will interfere with the other wave; for example, if the horizontal wave is slightly rotated so that a small vertical component appears, then it interferes with the vertical wave. As long as this interfering component is kept low with respect to the main signal (i.e., less than 30 dB), the interference is negligible.

If the components E_y and E_z are not in phase, then the direction of the resultant vector varies with time. In the case in which the locus of the tip of **E** describes an ellipse, the wave is said to be elliptically polarized. In addition, if the tip rotates clockwise, we obtain a right-hand (RH) elliptical polarization. For the opposite rotation we obtain a left-hand (LH) elliptical polarization. In the special case in which the tip of **E** describes a circle, we obtain a circular polarization (CP).

Right-handed circular polarization (RHCP) does not interfere with left-handed circular polarization (LHCP). These two polarizations could also be employed in a frequency-reuse system.

In the case of elliptical polarization the main axis of the ellipse determines the polarization orientation. The ratio of the major axis to the minor axis in decibels is called the "axial ratio." If the axial ratio is 0, we have circular polarization; if the axial ratio is ∞, we have linear polarization.

Depolarization producing interference in frequency-reuse systems has a number of causes, some of which are

- Impedance mismatch at the antenna ports
- Differential phase shift introduced by the polarizer
- Differential amplitude effects on the wave components
- Waveguide anomalies such as dispersion of beam and differential phase shift
- Reflector asymmetries and subreflector misalignment
- Faraday rotation and multipath effects

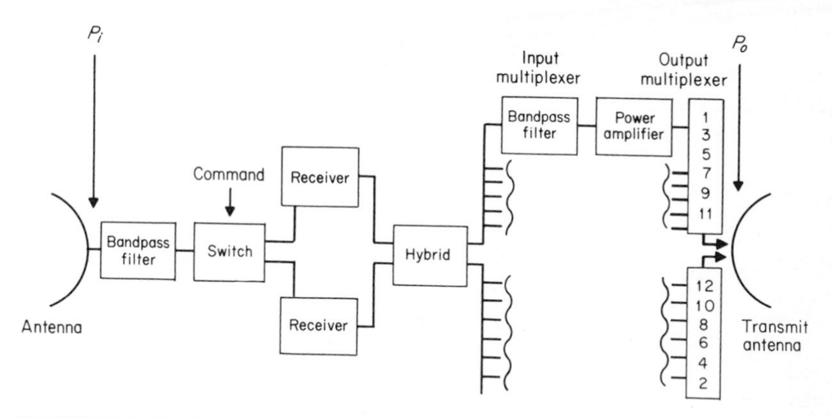

FIGURE 5-2 Twelve-transponder repeater.

TRANSPONDERS AND REPEATER CHANNELIZATION

A repeater is usually channelized both in frequency and power. For example, a repeater may utilize the 500-MHz bandwidth allowed to derive 12 channels of 36-MHz bandwidth per channel and 4-MHz separation between channels. With dual polarization, the repeater can provide 24 channels over the same 500-MHz bandwidth. Each channel utilizes its own TWT amplifier and is commonly referred to as a "transponder." At C band, 5- to 10-W TWTs have been used extensively. One of the above-described transponders is capable of transmitting an analog color television channel or approximately 1000 one-way analog telephone circuits. For a digital bit stream, this type of transponder is rated at slightly more than 60 Mb/s (single-carrier operation, QPSK).

A generalized block diagram of a repeater with redundant receivers and 12 power amplifiers—one for each transponder—is depicted in Fig. 5-2. The input power is P_i and the output power is P_o. The power radiated out of the transmit antenna is expressed in decibels referred to 1 W (dBW) as effective isotropic radiated power (EIRP).°

The wideband receiver consists of three elements: a low-noise amplifier, a frequency translator, and a driver amplifier, as depicted in Fig. 5-3. Low-noise amplifiers usually employ tunnel diodes or other low-noise semiconductor devices. Of course, there are numerous ways in which one can combine the various blocks in Fig. 5-2; for example, one can assume that the bandpass filters are combined with the amplifiers. For repeaters operating in the C band (4-GHz/6-GHz frequency band), which allows a 500-MHz bandwidth for satellite transmission, the frequency channelization in a 12-transponder repeater (such as the one in Fig. 5-2) would be as depicted in Fig. 5-4.

°International standards limit the power per unit area allowed to reach the earth's surface.

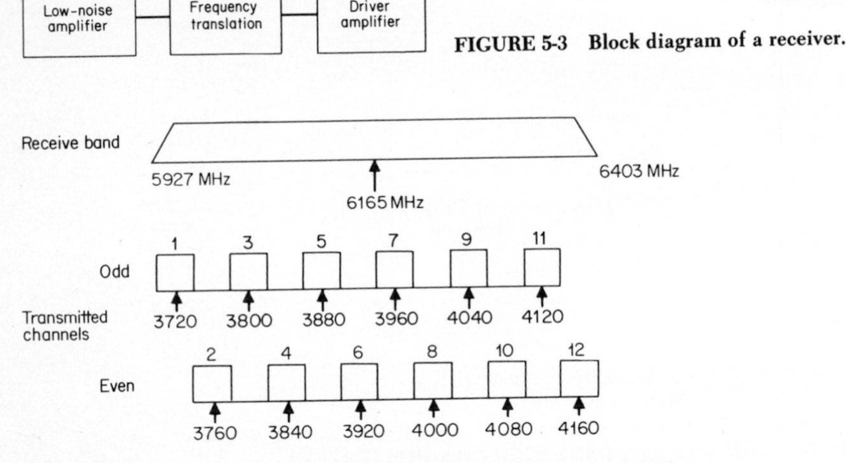

FIGURE 5-3 Block diagram of a receiver.

FIGURE 5-4 Receiving and transmitting bands of a 12-transponder C-band repeater.

In the case of a 24-transponder repeater, the same channelization occurs for each polarization; that is, 12 channels (1, 5, 9, 13, 17, 21/2, 6, 10, 14, 18, 22) are placed in horizontal polarization and 12 channels (3, 7, 11, 15, 19, 23/4, 8, 12, 16, 20, 24) are placed in vertical polarization. This interleaving arrangement enhances channel isolation in the input and output multiplexers by allowing much wider guard frequencies. In the case of frequency reuse, additional isolation is obtained by utilizing regional beams (more than one transmitting antenna) and/or spatially interleaving beams.

POWER AMPLIFIERS AND NONLINEARITIES

In most cases the power amplifier employs a traveling-wave tube which is the main source of nonlinearities in the repeater. As a result, in multicarrier operation, intermodulation products would be produced. A typical transfer characteristic of such a nonlinear repeater is depicted in Fig. 5-5.

The input power $P_{i,\max}$ that saturates the repeater amplifier is called maximum saturation power. For multicarrier operations, one must "back off" the operating point in order to restrict the operation in the linear region. In Fig. 5-5 the input power is backed off by BO_i, which corresponds to an output back-off BO_o, in order to achieve relative linearity for multicarrier operation. The magnitude of BO_i depends on the number of operating carriers.

The repeater does not usually have a power-control capability. Power control must be exercised from the transmitting earth stations. The efficiency of TWT amplifiers is about 30 percent. This efficiency is defined as the percentage equivalent of the ratio of the total output RF power at the fundamental frequency to the dc power provided by the power supply to the TWT amplifier. In addition to intermodulation products, a nonlinear amplifier of this kind could also introduce amplitude and phase distortion.

Quite often, all—or at least several—of the transponders in a repeater are identical. This allows flexibility for backup arrangements in case of failure as well as easy reconfiguration of the total system. Hybrid satellites with transponders operating in both C and K_u bands have also been deployed.

High-power amplifiers (HPAs) using solid-state devices rather than TWTs have been developed and introduced in operational satellites. Improvements in life and performance can be expected with the use of solid-state devices. The present-day limitation of these devices is the total output power that can be provided at microwave frequencies. Significant progress has been made in the development of gallium arsenide and high-power field-effect transistors (FETs) in the microwave range. Solid-state devices at 4 GHz have provided several watts output with gain in the range of 15 to 17 dB. Multistage solid-state power amplifiers with as high as 15 W output power in the 4-GHz frequency band were operational around 1981.

Traveling-Wave Tubes

The TWT is the key component of the microwave transmitter. It is the transponder output power amplifier, and very often it is also the intermediate driver amplifier. Because of its saturation characteristics, it is the largest contributor to repeater and overall transmission nonlinearities. In addition, having a cathode with limited lifetime, the TWT is another life-limiting component of the spacecraft.

A TWT consists of an electron gun with the appropriate beam-focusing magnets. The electron beam is guided at a beam collector in a glass envelope. A

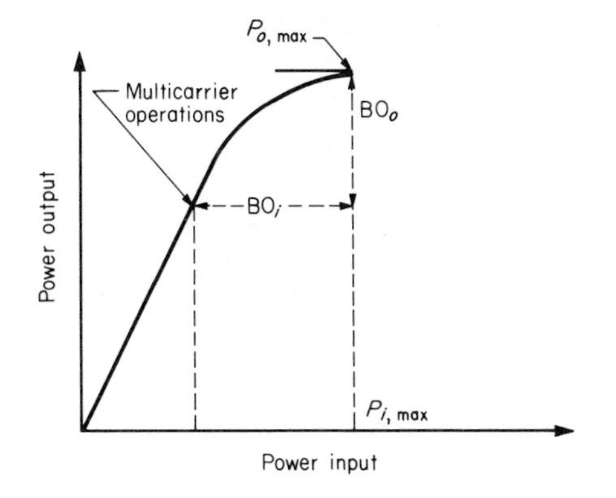

FIGURE 5-5 Nonlinear power characteristics of a TWT-amplifier repeater.

helix delay line encircles the beam, and its current interacts with the moving electrons of the beam, causing amplification of the RF signal applied to the TWT. Typical voltage requirements for efficient helix operation are in the range of 3 to 4 kV. The schematic in Fig. 5-6 depicts the main elements of a TWT.

As the result of an interaction process, part of the kinetic energy of the beam electrons is converted to RF energy propagated through the output coaxial cable. Two types of cathodes have been used in the design of electron guns for TWTs, namely, oxide-coated cathodes and dispenser-type cathodes. Oxide-coated cathodes utilize a base of nickel doped with small fractions of zirconium or magnesium and coated with an alkaline oxide such as barium oxide or strontium oxide. Oxide-coated cathodes operate at approximately 700°C and provide a current density in the range of 100 to 250 mA/cm^2 with a maximum lifetime in the neighborhood of 100,000 h. Dispenser-type cathodes utilize porous tungsten impregnated with an alkaline earth such as barium aluminates. These cathodes operate at temperatures on the order of 1000°C, providing a current density of up to 800 mA/cm^2. The life expectancy of the dispenser-type cathodes is projected to be 150,000 h. Improved technology for metal-to-ceramic vacuum seals has resulted in a better-controlled environment within the tube and has also contributed to longer TWT life.

Overall tube efficiency is defined as the ratio of maximum RF power output at the fundamental frequency to the electric power required by the TWT. This efficiency depends on the beam efficiency (i.e., the efficiency of the coupling of helix and electron current) and the collector efficiency (i.e., the efficiency of the energy recovery process). With the application of advanced collector configurations, the overall tube saturation efficiency has been significantly improved. In addition, with improved electron-beam focusing techniques, such as the utilization of periodic permanent magnet (PPM) structures, further overall efficiencies and weight reductions have been achieved in TWT technology. The beam efficiency (helix-to-beam current coupling) depends on the beam current, the helix voltage, and the beam-to-helix coupling impedance. The design of the ceramic rods supporting the helix is also important in achieving high beam effi-

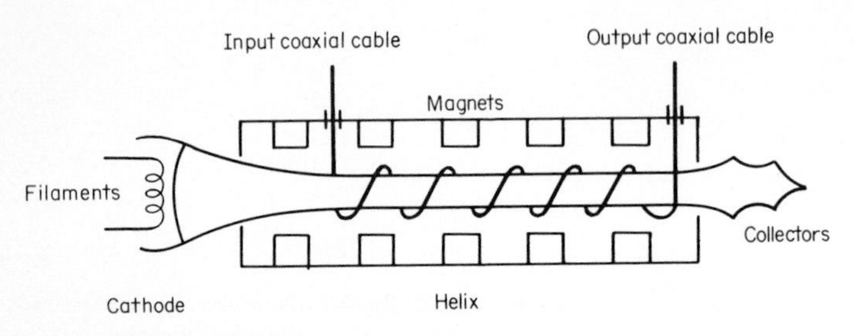

FIGURE 5-6 Schematic of a traveling-wave tube.

ciencies. Typical overall efficiencies range from 15 to 45 percent, depending on total output power, bandwidth, frequency, and design characteristics.

Typical saturated gain ranges from 40 to 70 dB and typical saturated output power from 5 to 30 W.

Because of their nonlinear operation, TWTs exhibit an amplitude modulation (AM) to phase modulation (PM) conversion characteristic which is measured by the AM/PM coefficient K_p in degrees per decibel. Typical values for K_p range from 2 to 7° per decibel. At saturation, typical phase-shift delay ranges from 15 to 50°.

The term TWT amplifier usually implies a TWT with its power supply. Power supplies for TWTs require careful and meticulous design. Many TWT amplifiers have failed because of power-supply failure.

TWT amplifiers for space applications around 1982 achieved gains up to 70 dB with an overall saturation efficiency of up to 40 percent for output powers of 10 to 20 W. However, TWT amplifiers for terrestrial use can reach much higher power, i.e., in the hundreds of watts.

Intermodulation Products

As an illustration of the effect of power amplifier nonlinearities and the creation of intermodulation product interference, an elementary analysis of a simple problem is given in this section.

The output voltage e_o of a nonlinear amplifier in terms of the input voltage e_1 is given by the relationship

$$e_o = k_1 e_1 + k_2 e_1^2 + k_3 e_1^3 \tag{5-1}$$

The input consists of two sinusoidals, i.e.,

$$e_1 = A \cos \omega_1 t + B \cos \omega_2 t \tag{5-2}$$

The intermodulation products and the suppression coefficient will be derived.

We have

$$e_1^2 = A^2 \cos^2 \omega_1 t + B^2 \cos^2 \omega_2 t + 2AB \cos \omega_1 t \cos \omega_2 t \tag{5-3}$$

where

$$\begin{aligned} \cos^2 \omega_1 t &= \tfrac{1}{2}(1 + \cos 2\omega_1 t) \\ \cos^2 \omega_2 t &= \tfrac{1}{2}(1 + \cos 2\omega_2 t) \\ \cos \omega_1 t \cos \omega_2 t &= \tfrac{1}{2} \cos (\omega_1 - \omega_2)t + \tfrac{1}{2} \cos (\omega_1 + \omega_2) \end{aligned} \tag{5-4}$$

We also have

$$\begin{aligned} e_1^3 = A^3 \cos^3 \omega_1 t + B^3 \cos^3 \omega_2 t + 3A^2B \cos^2 \omega_1 t \cos \omega_2 t \\ + 3AB^2 \cos \omega_1 t \cos^2 \omega_2 t \end{aligned} \tag{5-5}$$

where

$$\cos^3 \omega_1 t = \tfrac{1}{4}(\cos 3\omega_1 t + 3 \cos \omega_1 t) \qquad (5\text{-}6)$$
$$\cos^3 \omega_2 t = \tfrac{1}{4}(\cos 3\omega_2 t + 3 \cos \omega_2 t)$$

By introducing Eqs. (5-3) and (5-5) into Eq. (5-1) and making use of the identities in Eqs. (5-4) and (5-6), we obtain

$$\text{Row 1} \begin{cases} e_o = Ak_1\left(1 + \dfrac{3}{4}\dfrac{k_3}{k_1}A^2 + \dfrac{3}{2}\dfrac{k_3}{k_1}B^2\right)\cos \omega_1 t \\ \quad + Bk_1\left(1 + \dfrac{3}{4}\dfrac{k_3}{k_1}B^2 + \dfrac{3}{2}\dfrac{k_3}{k_1}A^2\right)\cos \omega_2 t \end{cases}$$

$$\text{Row 2} \qquad + \tfrac{3}{4}ABk_3[A \cos (2\omega_1 - \omega_2)t + B \cos (2\omega_2 - \omega_1)t]$$

$$\text{Row 3} \qquad + \frac{k_2}{2}[A^2 + B^2 + 2AB \cos (\omega_1 - \omega_2)t]$$

$$\text{Row 4} \qquad + \frac{k_2}{2}[A^2 \cos 2\omega_1 t + B^2 \cos 2\omega_2 t + 2AB \cos (\omega_1 + \omega_2)t]$$

$$\text{Row 5} \begin{cases} + \dfrac{k_3}{4}[A^3 \cos 3\omega_1 t + B^3 \cos 3\omega_2 t + 3A^2B \cos (2\omega_1 + \omega_2)t \\ + 3AB^2 \cos (2\omega_2 + \omega_1)t] \end{cases}$$

Now if we assume ω_1 to be very close to ω_2, then we have

$$2\omega_1 - \omega_2 \approx \omega_1 \qquad 2\omega_2 - \omega_1 \approx \omega_2$$
$$\omega_1 - \omega_2 \approx 0$$
$$2\omega_1 + \omega_2 \approx 3\omega_1 \qquad 2\omega_2 + \omega_1 \approx 3\omega_2$$

and, by passing e_o through a bandpass filter with cutoffs ω_o and ω_o' such that

$$\omega_o < \omega_1 \qquad \omega_2 < \omega_o'$$

and

$$\omega_o > 0 \qquad \omega_o' < 2\omega_1, 2\omega_2$$

we get

$$\begin{aligned} e_o = {} & Ak_1\left(1 + \frac{3}{4}\frac{k_3}{k_1}A^2 + \frac{3}{2}\frac{k_3}{k_1}B^2\right)\cos \omega_1 t \\ & + Bk_1\left(1 + \frac{3}{4}\frac{k_3}{k_1}B^2 + \frac{3}{2}\frac{k_3}{k_1}A^2\right)\cos \omega_2 t \\ & + \tfrac{3}{4}ABk_3[A \cos (2\omega_1 - \omega_2)t + B \cos (2\omega_2 - \omega_1)t] \end{aligned} \qquad (5\text{-}7)$$

In the first and second rows Eq. (5-7) gives the fundamentals suppressed, and in the third row it gives the third-order intermodulation products. Usually a detailed and complete analysis for intermodulation products of modulated carriers requires the utilization of electronic computers with sophisticated programs.

RECEIVERS AND NOISE CONSIDERATIONS

The typical receiver employed by a satellite repeater is a wideband receiver with a single frequency conversion. Double conversion has not been widely used because of many disadvantages such as added weight and complexity.

The most critical element of a receiver is the front-end low-noise amplifier (LNA). The application of tunnel-diode amplifiers and low-noise field-effect transistors has appreciably improved the design characteristics of LNAs. The amplification of a very low level wideband signal (most often 500-MHz bandwidth) without the introduction of appreciable noise is the main performance objective of the low-noise amplifier.

Noise Generated by Electronic Devices

A summary review of the mathematical representation of noise generated from some typical devices will clarify this section as well as subsequent chapters.

Thermal Noise Thermal noise (also often called "Johnson noise") across a resistor load is generated because of the thermal agitation of electrons within the resistor. This completely random motion of the electrons occurs with velocities whose mean value depends on the absolute temperature of the resistor. A voltage is generated across the resistor as the result of a large number of random components; the voltage therefore obeys a gaussian distribution (central limit theorem). Since no net charge is accumulated at either end of the resistor, the average value of the thermal-noise-generated voltage is zero. However, the rms value of the thermal-noise voltage equals the standard deviation σ_n of the gaussian distribution function that describes the probability density function $p(V_n)$ of this voltage V_n with zero mean, namely,

$$p(V_n) = \frac{\exp(-V_n^2/2\sigma_n^2)}{\sqrt{2\pi}\sigma_n} \tag{5-8}$$

and the probability distribution function is

$$P(V_n) = \frac{1}{\sqrt{2\pi}\sigma_n} \int_{-\infty}^{V_n} \exp\left(-\frac{x^2}{2\sigma_n^2}\right) dx \tag{5-9}$$

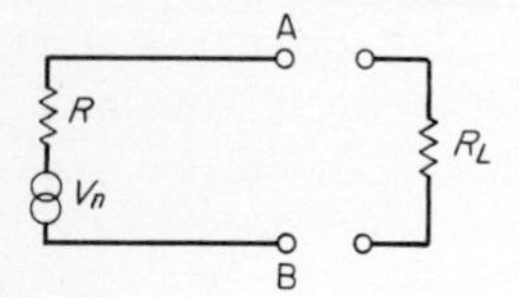

FIGURE 5-7 Voltage equivalent circuit for resistor noise.

The average voltage $\overline{V}_n$ obtained by fully rectifying the noise voltage $|V_n|$ is

$$\overline{V}_n = \sigma_n \sqrt{\frac{2}{\pi}} \tag{5-10}$$

and the form factor, i.e., ratio of rms to average of absolute voltage, is

$$f_n = \sqrt{\frac{\pi}{2}} = 1.253 \tag{5-11}$$

Since the noise voltage as a function of time is not time-limited, a Fourier transform of this voltage cannot be found. The rms noise voltage results in a power level with a power density function depending on frequency. On the basis of the equipartition theory,* the power density function $n(f)$ is given as

$$n(f) = kT \qquad \text{W/Hz} \tag{5-12}$$

where k is Boltzmann's constant, 1.3805×10^{-23} J/K, and T is the absolute temperature of the resistor. This noise, with power density constant as a function of frequency, is called "white noise."

The above approximate relationship for the thermal-noise power density can be refined by applying quantum-mechanics principles. It can be shown that

$$n(f) = \frac{hf}{\exp(hf/kT) - 1} \qquad \text{W/Hz} \tag{5-13}$$

where h is Planck's constant, 6.625×10^{-34} J/s. This function is used at high frequencies. For example, at room temperature ($T = 290$ K), the factor hf becomes significant for frequencies above 4000 GHz. However, in most practical cases the expression $n(f) = kT$ is adequate, and the noise power N over a bandwidth B in hertz will be

$$N = Bn(f) = kTB \tag{5-14}$$

Pure reactive components, such as capacitors or inductors, do not generate thermal noise. However, since they generate frequency-dependent impedances, they influence the total noise power of a system.

*Equipartition theory: In a system at thermal equilibrium, the energy per degree of freedom (degree of energy storage) is $\frac{1}{2}kT$.

The equivalent circuit for a noisy resistor R is given in Fig. 5-7, where the resistor R is noiseless and the noise source $V_n(t)$ furnishes gaussian white noise with an average value $\overline{V}_n(t) = 0$.

The available power to a load across terminals A and B from thermal noise of a resistor R at temperature T, in kelvins, over a bandwidth B will be

$$N = kTB \tag{5-15}$$

However, the power that a voltage source V_n with an rms value $V_{n,\mathrm{rms}}$ can deliver to a load across A and B is maximum when $R_L = R$, and this maximum power will be

$$P_{\mathrm{AB,max}} = \frac{V^2_{n,\mathrm{rms}}}{4R} \tag{5-16}$$

Consequently,

$$\frac{V^2_{n,\mathrm{rms}}}{4R} = kTB \tag{5-17}$$

and

$$V_{n,\mathrm{rms}} = \sqrt{4kTRB} \tag{5-18}$$

The current equivalent circuit for a noisy resistor R is depicted in Fig. 5-8. Again utilizing the concept of the available noise power at terminals A and B, one can show that

$$i_{n,\mathrm{rms}} = \sqrt{\frac{4kTB}{R}} \tag{5-19}$$

Reactive elements do not generate noise, but since an impedance $Z(f)$ can be written as

$$Z(f) = R(f) + jX(f) \tag{5-20}$$

the rms open-circuit noise voltage over a small frequency band Δf of this impedance will be

$$V_{n,\mathrm{rms}} = \sqrt{4kTR(f)\,\Delta f} \tag{5-21}$$

This relation clearly indicates that the noise rms voltage is a function of frequency f.

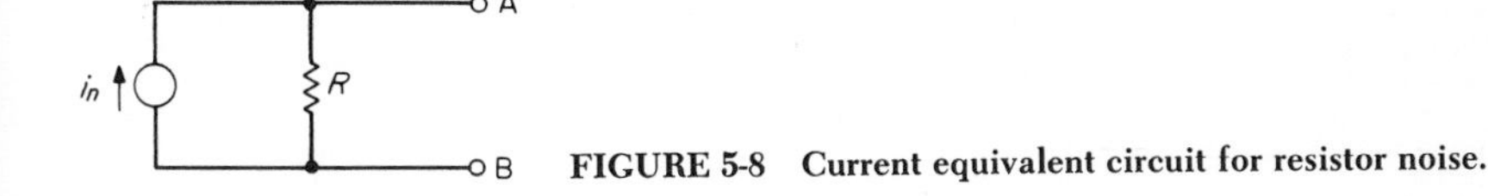

FIGURE 5-8 Current equivalent circuit for resistor noise.

The available noise power over the narrow band Δf will be

$$N(f) = \frac{V_{n,\text{rms}}^2}{4R(f)} = kT\,\Delta f \tag{5-22}$$

Shot Noise Shot noise, found in most active devices (such as transistors) was first observed in the anode current of vacuum tubes. It results from the fact that direct electric current consists of a flow of electrons with discrete nature. Since shot noise is made up of a very large number of contributions (from each flowing electron) which are independent of each other, it has a gaussian probability distribution.

The rms noise current per hertz is

$$i_{n,\text{rms}}^2 = 2qI \tag{5-23}$$

where I is the direct current in amperes and q the charge of an electron in coulombs ($q = 1.6 \times 10^{-19}$ C). Shot noise is not related to the random motions of electrons, and, as a result, it is not related to temperature. The variance of the gaussian distribution is $i_{n,\text{rms}}$. In addition, shot noise contains frequency components with uniform distribution, and as a result it is "white noise."

Low-Frequency Noise The $1/f$ noise, or low-frequency noise, also has gaussian distribution. Surface and contact irregularities in semiconductors are what cause the generation of this noise. Fluctuations in the conductivity of the medium which result from contact irregularities can be reduced by proper treatment of semiconductor surfaces, and, as a result, the $1/f$ noise can also be appreciably reduced.

The spectral density obeys the relationship

$$n(f) = \frac{C}{f^m} \tag{5-24}$$

where m ranges from 0.8 to 1.5. This relationship is valid for frequencies as low as a fraction of a hertz, but it becomes less accurate as f approaches zero. Similarly, the relationship is not valid for very high frequencies.

Impulse Noise This noise arises from various transients and discharges and consists of spikes of energy having approximately a flat frequency spectrum. Assuming that the pulses occur at random intervals and are independent of each other, the number of pulses arriving in any fixed interval will follow a Poisson distribution:

$$P(n) = \frac{(\bar{n}T)^n \exp(-\bar{n}T)}{n!} \tag{5-25}$$

where $P(n)$ is the probability that exactly n pulses occur at a time interval T and n is the average number of pulses per unit time.

The LNA and Receiver as Two-Port Networks

The LNA section of the receiver could be considered as a two-port network and consequently could be analyzed—from the noise standpoint—by application of the well-known definitions and techniques of two-port network theory. Of course, the definitions and analysis which follow are applicable to any two-port network; for example, they are applicable to the total receiver when it is appropriately depicted as a two-port network.

A typical input-output representation of such a network is shown in Fig. 5-9. In general this network internally generates noise which contributes to the output noise power. In addition, the input source noise will also contribute to the output noise power.

Available power gain $G_a(f)$ is defined as the ratio of the available signal power at the output to the available source power, but since

$$\text{Available signal power at output} = \frac{|V_o|^2}{4R_o}$$

and

$$\text{Available source power at input} = \frac{|V_s|^2}{4R_s}$$

we have

$$G_a(f) = \frac{|V_o|^2/4R_o}{|V_s|^2/4R_s} = \frac{|V_o|^2}{|V_s|^2}\frac{R_s}{R_o}$$

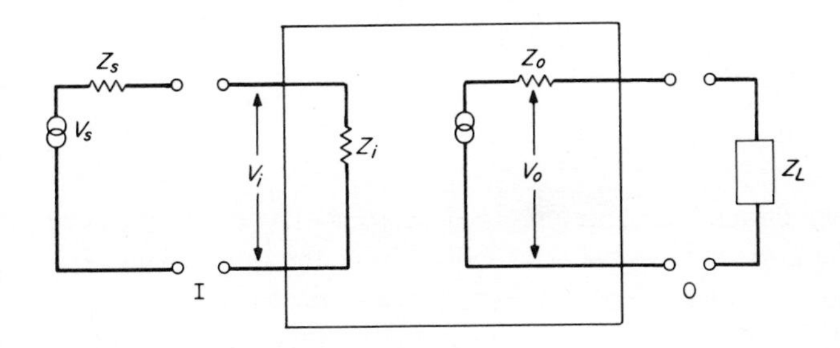

FIGURE 5-9 Two-port noisy network.

V_s = input source voltage

$Z_s = R_s + jX_s$ = input source impedance

$Z_i = R_1 + jX_1$ = network input impedance

$Z_o = R_o + jX_o$ = network output impedance

$Z_L = R_L + jX_L$ = load impedance

V_o = output open-circuit voltage

V_i = network input voltage

$H(f) = V_o/V_i$ = transmittance function

But from the input equivalent circuit we have

$$V_i = V_s \frac{Z_i}{Z_i + Z_s}$$

and therefore

$$G_a(f) = \frac{|V_o|^2}{|V_i|^2} \left| \frac{Z_i}{Z_i + Z_s} \right|^2 \frac{R_s}{R_o}$$

or

$$G_a(f) = \left| \frac{H(f)Z_i}{Z_i + Z_s} \right|^2 \frac{R_s}{R_o} \tag{5-26}$$

Noise Temperature The available noise power of a thermal-noise source is given by the relationship

$$N(f) = kT\Delta f \tag{5-27}$$

The temperature $T = N(f)/(k\,\Delta f)$ is called "noise temperature." In the case of resistor thermal noise, the noise temperature is the physical temperature of the resistor.

It is convenient to generalize the noise temperature concept for sources of noise other than thermal sources. If the available noise power of a source is $N(f)$, then noise temperature is defined as

$$T = \frac{N(f)}{k\,\Delta f} \tag{5-28}$$

This noise temperature is not necessarily the physical temperature of the source, and it is a function of the frequency.

Noise Figure Noise figure F (or, more precisely, spot noise figure over a very narrow frequency band Δf) of a network is defined to be the ratio having, as a numerator, the ratio of available signal power S_1 to the available noise power N_1 at the input I of the network and, as a denominator, the ratio of available signal power S_o to the available noise power N_o at the output O (for a very narrow frequency band Δf and the source noise temperature at the *standard temperature* $T_o = 290$ K).

According to the definition,

$$F = \frac{S_1/N_1}{S_o/N_o} = \frac{S_1}{S_o}\frac{N_o}{N_1} \tag{5-29}$$

But $N_1 = kT_o\,\Delta f$ and $G_a(f) = S_o/S_1$; therefore,

$$F = \frac{N_o}{G_a(f)kT_o\,\Delta f} \tag{5-30}$$

$G_a(f)kT_o\,\Delta f$ is the available noise power at the output for a network that does not itself generate any internal noise, when the signal noise source is at standard temperature T_o.

Equation (5-30) could also be used for defining the noise figure F. The noise figure F gives a measure of the network's contribution to the total output noise from its internal noise sources, which may be caused by various elements such as active semiconductor devices and passive resistive elements.

Noise Bandwidth Noise bandwidth B_n of a given transmittance function $H(f)$ is defined as the bandwidth of an ideal bandpass filter which has a constant transmittance value over B_n equal to the maximum absolute value of the transmittance $H(f)$ and which delivers the same average power from a white-noise source as the given transmittance function.

In Fig. 5-10, the areas under the solid curve and in the dashed rectangle are equal.

Average Noise Figure From Eq. (5-30) and the definition of the noise bandwidth, the average noise figure $\overline{F}$ is defined as

$$\overline{F} = \frac{\overline{N}_o}{G_{\max} kT_o B_n} \tag{5-31}$$

where $\overline{N}_o$ is the ratio of the total available noise power at the output to the total noise bandwidth, and $G_{\max}kT_oB_n$ is the total noise power at the output due only to the input source available noise power.

It can easily be shown that

$$F = \frac{\int_0^{\infty} G_a(f)F\,df}{\int_0^{\infty} G_a(f)\,df} \tag{5-32}$$

Effective Input Noise Temperature T_e Instead of using the noise figure F for describing a network's internal contribution to the output noise power, the effective input noise temperature can be used to achieve the same result.

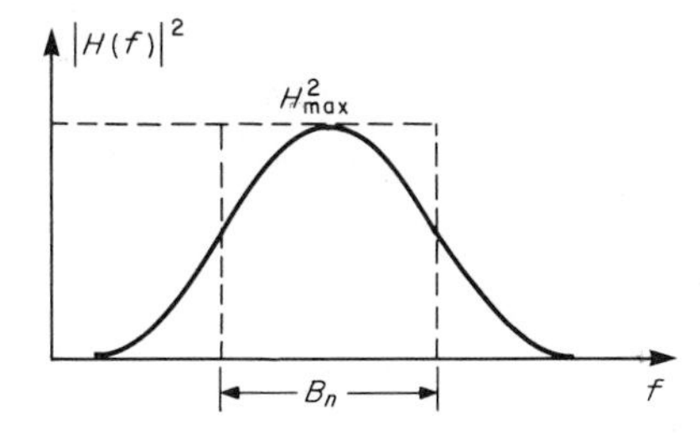

FIGURE 5-10 Noise bandwidth.

If the noisy network were replaced by a hypothetical one—identical in all respects except that it is noiseless and it also has an additional input noise source—at temperature T_e, contributing the same output available noise power N_{oe} as the noisy network, we would have

$$T_e = \frac{N_{oe}}{G_a(f)k\,\Delta f} \tag{5-33}$$

and the total output available noise power, with the source noise at temperature T taken into account, would be

$$N_o = G_a(f)k(T + T_e)\,\Delta f \tag{5-34}$$

Since T_e and F express in different terms the measure of one and the same effect, they must be related through an explicit unique relationship.

In fact, using Eq. (5-30), we can express the output noise in terms of the noise figure as

$$N_o = FG_a(f)kT_o\,\Delta f$$

Comparing this relationship with Eq. (5-34), when $T = T_o$, we get

$$FG_a(f)kT_o\,\Delta f = G_z(f)k(T_o + T_e)\,\Delta f$$

or

$$FT_o = T_o + T_e$$

and

$$F = 1 + \frac{T_e}{T_o} \tag{5-35}$$

Attenuators (Passive) Consider the network in Fig. 5-9 to be a passive attenuator with its lossy elements at temperature T, while the temperature of the noise source at the input is T_S. If the temperature $T_S = T$, then the output load will see only one noise source at temperature T and the output available noise power will be simply $N_o = kT\,\Delta f$. If $T_S \neq T$, then the input noise source will contribute only with its excess temperature $T_S - T$. The output noise power will be

$$N_o = G_a(f)k(T_S - T)\,\Delta f + kT\,\Delta f$$

or

$$N_o = k\{G_a(f)T_S + [1 - G_a(f)]T\}\,\Delta f \tag{5-36}$$

The effective input noise temperature will be

$$T_e = \frac{1 - G_a(f)}{G_a(f)}\,T \tag{5-37}$$

By defining $L_a = 1/G_a(f)$ as the loss ratio of the attenuator, we have

$$T_e = (L_a - 1)T \tag{5-38}$$

and because of Eq. (5-35),

$$F = 1 + (L_a - 1)\frac{T}{T_o} \tag{5-39}$$

An attenuator designed exclusively with reactive elements will have an effective input temperature of zero. If $T = T_o$ (the internal temperature of the attenuator is standard), then

$$F = L_a \tag{5-40}$$

Cascaded Networks When a network like the LNA has many stages, it can be considered to be a cascaded network. In the case of two cascaded networks (see Fig. 5-11), with network 1 having available power gain G_1, effective input noise temperature T_{e1}, and noise figure F_1 and network 2 having corresponding values G_2, T_{e2}, and F_2, the total noise power N_o at the output will be

$$N_o = G_1G_2kT\,\Delta f + G_1G_2kT_{e1}\,\Delta f + G_2kT_{e2}\,\Delta f$$

But, if $T_{e1,2}$ is the effective input noise temperature of the combination,

$$N_o = G_1G_2K(T + T_{e1,2})\,\Delta f$$

and, by equating the right sides of these relationships, we get

$$T_{e1,2} = T_{e1} + \frac{T_{e2}}{G_1}$$

By repeating the same process for n cascaded networks we get

$$T_{e1,\ldots,n} = T_{e1} + \frac{T_{e2}}{G_1} + \frac{T_{e3}}{G_1G_2} + \cdots + \frac{T_{en}}{G_1G_2\cdots G_{n-1}} \tag{5-41}$$

and, by applying Eq. (5-35), we get

$$F_{1,\ldots,n} = F_1 + \frac{F_2 - 1}{G_1} + \frac{F_3 - 1}{G_1G_2} + \cdots + \frac{F_n - 1}{G_1G_2\cdots G_{n-1}} \tag{5-42}$$

If one of the above networks is an attenuator, then the gain of the attenuator $G = 1/L$ will enter Eq. (5-41) or the noise figure $F = L$ will enter Eq. (5-42).

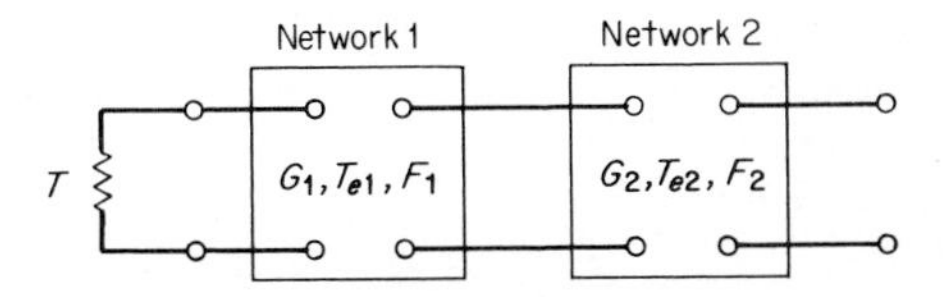

FIGURE 5-11 Cascaded networks.

If the attenuator elements are not at standard temperature T_o, then the noise figure of Eq. (5-39) must be used.

EXAMPLE 5-1 SATELLITE-RECEIVER-SYSTEM NOISE TEMPERATURE

Consider the satellite receiver front end depicted in Fig. 5-12. The effective system noise temperature T_S at the receiver input consists of two temperatures, T_{S1} due to the antenna and lossy elements and T_{S2} due to the amplifier stages.

The noise at receiver input due only to antenna and lossy elements is

$$kT_{S1}\,\Delta f = kT_A \frac{1}{L}\Delta f + k(L-1)T_o \frac{1}{L}\Delta f$$

since $1/L$ is the gain of the lossy section and $(L-1)T_o$ is the lossy-section effective noise temperature. Consequently

$$T_{S1} = \frac{T_A}{L} + \frac{L-1}{L} T_o$$

When Eq. (5-41) is applied, the effective noise temperature due to the two amplifiers becomes

$$T_{S2} = T_{R1} + \frac{T_{R2}}{G}$$

and consequently

$$T_S = T_{S1} + T_{S2} = \frac{T_A}{L} + \frac{L-1}{L} T_o + T_{R1} + \frac{T_{R2}}{G} \qquad (5\text{-}43)$$

This relationship applies both for an earth-station receiver and for a satellite receiver. In the case of the earth station, the antenna noise temper-

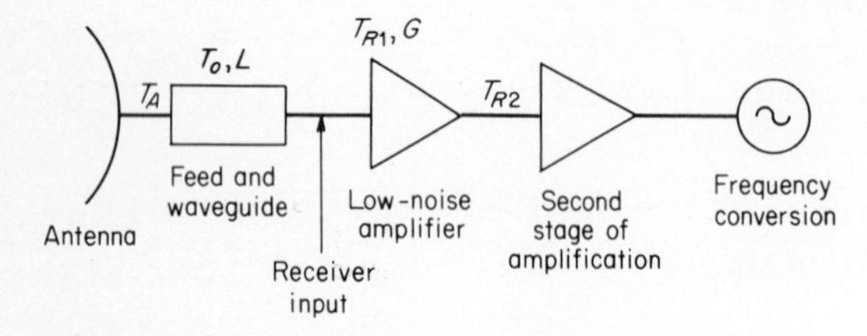

FIGURE 5-12 Example of a satellite receiver. T_A = antenna effective noise temperature; L = feed and waveguide loss at standard temperature T_o; T_{R1} = LNA effective noise temperature; T_{R2} = effective noise temperature of second stage and beyond; G = gain of LNA.

ature T_A must include the contribution of the dish itself, the sky noise, atmospheric absorption effects, and any precipitation or heavy-fog effects. In the case of the satellite receiver, the antenna effective temperature T_A must also include all the appropriate effects from space influencing this T_A at the input of the lossy section (see Chap. 6).

Amplification Devices with Low-Noise Characteristics

Microwave integrated circuit (MIC) technology has produced a number of devices for use in the front end of communications satellite LNAs. The frequency response of these amplifiers must be flat to within 0.2 dB over at least a 500-MHz bandwidth. The front-end first stage must have a low noise figure (less than 5 dB) when operated in an uncooled mode and must provide adequate gain, at least 5 dB.

Low-noise bipolar transistors for frequencies reaching up to 10 GHz were first used in the late 1960s. At the same time tunnel-diode amplifiers (TDAs) were successfully developed as low-noise amplifiers for communications satellite applications at C band. By 1971, however, developments in field-effect transistor (FET) technology had progressed enough to produce the first units for microwave frequencies. By 1980 gallium arsenide FETs had surpassed the performance of bipolar transistors in the higher frequency ranges.

A number of FET structures have been developed. More specifically, some of the structures that have prevailed are

- Metal-semiconductor field-effect transistors (MESFETs)
- Insulated-gate FETs (IGFETs)
- *pn* junction FETs (JFETs)

FETs utilize submicron structures and monolithic integration. Two of the terminals of the FET, the "source" and the "drain," are usually alloyed ohmic contacts to the basic conductive layer. The third terminal, the "gate" between source and drain, is a Schottky contact (metal to semiconductor), and the width of the free-electron carrier depletion layer under the gate controls the current from drain to source. A variety of semiconductor materials such as gallium arsenide have been used that are superior to silicon with respect to majority-carrier transport properties. In addition, high-output-power FETs, in the range of several watts, have been developed as microwave power amplifiers. Low-noise FETs operating at microwave frequencies can provide a gain of over 10 dB and a noise figure between 2.5 and 5 dB.

PERFORMANCE CRITERIA

An important element in the overall performance of the spacecraft communications system is the antenna. Spacecraft antennas were discussed in Chap. 4

with the other spacecraft subsystems. A more detailed treatment of the antenna influence on the overall communications link performance will be discussed later, in Chap. 7. In addition, in Chap. 6, the most important parameters determining antenna performance are discussed. After a signal is received by the antenna, it is processed in the repeater electronics. The channelized transparent repeater consists of a number of transponders sized appropriately to optimize link performance.

Transponder Gain and Output Power

Communications transponder gains are in the range of 100 to 130 dB. The transponder gain is the sum of the receiver and transmitter gains reduced by the losses of the signal-processing elements. Examples of these loss-contributing elements are the channelization filters, equalizers, and channel interconnection switches. These elements contribute 10 to 20 dB of loss. The receiver gain is approximately equal to the transmitter gain, in the range of 50 to 70 dB. The transmitter saturation power is in the range of 5 to 25 W. Solid-state transponders at the present time are limited to the 4- to 10-W range. TWT amplifiers operate efficiently in the 7.5- to 20-W range.

Effective Isotropic Radiated Power

The power radiated by the transmitting antenna is not uniformly distributed. The power distribution depends on the directivity of the antenna. The radiated power P_T of the transmitting antenna multiplied by the gain G_T of the same antenna is defined as the effective isotropic radiated power (EIRP), which is usually measured in decibels referred to 1 W (dBW). Typical satellite EIRPs range from 30 to 45 dBW. (See Chap. 7 for more detailed discussion.)

Receiver *G/T*

The receiver noise figure and receiving antenna gain will determine the input G/T figure of merit. The first stage of amplification is critical, requiring a low noise figure and a relatively high gain. The antenna gain is also influenced by the attitude stabilization accuracy. A typical value of G/T is about -6 dB/K.

Bandwidth and Passband

The overall bandwidth allowable, as previously discussed, is limited by international frequency-allocation regulations. In general the allowed bands in C band and K_u band have a 500-MHz total link bandwidth. Frequency reuse techniques utilizing cross-polarization and/or spot and regional beams with spatial separation could more than double the channels normally available within this bandwidth limitation.

The wideband receivers are usually designed with a passband wider than the total link bandwidth; the receiver input filter is designed to satisfy the out-of-band rejection requirement. Other elements such as mixers have wider bandwidths. The overall design must also meet the requirements of signal intermodulation and harmonic rejection.

Linearity and Intermodulation Products

The receiver nonlinearity is usually due to the input filters and the mixer. It may result in both amplitude and phase distortion. The receiver is usually designed so that its contribution to distortion is only a small fraction of the distortion caused by the output power amplifier. The nonlinearities of TWTs and power amplifiers have been discussed in a previous section.

In the case of a single saturated signal per high-power amplifier, cross talk between channels will be determined by the receiver nonlinearities. In general the overall repeater nonlinearities will produce intermodulation interference, intelligible cross talk, and increased bit error rates in digital transmission. By selecting the appropriate operating point within the linear region, these effects can be minimized.

Spurious Outputs

Local oscillator (LO) harmonics produced by the crystal oscillator as well as harmonics in the microwave frequencies are the main source of spurious outputs. Coupling of the transmitter output into the receiver could also be a source of spurious outputs. Suppression of the harmonics through appropriate filtering and isolation of transmitter output from the receiver circuitry is usually required to meet the spurious-output specification.

Frequency Stability

Frequency stability is usually specified in two distinct terms: "short-term stability" and "long-term stability." The receiver LO and the crystal in the LO are the main sources of frequency instability. Aging of crystals is used to reduce long-term effects.

REFERENCES

Haykin, Simon: *Communication Systems*, John Wiley & Sons, New York, 1978.

Liechti, C. A.: "Microwave Field Effect Transistors," *IEEE Transactions on Microwave Theory and Techniques*, vol. MTT-24, 1976, pp. 279–300.

Members of the Technical Staff of Bell Telephone Laboratories: *Transmission Systems for Communications*, rev. 4th ed., Bell Telephone Laboratories, Winston Salem, N.C., 1971.

Ozaki, H., M. Eick, and N. Silence: *Communications Satellite Receiver Design—Satellite Communications*, IEEE Press, New York, 1979.

Reference Data for Satellite Communications and Earth Stations, ITT Space Communications, Ramsey, N.J., 1972.

Strauss, R., J. Bretting, and R. Metivier: *Travelling Wave Tubes for Communication Satellites*, IEEE Press, New York, 1979.

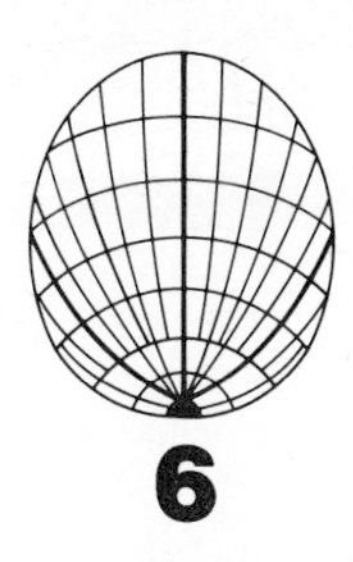

6 EARTH STATIONS AND TERRESTRIAL LINKS

There are two categories of terrestrial stations: telemetry, tracking, and command stations utilized for the control of the spacecraft, and stations utilized mainly for the purpose of communicating with similar stations. By the term "earth station" we mean the latter category, that is, the communicating stations utilizing the satellite as a repeater.

An earth station consists of the antenna dish, the antenna feed, the uplink (transmitting electronics), the downlink (receiving electronics), the tracking system, and the auxiliary subsystems such as those for power, temperature control, and deicing. Earth stations communicating with station-keeping satellites in geostationary orbit do not have to track a moving satellite and, as a result, do not need a sophisticated tracking system. The mount allows the dish to move in azimuth and elevation in order to lock on the spacecraft. Once this has been accomplished, no additional motion is necessary. Therefore, a continuous automatic tracking system is not required. The station is usually powered by the local electric power system. In cases in which uninterrupted operation is required, a diesel-generator set with storage batteries is provided to take up the transition from one power system to the other in case of a failure. Figure 6-1 depicts a commercial earth station for bulk, long-haul (trunk) satellite communications, with a parabolic dish. The block diagram of an earth station is depicted in Fig. 6-2. The uplink interfaces with the diplexer through the high-power amplifier. The downlink interfaces with the same diplexer through the low-noise receiver.

The discussion in Chap. 5 regarding components such as TWT amplifiers and low-noise devices for LNA design is also applicable to earth-station technology.

FIGURE 6-1 **Commercial earth station in San Francisco.** ***(Courtesy of American Satellite Company.)***

ANTENNA

The gain G of a parabolic antenna is directly proportional to the area of the aperture A_o or the square of the diameter D of the dish:

$$G \propto A_o \qquad (A_o \propto D^2)$$

The efficiency of the aperture depends on many factors such as the efficiency of the reflector system (which also depends on the reflector's geometric accuracy) and the efficiency of the feed system. A typical figure for this efficiency for parabolic dishes is 0.54.

When multiple beams with different polarization are used, the polarization isolation capability of the feed system is very important. An isolation of better than 30 dB for signals at the same frequency but opposite polarization is usually achieved for well-designed systems. Axial ratio is defined as the ratio of the two axes of the ellipse of an elliptical polarization. The major axis defines the polarization orientation. Linear and circular polarizations (which correspond to axial ratios of ∞ and 0 dB, respectively) result in elliptical polarizations due to imperfections.

There is a great difference between the transmitted and the received power because of the weakness of the received signal. To prevent interference with the sensitive receiver in the neighborhood of a high-power transmitter, a high-performance diplexer and receiving and transmitting filters are required. Of course, the higher the filter attenuation, the higher the signal distortion due to group delay. As a result, the filter design tends to be quite sophisticated in order to avoid excessive distortion.

Power gain, or simply gain, of an antenna in a given direction is defined as the ratio of the maximum radiation intensity produced in that direction to the maximum radiation intensity produced in the same direction from a reference antenna with the same power input. The reference antenna is a hypothetical lossless antenna which radiates uniformly in all directions. The radiation intensity ϕ in a given direction of an antenna radiating a total power W is the power per unit solid angle radiated in that direction. An isotropic radiator with total radiating power W will have a radiating intensity $\phi_0 = W/4\pi$. The gain of the antenna therefore will be

$$G = \frac{\phi}{\phi_0} = \frac{4\pi\phi}{W} \tag{6-1}$$

The directivity, or maximum directive gain, is

$$G_{\max} = \frac{4\pi\phi_{\max}}{W} \tag{6-2}$$

The gain of an antenna when it is receiving is the same as the gain when it is transmitting. Of course, this gain can be realized by a receiving antenna only in the presence of a properly polarized field.

The effective area A of an antenna is defined as a function of the gain by the relation

$$A = \frac{\lambda^2 G}{4\pi} \tag{6-3}$$

where λ is the wavelength of the radiated wave.

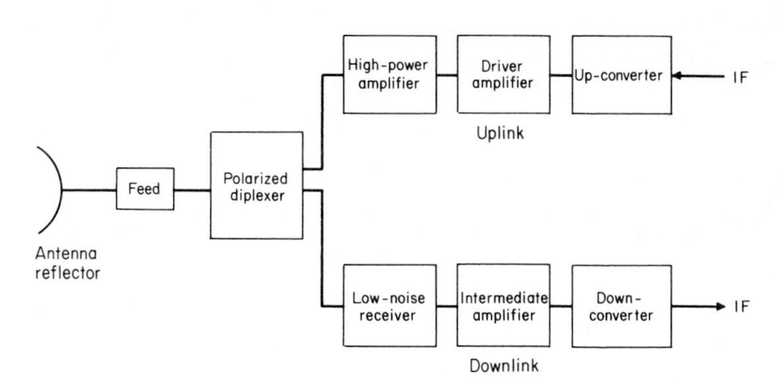

FIGURE 6-2 Block diagram of an earth station.

The total power received by a receiving antenna when p is the power density at the antenna terminals can be shown to be

$$W = pA \tag{6-4}$$

Again, the received wave must be appropriately polarized.

The effective area of an antenna is related to the aperture area A_o through the efficiency factor k:

$$A = kA_o \tag{6-5}$$

The factor k varies from 0.54 for a parabolic reflector to 0.81 for a high-efficiency shaped reflector.

Closely associated with antenna gain is beam width, since, in order to concentrate the power, one should decrease the radiation angle. The square of the beam width is inversely proportional to the antenna aperture. For a given aperture the beam width is also inversely proportional to frequency. If the beam width is θ,

$$\theta^2 = k_1 \frac{1}{G} \qquad \text{or} \qquad \theta^2 = k_2 \frac{\lambda^2}{A_o} = k_3 \frac{1}{f^2 A_o} \tag{6-6}$$

An approximate relationship for the beam width θ of a parabolic reflector is

$$\theta \approx \sqrt{\frac{27{,}000}{G}} \approx \frac{233\lambda}{\pi D} \tag{6-7}$$

In the case of larger parabolic dishes, the design imperfections of the dish degrade the axial gain of the antenna so that the actual gain G_o is given as a fraction g of the theoretical gain G:

$$G_o = gG \tag{6-8}$$

where

$$g = \exp\left[-A\left(\frac{4\pi E_{rms}}{2}\right)^2\right]$$

and E_{rms} is the rms error of the dish in the direction of the main axis of the antenna. A is a correction factor given by the relation

$$A = \frac{1}{1 + (D/4F)^2} \tag{6-9}$$

where D is the antenna dish diameter and F is the focal length. For shallow reflectors in which $4F \gg D$, $A \approx 1$. The rms error E_{rms} depends on the dish manufacturing process and could be from a thousandth of an inch up to about 1 in. Equation (6-9) is an empirical relation derived from experimental data.

LOW-NOISE RECEIVER: SYSTEM NOISE TEMPERATURE

One of the most critical components in the space communications link is the earth-station low-noise receiver—more specifically, the first stage of the receiver, the low-noise preamplifier. Because of the very low signal power at the receiving antenna, this preamplifier must be extremely sensitive and introduce as little noise as possible. Most of the time, in order to avoid the losses and noise introduced by waveguides, the preamplifier is mounted directly on the antenna dish, receiving the energy from the focusing element.

Cryogenically cooled parametric amplifiers and solid-state components such as tunnel diodes or gallium arsenide field-effect transistors (FETs) are used in the preamplifier chain to obtain high gain with low-noise performance. The noise introduced by the preamplifier is measured by the effective input noise temperature T of this amplifier in kelvins, related to the equivalent thermal noise power by the equation

$$N = kBT \tag{6-10}$$

where k is Boltzmann's constant, 1.3805×10^{-23} J/K, and B is the bandwidth in hertz. The receiver is usually wideband, designed for receiving all the transmitted channels. Cooled preamplifiers have been designed with noise temperatures as low as 20 K. Uncooled GaAs FET amplifiers have noise temperatures of the order of magnitude of 100 K. Because of high reliability requirements, the earth station often utilizes dual receivers.

In addition to the receiver electronic equipment noise, the antenna noise, as well as noise due to atmospheric and cosmic sources, must be added at the receiver input. Noise is also introduced from the losses in the section from the antenna feed system to the receiver. The discussion in Chap. 5 regarding device noise and the analysis of two-port networks is directly applicable in the analysis that follows this section.

The total noise temperature that limits the sensitivity of the earth station is a combination of the following factors:

- Sky noise, atmospheric absorption, and rain
- Transmission-line losses including the feed losses
- Receiver system temperature T_R
- Antenna noise temperature

The elements that contribute to the antenna noise temperature are spillover of radiation, blockage, and, in general, various inefficiencies.

Taking into account only the sky noise under clear conditions and the spillover characteristics, we find the noise temperature T_A for the antenna to be

$$T_A = K_1T_s + K_2T_s + K_2T_g \tag{6-11}$$

where T_s is the sky temperature, T_g is the ground temperature, K_1 is the efficiency of the main beam, and

$$K_2 = \frac{1 - K_1}{2} \tag{6-12}$$

if one assumes that half of the spurious radiation is directed to the sky and half to the ground. The ground temperature can be approximated as

$$T_g = T_o = 290 \text{ K} \qquad (17°\text{C}) \tag{6-13}$$

In Fig. 6-3 the sky temperature is given by curves that include atmospheric absorption and cosmic noise under clear conditions.

Rainfall causes substantial degradation, which is frequency-dependent, by attenuating the signal and raising the noise temperature of the system. If L_R is the power loss ratio due to rain, then the rain attenuation A_R in decibels is given as

$$A_R = 10 \log L_R \tag{6-14}$$

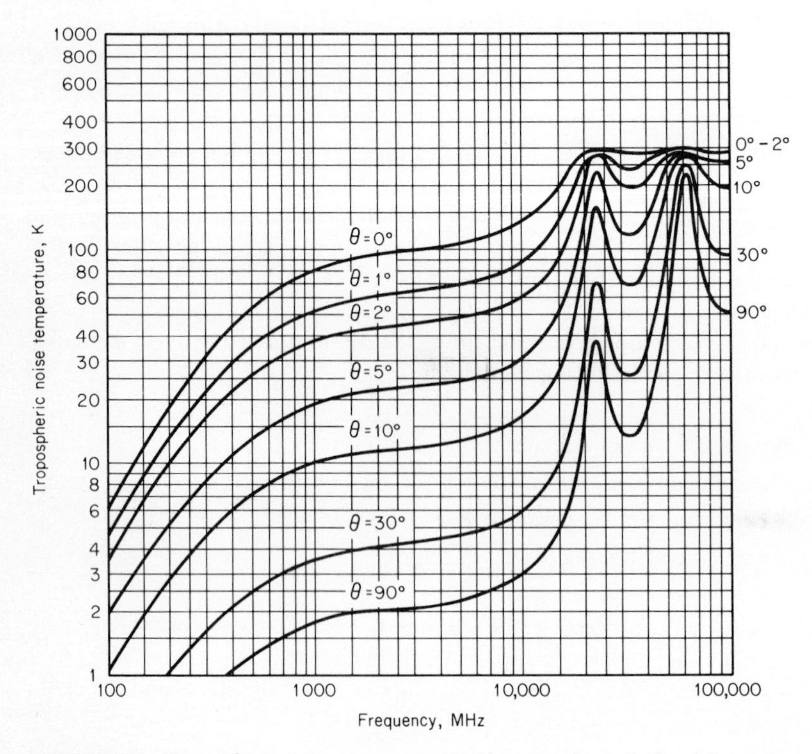

FIGURE 6-3 Tropospheric noise temperature in clear weather calculated for various elevation angles.

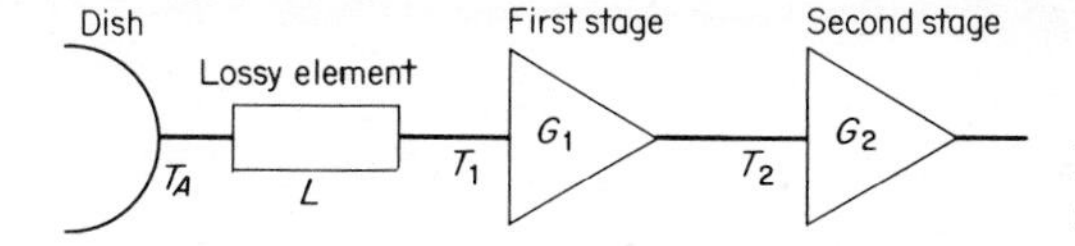

FIGURE 6-4 Earth-station receiving section (block diagram).

and the antenna system temperature T_A is

$$T_A = T_{AC} + \frac{L_R - 1}{L_R} T_o \tag{6-15}$$

where T_{AC} is the antenna noise temperature for clear weather.

The noise figure F of a lossy element with loss L is equal to L, and the noise temperature of an element at its input is $T_{in} = (F - 1)T_o$. In general, the noise temperature of a system with n components in series will be

$$T_S = T + \frac{T_2}{G_1} + \frac{T_3}{G_1 G_2} + \cdots + \frac{T_n}{G_1 G_2 \cdots G_{n-1}} \tag{6-16}$$

where G_i and T_i are the gain and noise temperature of the ith component. Consequently, the total noise temperature of the earth station, including antenna temperature T_A, transmission losses L, and receiver first-stage gain G_1 as depicted in Fig. 6-4 at the input of the first stage will be

$$T_S = \frac{T_A}{L} + \frac{L - 1}{L} T_o + T_1 + \frac{T_2}{G_1} \tag{6-17}$$

where

T_A = equivalent antenna noise temperature at antenna output
L = transmission loss between output of antenna and input of receiver
$= \dfrac{\text{power at antenna}}{\text{power at receiver}}$
T_o = ambient temperature
T_1 = noise temperature of the receiver first stage
T_2 = noise temperature at the input of the receiver second stage
G_1 = receiver first-stage gain

The equivalent system noise temperature at the output of the antenna will be

$$T_{SA} = T_A + (L - 1)T_o + LT_1 + \frac{LT_2}{G_1}$$

The ratio G/T_S, i.e., gain over system temperature at receiver input of an earth station, is very often used to characterize the receiving sensitivity of the earth-station antenna system.

A typical G/T for a 10-m earth station is 33 dB/K. A typical G/T for a satellite receiver system is -6 dB/K.

EXAMPLE 6-1 ANTENNA GAIN AND NOISE TEMPERATURE

Given an earth station operating at 4 GHz and an elevation angle θ = 30°, employing a parabolic dish with diameter D = 10 m and aperture efficiency k = 0.54, determine the clear-weather antenna noise temperature T_{AC} and the antenna gain G.

From Fig. 6-3 for frequency 4 GHz and θ = 30°, the sky temperature T_s = 4.1 K.

The coefficients of Eq. (6-11) are

$$K_1 = K = 0.54$$

$$K_2 = \frac{1 - K_1}{2} = \frac{1 - 0.54}{2} = 0.23$$

and T_{AC}, the antenna noise temperature for clear weather, is

$$T_{AC} = 0.54 \times 4.1 \text{ K} + 0.23 \times 4.1 \text{ K} + 0.23 \times 290 \text{ K} = 69.8 \text{ K}$$

The wavelength λ at 4 GHz for $c = 3 \times 10^8$ m/s (the velocity of light) is

$$\lambda = \frac{c}{\lambda} = \frac{3 \times 10^8 \text{ m/s}}{4 \times 10^9 \text{ s}^{-1}} = 0.075 \text{ m} = 7.5 \text{ cm}$$

The antenna aperture A_o is

$$A_o = \frac{\pi D^2}{4} = \frac{\pi \times 10^2 \text{ m}^2}{4} = 78.5 \text{ m}^2$$

and the antenna effective area A is

$$A = kA_o = 0.54 \times 78.5 \text{ m}^2 = 42.39 \text{ m}^2$$

The gain G is

$$G = \frac{4\pi A}{\lambda^2} = \frac{4\pi \times 42.39 \text{ m}^2}{(0.075\ )^2 \text{ m}^2} = 94{,}652.16$$

or

$$G = 49.76 \text{ dB}$$

EXAMPLE 6-2 FIGURE OF MERIT G/T

For a rainy day when the signal experiences a 1-dB total attenuation, what is the earth-station $(G/T)_S$ if the antenna lossy element results in a signal loss of 0.5 dB and the receiver noise temperature is T_R = 120 K?

The power attenuation L_R due to rain is, from Eq. (6-14),

$$L_R = \log^{-1} \frac{A_R}{10} = \log^{-1} \frac{1}{10} = 1.259$$

and the antenna noise temperature is

$$T_A = T_{AC} + \frac{L_R - 1}{L_R} T_o = 69.8 \text{ K} + \left(\frac{1.259 - 1}{1.259}\right)(290 \text{ K}) = 129.5 \text{ K}$$

The power loss ratio at the antenna feed output is

$$L = \log^{-1} \frac{0.5}{10} = 1.12$$

The system noise temperature T_S can now be calculated:

$$\begin{aligned} T_S &= \frac{T_A}{L} + \frac{L - 1}{L} T_o + T_R \\ &= \frac{129.5 \text{ K}}{1.12} + \left(\frac{1.12 - 1}{1.12}\right)(290 \text{ K}) + 120 \text{ K} \\ &= 266.7 \text{ K} \end{aligned}$$

or by taking 10 log 266.7 we obtain

$$T_S = 24.26 \text{ dBK}$$

The gain G_S in front of the receiver is

$$G_S = 49.76 \text{ dB} - 0.5 \text{ dB} = 49.26 \text{ dB}$$

and

$$\left(\frac{G}{T}\right)_S = 49.76 - 24.26 = 25 \text{ dB/K}$$

The 49.26-dB gain corresponds to a power ratio of 84,333.475. An alternative calculation for $(G/T)_S$ is

$$\left(\frac{G}{T}\right)_S = 10 \log \frac{84{,}333.475}{266.7 \text{ K}} = 25 \text{ dB/K}$$

SUN-TRANSIT OUTAGES

If the sun happens to be positioned behind the satellite when the earth-station dish is directed toward the spacecraft, a sun outage occurs, and communications are interrupted, since the noise temperature of the sun is over 25,000 K. The sun appears as a disk with a look angle (from the earth station) of approximately

29 minutes of a degree. At sun-transit outage the satellite shadow falls on the earth terminal, and the outage lasts until the shadow traverses the earth terminal. The peak outage time is a few minutes. The outage occurs every day for approximately 6 days twice a year. The exact time of the year depends, of course, upon the position of the earth station with respect to the satellite orbital spot.

HIGH-POWER AMPLIFICATION

A high-power amplifier (HPA) usually employs a TWT or a klystron. If a multicarrier operation is anticipated, the total power of the amplifier must be substantially higher than the power per carrier in order to avoid intermodulation products. This is necessary because of the nonlinear characteristics of these devices when operated close to saturation level. A semiconductor HPA could be used for low-power applications.

The radiated power from the antenna is measured in decibels referred to 1 W (dBW) along the main beam, and it is expressed in terms of effective isotropic radiated power (EIRP).

Besides the main beam, the antenna will generate side lobes. The side lobes are measured at their maximum value relative to the main beam in decibels. A typical value of a first side lobe for a 10-m dish is −14 dB.

The power per carrier must be accurately controlled to achieve the proper operating point of the satellite transponder.

EARTH-STATION SUPPORTING SYSTEMS AND ELECTRONICS

The operation of an earth station requires electric power and very often an uninterruptable power supply (UPS). In addition, a certain amount of civil works is required for the installation of the antenna, to provide shielding against interfering radiation and housing for the electronics. Both the UPS and the civil works are briefly discussed in Chap. 2 under "Ground Segment."

The IF signal of the uplink is derived from the output of the modulator, while the output IF of the downlink feeds into the demodulator. The modem (modulator-demodulator), although not considered an element of the earth station, is housed with the earth-station electronics. In addition, when the earth station serves trunk routes, the high-level multiplexing equipment is also usually housed with the earth-station electronics.

TERRESTRIAL LINKS AND DISTRIBUTION

From the earth station the signal is carried to the customers' premises through a terrestrial transmission medium. Coaxial cable, fiber-optic cable, twisted-wire cable, microwave links, etc., can be used for carrying the signal to the user.

FIGURE 6-5 Central office interior. ***(Courtesy of American Satellite Company.)***

The nature of the signal and the distance to be carried will influence the economics and the selection of the appropriate method. Broadband signals over distances of a few kilometers could probably use fiber optics. Relatively narrowband signals over short distances will probably use a twisted-wire pair. Microwave links are normally used to carry multiplexed broadband signals over longer distances.

Quite often, when the signal is carrying messages to be widely distributed within a metropolitan area, a central office is located in the center of the city where the satellite signal is carried for demultiplexing and distributing through the public telephone distribution system. The connection between the central office and a customer's premises was defined in Chap. 2 as the local loop, and it is usually provided by the local telephone company. Figure 2-1 depicts a typical channel configuration.

Figure 6-5 shows a central office interior where satellite signals are demultiplexed, equalized, and interconnected to the local telephone company local-loop distribution network.

In Chap. 2 echo canceling and satellite delay compensation for data transmission were discussed under "Characteristic Advantages and Problems." These functions are usually performed at the central offices.

SIGNAL AND NOISE LEVELS IN TRANSMISSION SYSTEMS

In a transmission system it is customary to select a point as a reference and call it the zero transmission level point (0 TLP). The signal and noise levels at the various points of the transmission path are measured in decibels with respect to 0 TLP. For example, a signal of 20 dB at a given point on the transmission loop corresponds to an absolute power of 100 times the power of the signal at 0 TLP. The 0 TLP may or may not be physically accessible. The designation dBm0 is used to indicate the signal level in decibels referred to 1 mW (dBm) at 0 TLP. For measuring noise, it is also required to define a reference noise (rn) power. In telephone work this reference power is defined as 10^{-12} W, which corresponds to -90 dBm. Consequently, 0 dBrn means 10^{-12} W, or -90 dBm. A noise level of 10 dBrn0 means 10 dB above the reference noise power level at 0 TLP, or 10^{-11} W. At any other point in the transmission system, the noise is measured in decibels above the noise level at 0 TLP. The noise at 0 TLP is measured in dBrn0, that is, in decibels above the reference level.

Since the subjective effect of noise depends on the frequency response of the instrument that converts the electrical signal to voice as well as the frequency response of the ear of the listener, the frequency content of noise is important. For a 3-kHz channel the frequency characteristics of a standard telephone set (500 type) and those of a typical listener have been combined to create a standard which is referred to as the C curve. The C curve gives a weighting factor vs. frequency by which the noise-power spectrum is multiplied for a band of 3 kHz. This weighting factor is 1 for a 1000-Hz tone and less than 1 for all other frequencies. When noise is weighted with the C curve, the noise power of the previous example is written as 10 dBrnC0. This means -80 dBm of weighted noise power.

The International Telephone and Telegraph Consultative Committee (CCITT) has adopted a different weighting standard than that of the U.S. telephone system. When noise is weighted according to the CCITT standard, the noise is said to be psophometrically weighted and it is measured in picowatts of psophometrically weighted noise (pWp). For example, the CCITT allows a maximum noise interference in a telephone circuit of 10,000 pWp. Since the original psophometric definition refers to voltages, the conversion to power assumes a 600-Ω standard load. Consequently, the relation deriving psophometrically weighted power N_{pWp} in picowatts from psophometric voltage V_p is

$$N_{\mathrm{pWp}} = \frac{[V_p\,(\mathrm{mV})]^2}{600} \times 10^6$$

The psophometric weighting over a 3-kHz band decreases the noise power by approximately 2.5 dB, while the C-curve weighting decreases the noise power by 2.0 dB.

REFERENCES

Hogg, D. C., and Ta-shing Chu: "The Role of Rain in Satellite Communications," *Proceedings of the IEEE*, vol. 63, September 1975, pp. 1308–1331.

Members of the Technical Staff of Bell Telephone Laboratories: *Transmission Systems for Communications*, rev. 4th ed., Bell Telephone Laboratories, Winston Salem, N.C., 1971.

Reference Data for Satellite Communications and Earth Stations, ITT Space Communications, Ramsey, N.J., 1972.

Spilker, J. J., Jr.: *Digital Communications by Satellite*, Prentice-Hall, Englewood Cliffs, N.J., 1977.

7

THE SPACE COMMUNICATIONS LINK

The space communications link is usually only a portion of the overall communications link, that is, the link from end user to end user. In this chapter, the space link is defined as the earth station–to–earth station segment, including the satellite. The configuration of the rest of the link depends on specific applications. Subsequent chapters will cover some of the other segments of the link—the ones most frequently employed in useful applications.

The satellite link is a "carrier" system; that is, the actual message is carried by a high-frequency signal. The carrier could be a sinusoidal or square-shaped signal, transmitted as a continuous or bursting RF wave. The message information in analog systems is usually carried in the angle rather than the amplitude of a sinusoidal carrier. Digital systems in a similar fashion encode or "key" the actual message within some preselected characteristic of the carrier, for example, phase. The space-link analysis in this chapter will be restricted to the performance—in the presence of noise—of the communications carrier only and will not address the details involving the message signal which is modulating the carrier. The eventual quality of this message signal can be analyzed only after the discussion of the modulation, signaling, encoding, and detection processes, which are presented in subsequent chapters. By further applying classical communications techniques, the knowledge of the carrier-to-noise ratio (in which noise includes interference) in front of the receiving earth-station LNA is usually adequate for analyzing the postdetection quality of a message.

SPACE-LINK GEOMETRY

The terrestrial coordinates of the receiving and transmitting earth stations, together with the actual orbital location of the spacecraft in the geostationary arc, define the geometry of a space communications link.

The location of an earth station on the earth's surface is determined by two parameters, longitude and latitude. The location of the satellite on the geostationary arc is given by one parameter, the longitude. As a result, the generic problem to be solved in defining the geometric distances could be phrased as follows: Given a geostationary satellite S at an orbital station with longitude L_S and an earth station E with longitude L_E and latitude ϕ, find the distance d between the earth station and the satellite. Figure 7-1 depicts this geometry.

Assume the following:

- Earth's radius R_o ($R_o = GE$)
- Satellite distance from center of the earth R ($R = GS$)
- Earth station relative longitude w with respect to the satellite, where $w = |L_E - L_S|$

In addition, Fig. 7-1 depicts the positions of various significant points, with S′ being the subsatellite point on the equator and E′ the substation point on the equator where it is intersected by the plane of the major circle passing through the earth-station position E.

The spherical triangle EE′S′ is orthogonal at E′ since the equatorial plane and major-circle plane are orthogonal. Consequently, cos EGS = $\cos \phi \cos w$. But from the planar triangle EGS we have

$$d^2 = R^2 + R_o^2 - 2RR_o \cos \text{EGS}$$

and, because of the previous cosine relationship,

$$d^2 = R^2 + R_o^2 - 2RR_o \cos \phi \cos w \qquad (7\text{-}1)$$

Figure 7-2 represents the plane of the triangle EGS and its major-circle intersection with the earth's surface. By drawing the local horizontal (dashed line), we obtain from the triangle GSE

$$(\text{GS})^2 = (\text{GE})^2 + (\text{ES})^2 - 2(\text{GE})(\text{ES}) \cos (90° + \theta)$$

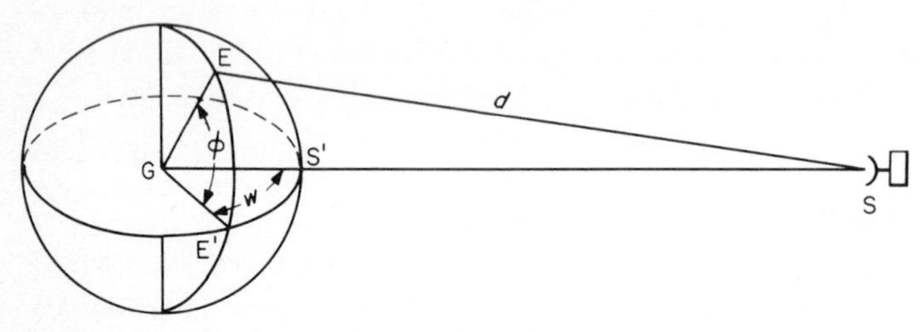

FIGURE 7-1 Earth station E with latitude ϕ and relative longitude w.

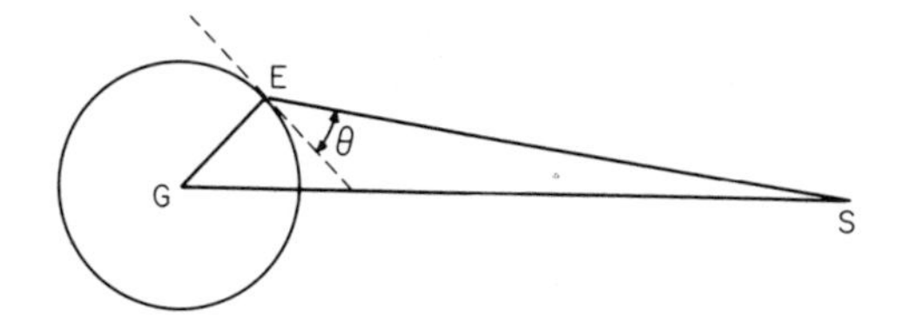

FIGURE 7-2 Earth station E and local horizon (elevation angle θ).

or

$$R^2 = R_o^2 + d^2 + 2R_o d \sin\theta$$

or

$$d^2 + 2R_o d \sin\theta - (R^2 - R_o^2) = 0$$

Solving for d, we obtain

$$d = R_o\left[\sqrt{\left(\frac{R}{R_o}\right)^2 - \cos^2\theta} - \sin\theta\right] \tag{7-2}$$

or

$$\sin\theta = \frac{R^2 - R_o^2 - d^2}{2R_o d} \tag{7-3}$$

Equation (7-2) gives the distance d in terms of the elevation angle θ, and Eq. (7-3) gives the elevation angle in terms of the distance d which could be computed from Eq. (7-1). The elevation angle is an important parameter because it determines the degree of atmospheric interference. For low elevation angles, the radiated wave must traverse a long distance through the atmosphere. As a result, low elevation angles correspond to high atmospheric interference. This is obvious from the representation in Fig. 7-2.

For adequate performance, elevation angles above 5° must be maintained. In some cases a 10° elevation angle is the minimum that can be tolerated.

Table 7-1 gives the extreme orbital spot locations for the U.S. orbital arc corresponding to 5 and 10° elevation angles.

TABLE 7-1 ORBITAL SPOT PERFORMANCE

	Degrees west longitude	
Location	For earth-station elevation 5°	For earth-station elevation 10°
Bangor, Maine	139	132
Honolulu, Hawaii	83	88
Anchorage, Alaska	89	91

Satellite Orbital Location

In the discussion "Space Segment and Launch Considerations" in Chap. 2, the significance of locating the spacecraft in an appropriate orbital location is illustrated with Figs. 2-4 and 2-5. In fact, Fig. 2-5 depicts one of the extreme locations (139°W longitude) for the U.S. orbital arc. As it can be seen from Table 7-1, the most eastward position of Conus, i.e., Bangor, Maine, will see the satellite with an elevation angle of 5°. This 5° elevation is just adequate for optimum link margins.

In Chap. 3, under "Station Keeping—Orbit Geometry," the subject of orbital locations is discussed again, with regard to north and south coverage limitations.

SATELLITE UPLINK

The satellite uplink where the transmitting earth-station signal is received by the satellite receiving antenna at a distance R_U is depicted in Fig. 7-3.

An isotropic reflector, transmitting the same power P_{TE} as the power radiated by the actual transmitting earth-station antenna, will result in a flux density ϕ_0 at the satellite receiving antenna, given by the relationship

$$\phi_0 = \frac{P_{TE}}{4\pi R_U^2}$$

If the transmitting antenna gain is G_{TE}, the flux density at the satellite receiving antenna will be

$$\phi_s = \frac{G_{TE} P_{TE}}{4\pi R_U^2} \tag{7-4}$$

The quantity $G_{TE}P_{TE}$ is the effective isotropic radiated power, usually measured in decibels referred to 1 W (dBW).

If the satellite receiving antenna has a gain G_{RS} and an effective aperture A_R and if L_U is the total power loss in the transmission path, the total power received by the satellite antenna is:

$$P_{RS} = \frac{G_{TE} P_{TE}}{4\pi R_U^2} A_R \frac{1}{L_U} \tag{7-5}$$

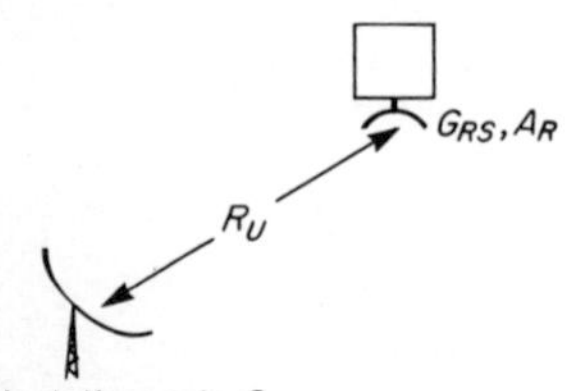

FIGURE 7-3 Uplink.

From the definition $A_R = \lambda^2 G_{RS}/4\pi$, where λ is the wavelength of the radiated wave,

$$P_{RS} = \frac{G_{TE}P_{TE}}{4\pi R_U^2}\frac{\lambda^2 G_{RS}}{4\pi}\frac{1}{L_U}$$

or

$$\frac{P_{RS}}{P_{TE}} = G_{TE}G_{RS}\left(\frac{\lambda}{4\pi R_U}\right)^2 \frac{1}{L_U} \tag{7-6}$$

We define

$$\text{Free space loss} = \left(\frac{4\pi R}{\lambda}\right)^2 \tag{7-7}$$

where R is the distance of transmission, in this case R_U.

It must be noted that in reality there is no loss of power from the propagation through free space. The factor $(4\pi R_U/\lambda)^2$ appears in Eq. (7-6) as a loss because of the specific way that antenna gain $G = 4\pi A/\lambda^2$ is defined.

However, the RF signal is attenuated when it is transmitted through the atmosphere because of absorption by the atmosphere, rain, fog, etc. It can also lose polarization because of effects such as Faraday rotation. The signal attenuation is included in the factor L_U and can be estimated from experimental data.

Equation (7-6) is a fundamental relationship, relating the power received by the satellite to the power transmitted by the earth station.

If the effective noise temperature of the satellite receiver is T_{RS}, then the noise power at the satellite receiver input will be $n_0 = kT_{RS}$, and if we replace P_{RS} by C, where C is the carrier power, Eq. (7-6) can be written as

$$\left(\frac{C}{n_0}\right)_U = P_{TE}G_{TE}\frac{G_{RS}}{T_{RS}}\left(\frac{\lambda}{4\pi R_U}\right)^2 \frac{1}{L_U}\frac{1}{k} \tag{7-8}$$

where

$\left(\frac{C}{n_0}\right)_U$ = uplink carrier-to-noise density ratio

$P_{TE}G_{TE}$ = earth-station EIRP

$\frac{G_{RS}}{T_{RS}}$ = satellite receiving sensitivity

$\left(\frac{4\pi R_U}{\lambda}\right)^2$ = free-space loss

L_U = uplink transmission medium losses

k = Boltzmann's constant

SATELLITE DOWNLINK

The reasoning applied for the uplink can also be applied for the downlink, and by simply replacing the subscript E with S (satellite for earth), the subscript S with E (earth for satellite), and the subscript D with U (up for down), we get for the downlink

$$\left(\frac{C}{n_0}\right)_D = P_{TS}G_{TS}\frac{G_{RE}}{T_{RE}}\left(\frac{\lambda}{4\pi R_D}\right)^2\frac{1}{L_D}\frac{1}{k} \tag{7-9}$$

where

$\left(\frac{C}{n_0}\right)_D$ = downlink carrier-to-noise density ratio

$P_{TS}G_{TS}$ = satellite EIRP

$\frac{G_{RE}}{T_{RE}}$ = earth-station receiving sensitivity

$\left(\frac{4\pi R_D}{\lambda}\right)^2$ = downlink space loss

L_D = downlink transmission-medium losses

TOTAL SPACE LINK

The schematic in Fig. 7-4 represents the total space-link configuration. The total noise power at the satellite receiver input per unit bandwidth is $n_S = kT_S$. If the transponder gain is G, we have, for the satellite noise at the satellite transmitter output,

$$n_{ST} = GkT_S$$

Therefore the noise power at the earth-station receiver input, due to satellite-generated noise, is

$$n_{SE} = G_D n_{ST}$$

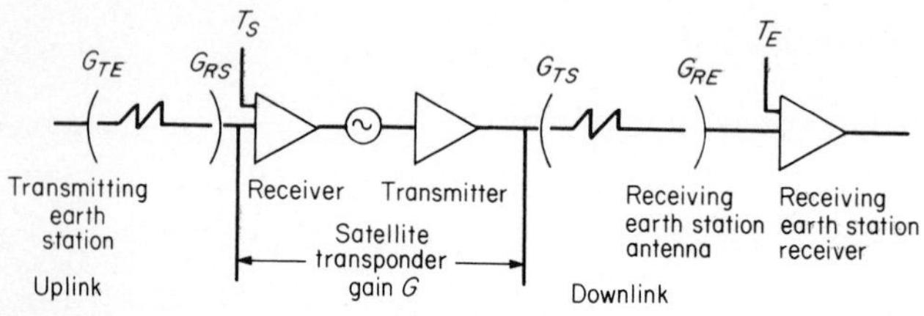

FIGURE 7-4 Total space-link configuration.

where

$$G_D = G_{TS} G_{RE} \left(\frac{\lambda}{4\pi R_D} \right)^2 \frac{1}{L_D}$$

[Note the analogy to Eq. (7-6).] Since kT_E is the additional noise power generated at the earth-station input, the total noise at the earth-station receiver input over a bandwidth Δf, that is, the total space-link noise, is

$$N_T = n_{SE}\,\Delta f + kT_E\,\Delta f$$

In regard to the carrier signal power, we have

$$C_{SR} = G_U P_{TE}$$

for the carrier power at the satellite receiver input, where

$$G_U = G_{TE} G_{RS} \left(\frac{\lambda}{2\pi R_U} \right)^2 \frac{1}{L_U}$$

[see Eq. (7-6)]. Therefore the carrier power at the satellite transmitter output is

$$C_{ST} = G_{SR} G$$

Consequently, the total carrier power at the earth-station receiver input is

$$C_{RE} = C_{ST} G_D$$

We call $C_{RE} = C_T$ the total link carrier power at the end of the space link, and therefore

$$\frac{C_T}{N_T} = \frac{C_{ST} G_D}{n_{SE}\,\Delta f + kT_E\,\Delta f}$$

or, by inversion,

$$\left(\frac{C}{N}\right)_T^{-1} = \left(\frac{C_T}{N_T}\right)^{-1} = \frac{n_{SE}\,\Delta f}{C_{ST} G_D} + \frac{kT_E\,\Delta f}{C_{ST} G_D}$$

$$= \frac{G_D G k T_S\,\Delta f}{C_{SR} G G_D} + \frac{kT_E\,\Delta f}{C_{RE}}$$

$$= \frac{kT_S\,\Delta f}{C_{SR}} + \frac{kT_E\,\Delta f}{C_{RE}}$$

Since

$$\frac{C_{SR}}{kT_S\,\Delta f} = \left(\frac{C}{N}\right)_U$$

which is the uplink carrier-to-noise ratio, and

$$\frac{C_{RE}}{kT_E\,\Delta f} = \left(\frac{C}{N}\right)_D$$

which is the downlink carrier-to-noise ratio,

$$\left(\frac{C}{N}\right)_T^{-1} = \left(\frac{C}{N}\right)_U^{-1} + \left(\frac{C}{N}\right)_D^{-1} \tag{7-10}$$

GENERAL LINK EQUATION

Equation (7-10) gives the total carrier-to-noise ratio as a function of the carrier-to-noise ratios of the uplink and downlink. With similar logic it can be proved that if $(C/N)_I$ is the carrier-to-noise ratio due to intermodulation products introduced by the nonlinear transponder, the total carrier-to-noise ratio will be

$$\left(\frac{C}{N}\right)_T^{-1} = \left(\frac{C}{N}\right)_U^{-1} + \left(\frac{C}{N}\right)_I^{-1} + \left(\frac{C}{N}\right)_D^{-1} \tag{7-11}$$

In a similar fashion it can also be deduced that

$$\left(\frac{C}{N}\right)_T^{-1} = \sum_1^n \left(\frac{C}{N}\right)_i^{-1} \tag{7-12}$$

when there are n contributing causes.

SOURCES OF INTERFERENCE

In addition to thermal noise, two more sources of noise causing a degradation of the carrier-to-noise ratio were previously discussed, namely, intermodulation products due to transponder nonlinearity and depolarization in orthogonally polarized transmissions. Several additional interfering factors will be pointed out in the discussion to follow.

Nonlinearities

Several of the devices and components involved in the implementation of the space link exhibit nonlinear characteristics resulting in the generation of intermodulation products within the band of the transmission (in-band intermodulation products). The more severe nonlinearity is the one exhibited by the transponder power amplifier operated at saturation. The example given in Chap. 6 provides an elementary treatment of this effect. In general, the total power of the in-band intermodulation products will appear similar to an interfering noise power N_I. A complete determination of this interfering power requires a complicated analysis and, in general, also requires the help of electronic computation.

In addition the various nonlinearities affect both the amplitude and phase of the modulated carrier. Because of the AM-PM conversion, amplitude effects will be translated into phase effects and will interfere with angle-modulated signals.

Polarization Interference

For transmissions with single polarization, a signal power loss will be experienced because of loss of polarization. In fact, if a rotation by an angle θ takes place, the electric-field component in the direction of the desired polarization will be decreased by the factor $\cos \theta$. The power of the signal will decrease by a factor of $\cos^2 \theta$. In the case of a dual-polarized transmission, an electric-field component will appear in the orthogonal direction. The magnitude of this component will vary as $\sin \theta$, and it will interfere with the main transmission in the other orthogonal mode. Again, the power of this component N_p should be considered as noise power for purposes of computing the total carrier-to-noise ratio. In addition, a reduction of total carrier power will take place because of the $\cos \theta$ effect discussed above.

In the case of broadband transmission, a complete and detailed analysis requires consideration of the total power spectrum.

Atmospheric Effects

Example 6-2 in Chap. 6 gives an illustration of the effect of rain on the transmission of high-frequency signals. The higher the frequency, the more severe the attenuation due to rain (see Fig. 5-1).

Adjacent Satellite Interference

The proximity of satellites in the orbital arc and the spillover of antenna patterns with side lobes cause interference problems from satellite to satellite.

Adjacent Channel Interference

The utilization of imperfect filters and other similar devices introduces cross talk. In addition, intermodulation products generated in one transponder may spill over in a neighboring transponder.

Frequency Jitter

Imperfect frequency translators and converters introduce jitter effects.

EXAMPLE 7-1 ELEVATION ANGLE AND TIME DELAY

Consider an earth station at Washington, D.C., with the following coordinates: 77°W longitude, 39°N latitude. The elevation angle when this station is pointed at a satellite positioned in a geostationary orbit at 99°W longitude will be determined. In addition, the exact time delay of the signal traveling from the earth station to the satellite will be derived.

From Fig. 7-1, the values of the key geometric parameters are

Earth station relative longitude $w = |L_E - L_S| = 99° - 77° = 22°$
Earth station latitude $\phi = 39°$
Earth radius $R_o = 6370$ km
Satellite distance from center of earth $R = 42{,}230$ km

From Eq. (7-1) the distance of the satellite from the earth station is

$$d^2 = R^2 + R_o^2 - 2RR_o \cos\phi \cos w = (42{,}230\text{ km})^2 + (6370\text{ km})^2 - 2(42{,}230\text{ km})(6370\text{ km}) \cos 39° \cos 22°$$

and

$$d = 37{,}900\text{ km}$$

The signal travel time will be

$$t = \frac{d}{c} = \frac{379 \times 10^5\text{ m}}{3 \times 10^8\text{ m/s}} = 126.3\text{ ms}$$

where c is the velocity of light. The elevation angle θ is given by Eq. (7-3), and therefore

$$\sin\theta = \frac{(42{,}230\text{ km})^2 - (6370\text{ km})^2 - (37{,}900\text{ km})^2}{2(6370\text{ km})(37{,}900\text{ km})} = 0.6346$$

and

$$\theta = 39.4°$$

EXAMPLE 7-2 SPACE-LINK COMPUTATIONS

The carrier frequency of an uplink is $f = 6000$ MHz, and the transmitting earth station EIRP is 85 dBW. The satellite receiver G/T is -7 dB/K, and the transmission losses are 0.5 dB. We will determine the carrier-to-noise ratio at the satellite receiver input for an earth-station–to–satellite distance $R_U = 35{,}860$ km.

If the velocity of light $c = 3 \times 10^8$ m/s, then the uplink space loss is

$$\left(\frac{4\pi R_U}{\lambda}\right)^2 = \left(\frac{4\pi R_U}{c/f}\right)^2 = \left(\frac{4\pi R_U f}{c}\right)^2$$

$$= \left[\frac{4\pi(3.586 \times 10^7 \text{ m})(6 \times 10^9 \text{ s}^{-1})}{3 \times 10^8 \text{ m/s}}\right]^2$$

$$= (90{,}126 \times 10^5)^2 \doteq 199.1 \text{ dB}$$

The Boltzmann constant $k = 1.38 \times 10^{-23}$ can be expressed in decibels as $k \doteq -228.6$ dB. From Eq. (7-8) we obtain

$$\left(\frac{C}{N}\right)_U = \text{EIRP} + \left(\frac{G}{T}\right)_S - \text{space loss} - \text{medium loss} - k$$

$$= 85 - 7 - 199.1 - 0.5 + 228.6 = 107 \text{ dB}$$

If the satellite downlink utilizes a 4000-MHz frequency and the satellite EIRP is 33 dBW, we will derive the carrier-to-noise ratio in the receiving earth station input for an earth station $(G/T)_E = 31$ dB/K. Assume the same distance $R_D = 35{,}860$ km and a transmission medium loss of 0.5 dB.

The downlink space loss will be

$$\left(\frac{4\pi R_D f}{c}\right)^2 = \left[\frac{4\pi(3.586 \times 10^7 \text{ m})(4 \times 10^9 \text{ s}^{-1})}{3 \times 10^8 \text{ m/s}}\right]^2$$

$$= (6008 \times 10^6)^2 \doteq 195.6 \text{ dB}$$

Consequently, by utilizing Eq. (7-9) we obtain

$$\left(\frac{C}{N}\right)_D = 33 + 31 - 195.6 - 0.5 + 228.6 = 96.5 \text{ dB}$$

The total space link $(C/N)_T$ can be now determined from the relationship

$$\left(\frac{C}{N}\right)_T^{-1} = \left(\frac{C}{N}\right)_U^{-1} + \left(\frac{C}{N}\right)_D^{-1}$$

The power ratios for the uplink and downlink must first be computed:

$$\left(\frac{C}{N}\right)_U = 107 \text{ dB} \doteq 5.012 \times 10^{10}$$

$$\left(\frac{C}{N}\right)_D = 96.5 \text{ dB} \doteq 4.467 \times 10^9$$

Therefore

$$\left(\frac{C}{N}\right)_T^{-1} = (5.012 \times 10^{10})^{-1} + (4.467 \times 10^{9})^{-1}$$

$$= 0.2 \times 10^{-10} + 2.24 \times 10^{-10} = 2.44 \times 10^{-10}$$

and

$$\left(\frac{C}{N}\right)_T = 0.4098 \times 10^{10} \doteq 96.1 \text{ dB}$$

The values of the various parameters in this example were obtained from an actual system. The results of the link analysis indicate that for this system the downlink $(C/N)_D$ determines the overall performance of the link. This conclusion is in general true for most operating systems.

In most space links, the noise figure of the receiving earth station LNA, which in effect determines the G/T of the earth station, is a critical parameter. It can substantially improve the downlink performance, and, as a result, the total link performance. Of course the other controllable parameter is the gain G of the earth station, which, for a given frequency, depends on the diameter of the parabolic dish.

It is interesting to see the effect of the intermodulation interference on the overall link performance. Assume, as is the case most of the time, that the $(C/N)_I$ is about 30 dB, which corresponds to a power ratio $(C/N)^{-1} = 10^{-3}$. It can be easily seen for the specifics of Example 7-2 that since

$$\left(\frac{C}{N}\right)_T^{-1} = \left(\frac{C}{N}\right)_U^{-1} + \left(\frac{C}{N}\right)_D^{-1} + \left(\frac{C}{N}\right)_I^{-1}$$

$(C/N)_I$ will dominate the performance and that $(C/N)_T \approx 30$ dB.

It is now obvious why, in multicarrier operations, the required back-off in the transponder is a critical parameter. Because of transponder nonlinearities, the magnitude of the back-off will determine the magnitude of $(C/N)_I$. The resultant $(C/N)_I$ could easily dominate the performance of the overall link. In single-carrier operation, receiver nonlinearities will determine the magnitude of the intermodulation products, or HPA saturation will generate interfering harmonics, or both effects will combine to degrade the link margin. As previously discussed, the AM-PM conversion will also be a factor. Since for dual polarization axial ratios are about 30 dB, polarization interference could be a dominant factor also.

Where high frequencies are used, the effect of rain attenuation can be a significant factor. For example, at 14,000 MHz with 4-mm precipitation the signal could attenuate 4 to 5 dB/km. In the case of the present downlink example, a 10-km path will introduce 50-dB attenuation and $(C/N)_D$ will be reduced from 96.5 dB to 46.5 dB. This is a substantial loss of link margin. At low elevation angles with a long slant distance through the atmosphere and under heavy rain conditions, a substantial degradation of the link could result in unacceptable performance.

REFERENCES

Members of the Technical Staff of Bell Telephone Laboratories: *Transmission Systems for Communications*, rev. 4th ed., Bell Telephone Laboratories, Winston Salem, N.C., 1971.

Reference Data for Satellite Communications and Earth Stations, ITT Space Communications, Ramsey, N.J., 1972.

Spilker, J. J., Jr.: *Digital Communications by Satellite*, Prentice-Hall, Englewood Cliffs, N.J., 1977.

Van Trees, Harry L. (ed.): *Satellite Communications*, IEEE Press, New York, 1979.

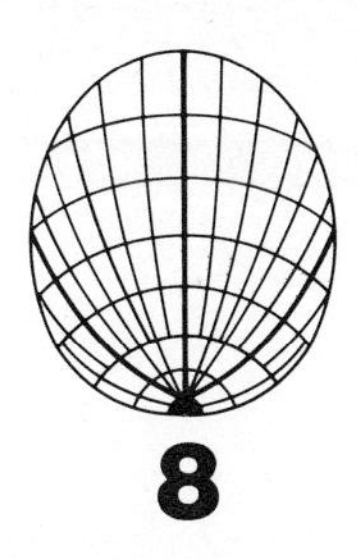

8

LINEAR SYSTEMS AND SIGNALS: SUMMARY OF FUNDAMENTAL PRINCIPLES

In the preceding chapter, the space-link computations resulted in the determination of the carrier-to-noise ratio C/N in front of the receiving earth station low-noise amplifier. The next step is to evaluate the effects of the communications channel and the various filters on the signal, carried by the modulated carrier, in the presence of noise. This evaluation can be carried out either in the time domain or in the frequency domain.

The frequency-domain analysis is performed with the aid of Fourier series and transforms, summarized in Appendix B. It is recommended that the reader become thoroughly familiar with the contents of Appendix B.

In this chapter we will study the various mathematical representations of the signals and noise, mainly in the frequency domain. Proper mathematical representation of the signal and noise is essential for carrying out the required analysis. In addition, we will study the mathematical representation of the communication channel and its elements for linear performance.

By assuming that the channel and its elements are linear systems, we simplify the analysis with the use of the superposition principle. In a linear system the superposition principle holds that the response of the system to a number of simultaneously applied inputs is the sum of the responses to each individual input.

TWO-PORT NETWORKS

Transfer Characteristic and Response to a Unit Impulse

Assume the two-port network of Fig. 8-1, where the input voltage V_1 is a simple sinusoid $V_1 = A \cos 2\pi ft$. We define the voltage transfer ratio $H(f)$ as the ratio of the output voltage V_o to the input voltage V_1, i.e.,

$$H(f) = \frac{V_o}{V_1}$$

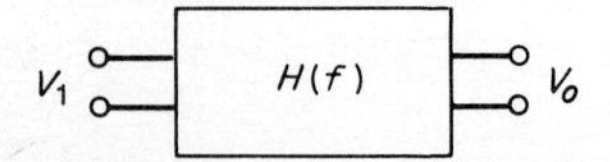

FIGURE 8-1 Two-port network.

If the two-port network consists of simple linear elements R, L, and C, the output voltage V_o will also be a sinusoid of the same frequency f. The voltage transfer function $H(f)$ uniquely characterizes the two-port network under consideration, from the standpoint of determining the output voltage V_o, given the input voltage V_1.

In the more general case a two-port network could be characterized by its response to a unit impulse function often known as a "delta function." A unit impulse function $\delta(t - t_0)$ is defined as the function that is zero everywhere except at time $t = t_0$, when it has an infinite amplitude but a unit area over the infinitesimal time interval dt (see "Delta Function" in Appendix B). When t_0 is located in the interval between A and B:

$$\int_A^B \delta(t - t_0)\, dt = 1 \tag{8-1}$$

Of course, we also have

$$\int_{-\infty}^{\infty} \delta(t - t_0)\, dt = 1 \tag{8-2}$$

A more precise definition of the unit impulse is given by the integral

$$\int_A^B v(t)\, \delta(t - t_0)\, dt = \begin{cases} v(t_0) & \text{for } A < t_0 < B \\ 0 & \text{for } t_0 \leq A \text{ or } t_0 \geq B \end{cases} \tag{8-3a}$$

Of course we also have

$$\int_{-\infty}^{\infty} v(t)\, \delta(t - t_0)\, dt = v(t_0) \tag{8-3b}$$

The response $h(t)$ of the two-port network of Fig. 8-1 to the unit impulse $\delta(t)$ is also uniquely characterizing this network. In fact it can be proved that the output $v_o(t)$ as a function of time resulting from an input excitation $v_1(t)$ is given by the relationship

$$v_o(t) = \int_{-\infty}^{\infty} h(\tau) v_1(t - \tau)\, d\tau \tag{8-4}$$

This relationship is self-evident when the input excitation is a unit impulse. In fact, in this case the value of the integral in Eq. (8-4) becomes $h(t)$, which is the correct answer. If we consider now the input to be a series of successive impulses with areas $v_1(t)$, instead of unity, for each value of t, then the superposition principle applied to the linear network will give the output response as the sum of the responses to these individual impulses as stated in the integral of Eq. (8-4).

The integral in Eq. (8-4) is called the convolution of the two functions $h(t)$ and $v_1(t)$. In general the convolution of two functions $v_1(t)$ and $v_2(t)$ is defined as

$$f(t) = v_1(t) \otimes v_2(t) = \int_{-\infty}^{\infty} v_1(\tau)v_2(t - \tau)\, d\tau$$
$$= \int_{-\infty}^{\infty} v_2(\tau)v_1(t - \tau)\, d\tau \tag{8-5}$$

(see "Convolution" in Appendix B).

We can now see that the system response $h(t)$ to the unit impulse $\delta(t)$ can be made to provide the response of the linear system for any input excitation by using the convolution integral in accordance with Eq. (8-4).

It is constructive to perform the same analysis in the frequency domain, by using the Fourier transforms of the same functions.

The Fourier transform of the unit impulse is given by the relationship

$$\Delta(f) = \int_{-\infty}^{\infty} \delta(t)e^{-j2\pi ft}\, dt$$

From Eq. (8-3b) with $t_0 = 0$ we obtain $\Delta(f) = 1$. The frequency spectrum of the unit impulse is a constant for all frequencies; in other words, it contains all frequencies equally. As a result, when we examine the response of a system to a unit impulse, in effect we examine the response of the system to all frequencies. Consequently, the unit impulse response $h(t)$ in fact uniquely characterizes the system.

The Fourier transform of the unit impulse $\delta(t - t_0)$ is given as

$$\Delta(f) = \int_{-\infty}^{\infty} \delta(t - t_0)e^{-j2\pi ft}\, dt = e^{-j2\pi ft_0}$$

The magnitude $|\Delta(f)| = |e^{-j2\pi ft_0}| = 1$ is a constant and the phase $\theta_\delta = -2\pi ft_0$ is a linear function of the frequency. The frequency spectrum of the impulse function $k\,\delta(t - t_0)$, where k is a constant, is $ke^{-j2\pi ft_0}$ with magnitude $|\Delta f| = k$ and phase $\theta_\delta = -2\pi ft_0$.

Now let us consider the Fourier transforms $H(f)$ of the function $h(t)$ and $V_1(f)$ of the function $v_1(t)$.

The system response is given by Eq. (8-4) as

$$v_o(t) = \int_{-\infty}^{\infty} h(\tau)v_1(t - \tau)\, d\tau = \int_{-\infty}^{\infty} v_1(\tau)h(t - \tau)\, d\tau$$
$$= \int_{-\infty}^{\infty} v_1(\tau) \int_{-\infty}^{\infty} H(f)e^{j2\pi f(t-\tau)}\, df\, d\tau$$

By exchanging the order of integration, we find

$$v_o(t) = \int_{-\infty}^{\infty} H(f)e^{j2\pi ft} \int_{-\infty}^{\infty} v_1(\tau)e^{-j2\pi f\tau}\, d\tau\, df$$

$$= \int_{-\infty}^{\infty} H(f)V_1(f)e^{j2\pi ft}\, df$$

If $V_o(f)$ is the Fourier transform of the output,

$$V_o(f) = H(f)V_1(f) \tag{8-6}$$

This relationship is significant. First of all it proves that $H(f)$, the Fourier transform of $h(t)$, is identical to the voltage transfer ratio as defined in Eq. (8-1). The function $H(f)$ is called the system transfer function. In addition, since in proving Eq. (8-6) we made no special assumptions about $h(t)$ and $v_1(t)$, we can similarly prove, starting with Eq. (8-5), that the Fourier transform of the convolution of two functions is the product of the Fourier transforms of the individual functions; that is,

$$v_1(t) \otimes v_2(t) \leftrightarrows V_1(f)V_2(f) \tag{8-7}$$

The inverse of Eq. (8-7) is also true as a result of the duality principle:

$$V_1(f) \otimes V_2(f) \leftrightarrows v_1(t)v_2(t) \tag{8-8}$$

That is, the product of two functions has as a Fourier transform the convolution of the Fourier transforms of the individual functions.

A similar duality exists between the Fourier transform of an impulse function and the impulse function itself. We proved that the Fourier transform of an impulse function is a constant. Similarly a constant function in time has as a Fourier transform an impulse in the time domain.

If $\delta(f)$ is considered to be the Fourier transform of the function $f(t)$, then

$$f(t) = \int_{-\infty}^{\infty} \delta(f)e^{j2\pi ft}\, df = 1$$

from Eq. (8-3*b*).

Ideal Low-Pass Filter and the Distortionless Channel

A time delay is introduced to a signal when it is passed through an ideal low-pass filter. An ideal low-pass filter is defined to have a system transfer function $H(f)$ given by the relationship

$$H(f) = \begin{cases} 1 \cdot e^{-j2\pi f t_0} & \text{for } |f| \leq B/2 \\ 0 & \text{for } |f| \geq B/2 \end{cases} \tag{8-9}$$

This transfer characteristic has constant amplitude $|H(f)| = 1$ for all frequencies within the passband and a phase spectrum $\phi = -2\pi f t_0$ that is linear with frequency, as depicted in Fig. 8-2.

The response of this filter to a unit impulse $\delta(t)$ will be

$$h(t) = \int_{-\infty}^{\infty} H(f) e^{j2\pi ft}\, df = \int_{-B/2}^{B/2} e^{-j2\pi f t_0}\, e^{j2\pi ft}\, df$$

and

$$h(t) = B\,\frac{\sin \pi B(t - t_0)}{\pi B(t - t_0)} \tag{8-10}$$

The response $v_o(t)$ to an excitation $v_1(t)$ which has no frequencies outside $\pm B/2$ will be

$$V_o(f) = V_1(f)H(f) = V_1(f)e^{-j2\pi f t_0}$$

Consequently,

$$V_o(f) \leftrightarrows v_1(t - t_0)$$

(see "Time Shift" in Appendix B), or

$$v_o(t) = v_1(t - t_0) \tag{8-11}$$

Equation (8-11) indicates that an ideal low-pass filter will simply introduce a time delay for all signals that have a frequency spectrum contained within the bandwidth of the filter.

In general, any channel with a transfer characteristic $H(f)$ satisfying the relationship

$$H(f) = Ae^{j(-2\pi f t_0 \pm n\pi)} \tag{8-12}$$

where n is an integer, will only introduce a time delay and possibly a sign change. A channel with a constant-amplitude response and a linear phase-shift characteristic, such as the one represented by Eq. (8-12), is called a distortionless

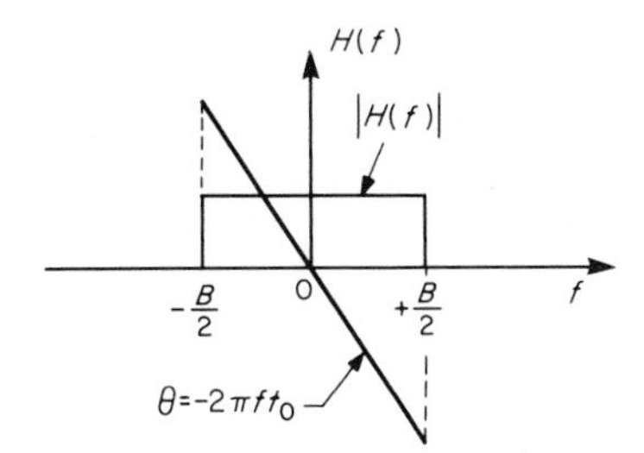

FIGURE 8-2 Ideal low-pass-filter transfer function.

channel. Such a channel is not physically realizable; it can only be approximated.

Let us express the transfer characteristic $H(f)$ of a linear system in terms of its magnitude and phase as shown in the following relationship:

$$H(f) = |H(f)|e^{j\theta(f)} \tag{8-13}$$

If $|H(f)|$ is not a constant, the response to an input signal will emerge with an amplitude distortion. If $\theta(f)$ is not a linear function of f, that is, if $d\theta(f)/df$ is not a constant, then the output signal will again be distorted. The quantity

$$\alpha = \frac{d\theta(f)}{dt} \tag{8-14}$$

is called "envelope delay." A constant envelope delay is a requirement for a distortionless channel. Envelope delay measures the time required to propagate a change in the envelope of the signal.

If we take the natural logarithm of each side of Eq. (8-13), we obtain

$$\ln H(f) = a(f) + jb(f) \tag{8-15}$$

where

$$a(f) = \ln |H(f)| \qquad \text{and} \qquad b(f) = \theta(f) \tag{8-16}$$

A "minimum-phase system" is a system in which

$$\begin{aligned} b(f) &= \frac{1}{\pi}\int_{-\infty}^{\infty} \frac{a(\tau)}{\tau - f}\,d\tau \\ a(f) &= a(\infty) - \frac{1}{\pi}\int_{-\infty}^{\infty} \frac{b(\tau)}{\tau - f}\,d\tau \end{aligned} \tag{8-17}$$

A minimum-phase system has the least possible phase shift for a given gain.* A minimum-phase network is realizable only if it satisfies certain gain and frequency conditions, namely,

$$\int_{-\infty}^{\infty} \frac{|a(f)|}{1 + (2\pi f)^2}\,df < \infty \tag{8-18}$$

The condition expressed in Eq. (8-18) is known as the Paley-Wiener condition.

EXAMPLE 8-1 RESPONSE OF THE LOW-PASS FILTER TO A SQUARE PULSE

The unit impulse response $h(t)$ of a low-pass filter is given by Eq. (8-10) as

$$h(t) = B \operatorname{sinc} B(t - t_0)$$

*See H. W. Bode, *Network Analysis and Feedback Amplifier Design*, Van Nostrand, New York, 1975.

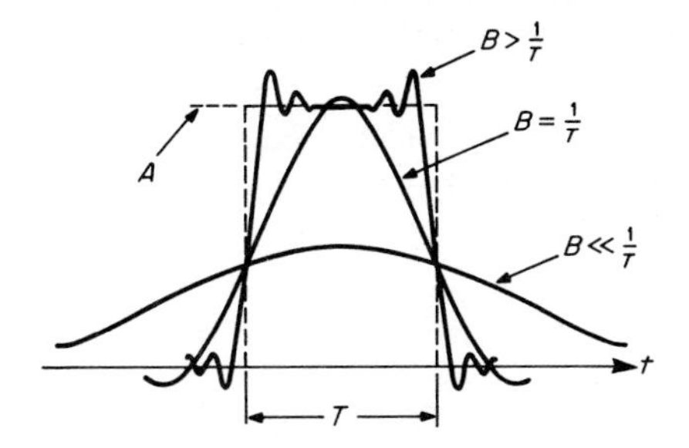

FIGURE 8-3 Response of a low-pass filter to a square pulse excitation.

where sinc $x = (\sin \pi x)/\pi x$. The expression for a rectangular pulse is given by Eq. (B-21) in Appendix B as

$$g(t) = \begin{cases} A & \text{for } |t| < T/2 \\ 0 & \text{for } |t| > T/2 \end{cases}$$

The response of the ideal low-pass filter to the rectangular pulse will be the convolution of $h(t)$ and $g(t)$:

$$y(t) = \int_{-\infty}^{\infty} g(\tau)h(t - \tau)\, d\tau \tag{8-19}$$

This can be written as

$$y(t) = \int_{-T/2}^{T/2} Ah(t - \tau)\, d\tau$$

$$= A \int_{-T/2}^{0} h(t - \tau)\, d\tau + A \int_{0}^{T/2} h(t - \tau)\, d\tau$$

By inserting the value of $h(t)$, we obtain

$$y(t) = \frac{A}{\pi}\left[\text{Si } 2\pi B\left(t + \frac{T}{2} - t_0\right) - \text{Si } 2\pi B\left(t - \frac{T}{2} - t_0\right)\right] \tag{8-20}$$

where Si x is defined as

$$\text{Si } x = \int_0^x \frac{\sin \theta}{\theta}\, d\theta \tag{8-21}$$

The function $y(t)$ is shown in Fig. 8-3. For relatively small B, the peak amplitude of the output increases linearly with the signal, attaining its maximum value for $B = 1/T$. Further increases in B will result in reproduction of the fine details of the pulse but will result in no further increase of amplitude.

NARROWBAND SIGNALS

Assume a signal of the form

$$m(t) = a(t) \cos [2\pi f_c t + \phi(t)] \tag{8-22}$$

where f_c is the carrier frequency. The amplitude $a(t)$ and the phase $\phi(t)$ are signals represented here as functions of time, possessing a certain frequency spectrum in the frequency domain. The bandwidth of these spectra is assumed to be limited in the sense that the maximum frequency f_m within each spectrum is much smaller than the carrier frequency f_c. For example, the functions $a(t)$ and $\phi(t)$ could be the signals representing the messages transmitted by the carrier system, $a(t)$ modulating the amplitude of the carrier and $\phi(t)$ modulating the phase.

The analysis of narrowband signals and systems can be greatly assisted by the use of Hilbert transforms. A short discussion of Hilbert transforms is given in Appendix C. It is recommended that the reader become familiar with their definition and properties.

Representation of a Bandpass Signal—Complex Envelope

The bandpass signal $m(t)$ with the frequency spectrum shown in Fig. 8-4*a* could be expressed as the real part of a complex signal:

$$m(t) = \text{Re}\,[a(t)e^{j[2\pi f_c t + \phi(t)]}] \tag{8-23}$$

or

$$m(t) = \text{Re}\,[a(t)e^{j\phi(t)}e^{j2\pi f_c t}] \tag{8-24}$$

It is therefore constructive to examine the properties of the function

$$g(t) = u(t)e^{j2\pi f_c t} \tag{8-25}$$

where

$$u(t) = a(t)e^{j\phi(t)} = a \cos \phi + ja \sin \phi \tag{8-26}$$

The function $u(t)$ is often called the complex envelope of the signal $m(t)$ satisfying the relationship

$$m(t) = \text{Re}\,[g(t)] = \text{Re}\,[u(t)e^{j2\pi f_c t}] \tag{8-27}$$

The complex envelope $u(t)$ can be expressed in terms of its real and imaginary parts:

$$u(t) = a(t) \cos \phi(t) + ja(t) \sin \phi(t) = u_r + ju_i \tag{8-28}$$

and therefore

$$\begin{aligned} g(t) &= u(t)e^{j2\pi f_c t} \\ &= (u_r + ju_i)(\cos 2\pi f_c t + j \sin 2\pi f_c t) \\ &= (u_r \cos 2\pi f_c t - u_i \sin 2\pi f_c t) \\ &\qquad + j(u_i \cos 2\pi f_c t + u_r \sin 2\pi f_c t) \end{aligned} \tag{8-29}$$

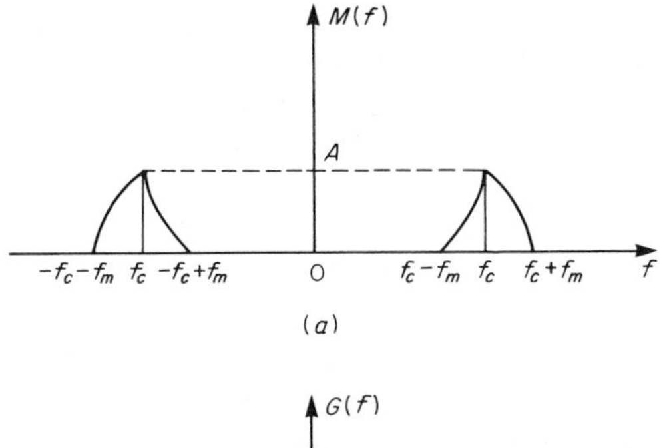

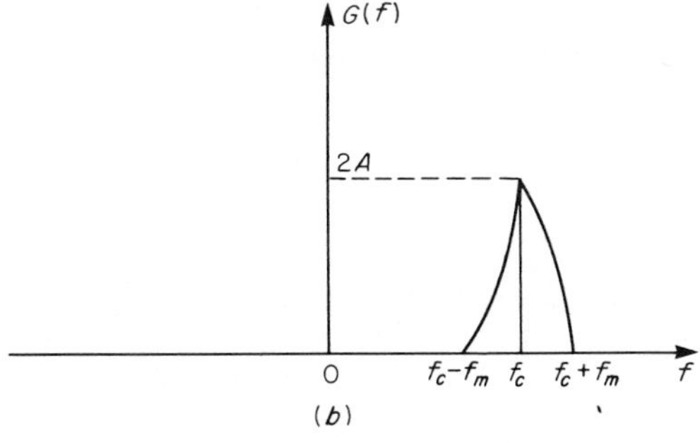

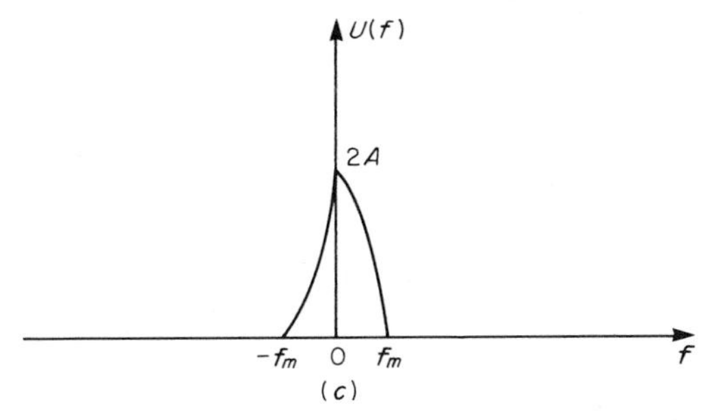

FIGURE 8-4 Frequency spectra for a narrowband signal and its preenvelope and complex envelope. (*a*) Frequency spectrum of $m(t)$. (*b*) Frequency spectrum of preenvelope $g(t) = m(t) + j\hat{m}(t)$. (*c*) Frequency spectrum of complex envelope $u(t)$.

and

$$m(t) = u_r \cos 2\pi f_c t - u_i \sin 2\pi f_c t \tag{8-30}$$

$$\hat{m}(t) = u_r \sin 2\pi f_c t + u_i \cos 2\pi f_c t \tag{8-31}$$

where $\hat{m}(t)$ is the Hilbert transform of $m(t)$. [See Appendix C, Eq. (C-10).] Consequently,

$$g(t) = m(t) + j\hat{m}(t) \tag{8-32}$$

Equation (8-32) indicates that $g(t)$ is the analytic signal, or "preenvelope," of the signal $m(t)$.

Equation (8-30) indicates how the signal $m(t)$ could be resolved in two components, the first ($u_r \cos 2\pi f_c t$) being the in-phase component and the second ($u_i \sin 2\pi f_c t$) being the quadrature component.

Equation (C-14) shows that the amplitude of the frequency spectrum of $g(t)$ will be twice that of $m(t)$ for positive frequencies only. This spectrum is depicted in Fig. 8-4*b*.

Since from Eq. (8-25) we obtain

$$u(t) = g(t)e^{-j2\pi f_c t} \tag{8-33}$$

the frequency spectrum of $u(t)$ shown in Fig. 8-4*c* will be the same as the one for $g(t)$ but translated in frequency by $-f_c$.

The complex envelope $u(t) = u_r(t) + ju_i(t)$ is a low-pass function as indicated by the frequency spectrum of Fig. 8-4*c*. In addition, both the in-phase component $u_r(t)$ and the quadrature component $u_i(t)$ are low-pass functions.

The significance of the analysis in this section is that the bandpass function $m(t)$ could be represented in terms of a low-pass function $u(t)$ and its components $u_r(t)$ and $u_i(t)$, which are also low-pass functions.

Operations with Bandpass Signals

Given two bandpass signals $m_1(t)$ and $m_2(t)$ of the same carrier frequency f_c, we can find the sum and the product by the following procedures.

Addition Introducing the complex envelopes $u_1(t)$ and $u_2(t)$ of the two bandpass signals, we have

$$m(t) = m_1(t) + m_2(t) = \text{Re}\,[u_1(t)e^{j2\pi f_c t}] + \text{Re}\,[u_2(t)e^{j2\pi f_c t}]$$

$$= \text{Re}\,[u(t)e^{j2\pi f_c t}]$$

where $u(t)$ is the complex envelope of $m(t)$. By considering the phasor representation of $m_1(t)$ and $m_2(t)$, it is easy to see that in general $u(t)$ will not be the sum of $u_1(t)$ and $u_2(t)$, but

$$|u(t)| = |u_1(t) + u_2(t)| = \sqrt{a_1^2 + a_2^2 + 2a_1a_2\cos(\phi_1 - \phi_2)}$$

where in general the a's and ϕ's are functions of time, and

$$u(t) = a(t)e^{j\phi(t)}$$

[see Eq. (8-26)].

Product The product of $m_1(t)$ and $m_2(t)$ with carrier frequencies f_1 and f_2 respectively is

$$m_1(t)m_2(t) = \text{Re}\,[u_1(t)e^{j2\pi f_1 t}]\,\text{Re}\,[u_2(t)e^{j2\pi f_2 t}]$$

But because, for any two complex functions g_1 and g_2, the following relationship is always true

$$\text{Re}\,g_1\,\text{Re}\,g_2 = \tfrac{1}{2}\,\text{Re}\,g_1g_2 + \tfrac{1}{2}\,\text{Re}\,g_1^*g_2$$

the previous relationship could be written as

$$m_1(t)m_2(t) = \text{Re}\,[\tfrac{1}{2}u_1(t)u_2(t)e^{j2\pi(f_1+f_2)t}] + \text{Re}\,[\tfrac{1}{2}u_1^*(t)u_2(t)e^{j2\pi(f_2-f_1)t}]$$

Fourier Transform Now if we want to convert $m(t)$ to the frequency domain, it is useful to express $m(t)$ as

$$m(t) = \tfrac{1}{2}[g(t) + g^*(t)] \tag{8-34}$$

and the Fourier transform $M(f)$ will be

$$\begin{aligned} M(f) &= \tfrac{1}{2}\int_{-\infty}^{\infty} [g(t) + g^*(t)]e^{-j2\pi ft}\,dt \\ &= \tfrac{1}{2}\int_{-\infty}^{\infty} g(t)e^{-j2\pi ft}\,dt + \tfrac{1}{2}\int_{-\infty}^{\infty} g^*(t)e^{-j2\pi ft}\,dt \\ &= \tfrac{1}{2}\int_{-\infty}^{\infty} u(t)e^{j2\pi f_c t}e^{-j2\pi ft}\,dt \\ &\quad + \tfrac{1}{2}\int_{-\infty}^{\infty} u^*(t)e^{-j2\pi f_c t}e^{-j2\pi ft}\,dt \end{aligned}$$

and

$$M(f) = \tfrac{1}{2}U(f - f_c) + \tfrac{1}{2}U^*(-f - f_c) \tag{8-35}$$

where $U(f)$ is the Fourier transform of the complex envelope $u(t)$.

Since the frequency f of the narrowband signal is such that $f \ll f_c$, we have two narrowband spectra, one the $\tfrac{1}{2}U(f)$ centered around f_c in the positive side of the frequency axis and a second one, the mirror image of $\tfrac{1}{2}U^*(f)$, centered around $-f_c$ (see Fig. 8-4).

BANDPASS SYSTEMS

General Representation: Response to a Bandpass Signal

Consider a linear system whose transfer function is band-limited and centered around a band-center frequency $\pm f_c$. For a bandpass system, the system transfer function will vanish for all frequencies except for a small band centered around $\pm f_c$. If the system unit impulse response is $g(t)$, the system transfer function $G(f)$ is the Fourier transform of $g(t)$. If we were to represent only the positive frequency part of $G(f)$ in terms of the transfer function $H(f)$ of an equivalent narrowband low-pass system centered around the zero frequency, we would obtain

$$G(f) = H(f - f_c) \qquad \text{for } f > 0 \tag{8-36}$$

where the total passband of $H(f)$ is small and close to zero. By our definition of $H(f)$, it is also true that

$$H(f - f_c) = 0 \qquad \text{for } f < 0 \tag{8-37}$$

(see Fig. 8-5). The response $g(t)$ of the original bandpass system to a unit impulse is given by the relationship

$$g(t) = \int_{-\infty}^{\infty} G(f)e^{j2\pi ft}\, df$$

By splitting the integral in two we have

$$g(t) = \int_{-\infty}^{0} G(f)e^{j2\pi ft}\, df + \int_{0}^{\infty} G(f)e^{j2\pi ft}\, df$$

or

$$g(t) = \int_{0}^{\infty} G(-f)e^{-j2\pi ft}\, df + \int_{0}^{\infty} G(f)e^{j2\pi ft}\, df$$

where f is always positive. But since $g(t)$ is a real function, $G(-f) = G^*(f)$, and the above relationship becomes

$$\begin{aligned} g(t) &= \int_{0}^{\infty} G^*(f)e^{-j2\pi ft}\, df + \int_{0}^{\infty} G(f)e^{j2\pi ft}\, df \\ &= \int_{0}^{\infty} \{[G(f)e^{j2\pi ft}] + [G(f)e^{j2\pi ft}]^*\}\, df \end{aligned}$$

and

$$g(t) = 2 \operatorname{Re} \int_{0}^{\infty} G(f)e^{j2\pi ft}\, df \tag{8-38}$$

But because of Eqs. (8-36) and (8-37),

$$g(t) = 2 \operatorname{Re} \int_{-\infty}^{\infty} H(f - f_c)e^{j2\pi ft}\, df$$

or

$$g(t) = 2 \operatorname{Re} e^{j2\pi f_c t} \int_{-\infty}^{\infty} H(f - f_c)e^{j2\pi (f-f_c)t}\, df$$

and

$$g(t) = 2 \operatorname{Re} [h(t)e^{j2\pi f_c t}] \tag{8-39}$$

where $h(t) \leftrightarrows H(f)$.

The function $h(t)$ represents the response of the low-pass system to a unit impulse. Both $h(t)$ and $H(f)$ may be complex functions since they do not correspond to an actual system. Equation (8-39) describes the response $g(t)$ of a

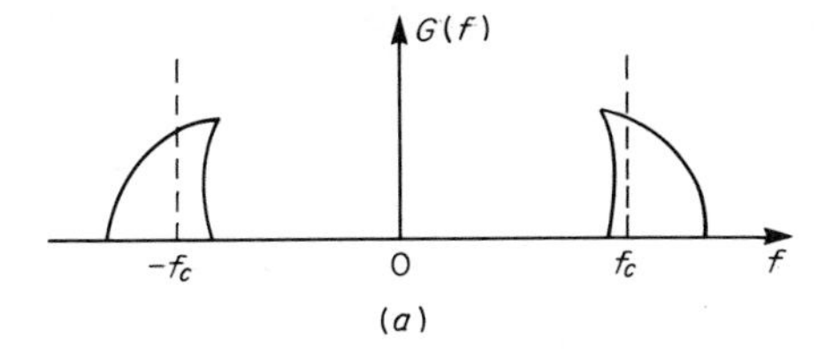

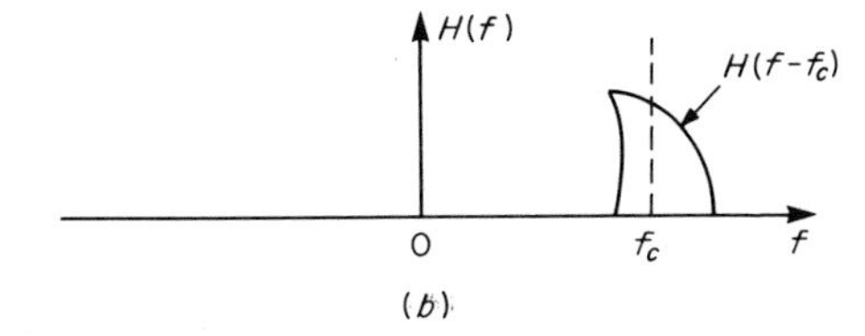

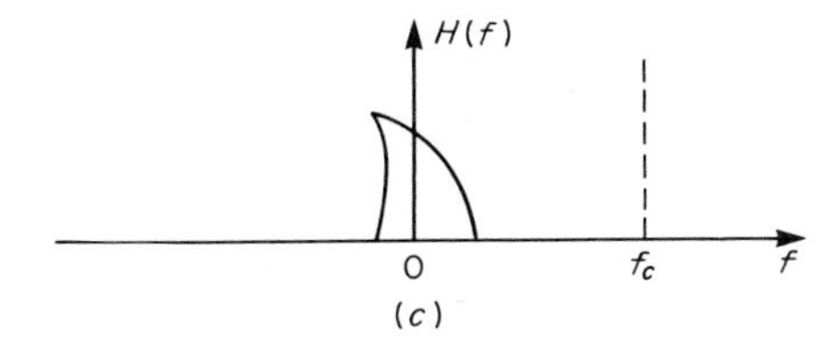

FIGURE 8-5 Low-pass-filter equivalent. (*a*) Bandpass filter. (*b*) Spectrum of $H(f - f_c)$. (*c*) Low-pass equivalent.

bandpass system with a center frequency f_c to a unit impulse excitation in terms of the response $h(t)$ of a low-pass system with a transfer function $H(f)$ defined by Eqs. (8-36) and (8-37).

Equation (8-39) can be written as

$$g(t) = h(t)e^{j2\pi f_c t} + h^*(t)e^{-j2\pi f_c t}$$

Taking the Fourier transform of both sides, we find

$$G(f) = H(f - f_c) + H^*(-f - f_c) \tag{8-40}$$

This relation is equivalent to Eq. (8-35), which refers to the Fourier transform of a narrowband signal.

The response of the bandpass system centered around the frequency f_c to the narrowband signal $m(t)$ of Eq. (8-22) is given by the convolution

$$v_o(t) = \int_{-\infty}^{\infty} m(\tau)g(t - \tau)\, d\tau$$

and therefore

$$V_o(f) = M(f)G(f)$$

By utilizing Eqs. (8-35) and (8-40) we obtain

$$V_o(f) = \tfrac{1}{2}[U(f - f_c) + U^*(-f - f_c)][H(f - f_c) + H^*(-f - f_c)]$$

or

$$V_o(f) = \tfrac{1}{2}U(f - f_c)H(f - f_c) + \tfrac{1}{2}U^*(-f - f_c)H^*(-f - f_c) \qquad (8\text{-}41)$$

since the products of nonoverlapping spectra vanish.

The output signal $v_o(t)$ is also a bandpass signal like $m(t)$, and, if $u_o(t)$ is its complex envelope,

$$v_o(t) = \operatorname{Re}\,[u_o(t)e^{j2\pi f_c t}] \qquad (8\text{-}42)$$

the analogy to Eq. (8-35) will consequently be

$$V_o(f) = \tfrac{1}{2}U_o(f - f_c) + \tfrac{1}{2}U_o^*(-f - f_c) \qquad (8\text{-}43)$$

By comparing Eqs. (8-43) and (8-41), we can see that

$$U_o(f) = U(f)H(f) \qquad (8\text{-}44)$$

and consequently

$$u_o(t) = \int_{-\infty}^{\infty} u(\tau)h(t - \tau)\,d\tau \qquad (8\text{-}45)$$

Thus the complex envelope of the output is given by the convolution of the complex envelope of the input and the unit impulse response of the hypothetical low-pass system.

The above analysis assumed that the center frequency f_c of the bandpass system coincided with the carrier frequency f_c. If in fact there is a difference $\pm\Delta f_c$, we have to multiply the complex envelope of the signal or the unit impulse response of the equivalent low-pass system by $e^{\pm j2\pi\,\Delta f_c t}$. Since all the functions involved in Eq. (8-45)—namely, $u_o(t)$, $u(t)$, and $h(t)$—are low-pass functions, the analysis of the bandpass problem was reduced to the analysis of a low-pass problem. The low-pass representation could substantially facilitate the analysis as well as the application of digital simulation by a digital computer. Numerical evaluation of Fourier transforms by application of fast Fourier transform algorithms substantially reduces the computation time.

EXAMPLE 8-2 COMPLEX ENVELOPE OF AN RF PULSE

An RF pulse is defined in Appendix B as a time function:

$$m_1(t) = \begin{cases} m(t)\cos 2\pi f_c t & \text{for } -T/2 < t < T/2 \\ 0 & \text{for } |t| > T/2 \end{cases}$$

where $m(t)$ is a rectangular pulse of duration T (see Fig. B-3).

By using the assumption $f_cT \gg 1$ we can simplify the Fourier transform $M_1(f)$

$$M_1(f) = \begin{cases} \dfrac{AT}{2} \operatorname{sinc}(f - f_c)T & \text{for } f > 0 \\ \\ \dfrac{AT}{2} \operatorname{sinc}(f + f_c)T & \text{for } f < 0 \end{cases}$$

(see "Frequency Shift," Appendix B). The Fourier transform $G(f)$ of the preenvelope is

$$G(f) = \begin{cases} AT \operatorname{sinc}(f - f_c)T & \text{for } f > 0 \\ 0 & \text{for } f < 0 \end{cases}$$

The Fourier transform $U(f)$ of the complex envelope is found by a frequency shift:

$$U(f) = G(f + f_c) = AT \operatorname{sinc} fT$$

and, as in Example B-1, $u(t)$ is a rectangular pulse with amplitude A and duration T.

EXAMPLE 8-3 RF PULSE THROUGH AN IDEAL BANDPASS FILTER

The transfer characteristic $G(f)$ of an ideal bandpass filter will be

$$G(f) = \begin{cases} e^{-j2\pi(f-f_c)t_0} & \text{for } f_c - W < f < f_c + W \\ e^{-j2\pi(f+f_c)t_0} & \text{for } -f_c - W < f < -f_c + W \\ 0 & \text{elsewhere} \end{cases}$$

where f_c is the midfrequency and $2W$ the bandwidth, and $|G(f)| = 1$ in the passband.

The equivalent low-pass filter $H(f)$ characteristic for which

$$H(f - f_c) = \begin{cases} G(f) & \text{for } f > 0 \\ 0 & \text{for } f < 0 \end{cases}$$

is given by

$$H(f) = \begin{cases} e^{-j2\pi f t_0} & \text{for } -W < f < W \\ 0 & \text{elsewhere} \end{cases}$$

The delta function response $h(t)$ of this filter will be

$$h(t) = \int_{-\infty}^{\infty} H(f)e^{j2\pi ft}\, df = 2W \operatorname{sinc} 2W(t - t_0)$$

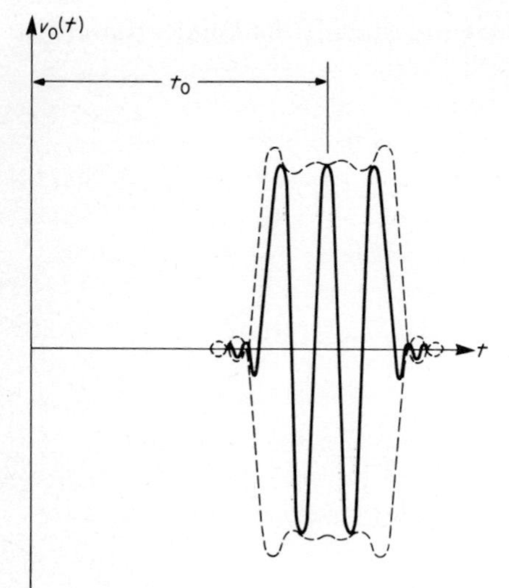

FIGURE 8-6 Bandpass filter response to an RF pulse.

The RF-pulse complex envelope in Example 8-3 was found to be

$$u(t) = \begin{cases} A & \text{for } T/2 < t < -T/2 \\ 0 & \text{elsewhere} \end{cases}$$

Consequently, the response $v_o(t)$ of the bandpass filter to the RF pulse will have a complex envelope:

$$u_o(t) = \int_{-\infty}^{\infty} u(\tau)h(t - \tau)\, d\tau$$

which yields

$$u_o(t) = \frac{A}{\pi}\left[\text{Si } 2\pi W\left(t + \frac{T}{2} - t_0\right) - \text{Si } 2\pi W\left(t - \frac{T}{2} - t_0\right)\right]$$

Since the complex envelope has only one real component,

$$v_o(t) = \frac{A}{\pi}\left[\text{Si } 2\pi W\left(t + \frac{T}{2} - t_0\right) - \text{Si } 2\pi W\left(t - \frac{T}{2} - t_0\right)\right] \times \cos 2\pi f_c t$$

This response is plotted in Fig. 8-6.

CORRELATION OF SIGNALS: ENERGY AND POWER DENSITIES

Autocorrelation—Energy and Power Density

Given two real aperiodic waveforms $v_1(t)$ and $v_2(t)$ with finite energy over the time domain $-\infty$ to $+\infty$, the cross-correlation function is defined as

$$R_{v1,v2}(\tau) = \int_{-\infty}^{\infty} v_1(t)v_2(t + \tau)\, dt \tag{8-46}$$

where τ is an arbitrary displacement in time. The multiplication of the two signals takes place after one is displaced by a distance τ in time. When this product is formed for all possible displacements τ, a process that corresponds to scanning the first signal with the displaced second signal, the cross-correlation is formed as a function of the displacement τ. For τ large and negative, the $v_2(t + \tau)$ signal is displaced to the extreme right of the t axis, and as τ increases toward positive values, $v_2(t + \tau)$ moves to the left from that extreme position. When $v_1(t) = v_2(t) = v(t)$, Eq. (8-46) becomes

$$R_v(\tau) = \int_{-\infty}^{\infty} v(t)v(t + \tau)\, dt \tag{8-47}$$

The function $R_v(\tau)$ is known as the "autocorrelation function" of signal $v(t)$.

The Fourier transform of $R_v(\tau)$ can be found by introducing the Fourier transform of $v(t + \tau)$ in Eq. (8-47),

$$R_v(\tau) = \int_{-\infty}^{\infty} v(t) \int_{-\infty}^{\infty} V(f)e^{j2\pi f(t+\tau)}\, df\, dt$$

and by changing the order of integration,

$$R_v(\tau) = \int_{-\infty}^{\infty} V(f)e^{j2\pi f\tau} \int_{-\infty}^{\infty} v(t)e^{j2\pi ft}\, dt\, df$$

$$= \int_{-\infty}^{\infty} V(f)e^{j2\pi f\tau}V^*(f)\, df$$

or

$$R_v(\tau) = \int_{-\infty}^{\infty} |V(f)|^2 e^{j2\pi f\tau}\, df \tag{8-48}$$

Therefore

$$R_v(\tau) \leftrightarrows |V(f)|^2 \tag{8-49}$$

The function $W_v(f) = |V(f)|^2$ is called the "energy density spectrum" of $v(t)$.

For $\tau = 0$, Eq. (8-48) becomes

$$R_v(0) = \int_{-\infty}^{\infty} |V(f)|^2\, df = \int_{-\infty}^{\infty} W_v(f)\, df \tag{8-50a}$$

and therefore $R_v(0)$ is the total energy of the signal $v(t)$.

It can be easily shown that

$$R_v(\tau) = R_v(-\tau) \tag{8-50b}$$

$$R_v(\tau) \leq R_v(0)$$

The Fourier transform of the cross-correlation function $R_{v1,v2}(\tau)$ can also be easily found by introducing the Fourier transform of $v_2(t + \tau)$ in Eq. (8-46) and then changing the sequence of integration. This process yields

$$W_{v1,v2}(f) = V_1^*(f)V_2(f) \leftrightarrows R_{v1,v2}(\tau) \tag{8-51}$$

It is important to pay attention in ordering the functions $v_1(t)$ and $v_2(t)$, because in general

$$R_{v1,v2}(\tau) \neq R_{v2,v1}(\tau)$$

In fact, it can be easily shown that

$$R_{v1,v2}(\tau) = R_{v2,v1}(-\tau) \tag{8-52}$$

and

$$W_{v1,v2}(f) = W^*{}_{v2,v1}(f) \tag{8-53}$$

Neither the correlation nor the cross-correlation functions uniquely determine the functions $v(t)$, $v_1(t)$, and $v_2(t)$ since they completely discard the phase spectrum.

In the case of periodic real functions, it can be shown that by adapting the definition

$$R_{v1,v2}(\tau) = \frac{1}{T_0}\int_{-T_0/2}^{T_0/2} v_1(t)v_2(t+\tau)\,dt \tag{8-54}$$

for cross-correlation when $v_1(t)$ and $v_2(t)$ have the same period T_0 and by adopting the definition

$$R_v(\tau) = \frac{1}{T_0}\int_{-T_0/2}^{T_0/2} v(t)v(t+\tau)\,dt \tag{8-55}$$

for autocorrelation when T_0 is the period of $v(t)$, we obtain $R_{v1,v2}(\tau)$ and $R_v(\tau)$ as periodic functions with period T_0.

The Fourier series of these two new periodic functions for $f_0 = 1/T_0$ are

$$R_{v1,v2} = \sum_{k=-\infty}^{\infty} a_1^*(k)a_2(k)e^{j2\pi kf_0\tau} \tag{8-56}$$

and

$$R_v(\tau) = \sum_{k=-\infty}^{\infty} |a(k)|^2 e^{j2\pi kf_0\tau} \tag{8-57}$$

where $a_1(k)$ and $a_2(k)$ are the Fourier series coefficients of $v_1(t)$ and $v_2(t)$ and $a(k)$ is the Fourier series coefficient of $v(t)$. From Eqs. (8-56) and (8-57), we can obtain the Fourier transforms

$$R_{v1,v2}(\tau) \leftrightarrows S_{v1,v2}(f) = \sum_{k=-\infty}^{\infty} a_1^*(k)a_2(k)\delta(f-kf_0) \tag{8-58}$$

and

$$R_v(\tau) \leftrightarrows S_v(f) = \sum_{k=-\infty}^{\infty} |a(k)|^2 \delta(f - kf_0) \tag{8-59}$$

The function $S_v(f)$ is called the "power density spectrum" of $v(t)$. We can also derive $R_v(0)$ as

$$R_v(0) = \int_{-\infty}^{\infty} S_v(f)\, df = \sum_{k=-\infty}^{\infty} |a(k)|^2 \tag{8-60}$$

where $R_v(0)$ represents the total power of the spectrum.

If we have a linear system with transfer function $H(f)$ and response $h(t)$ to a unit impulse, we can create the autocorrelation function of $h(t)$,

$$R_h(\tau) = \int_{-\infty}^{\infty} h(t)h(t + \tau)\, dt \tag{8-61}$$

and it is easy to show that

$$R_h(\tau) \leftrightarrows |H(f)|^2 \tag{8-62a}$$

But if $v_1(t)$ is an input signal to the linear system and $v_o(t)$ the output, we have

$$V_o(f) = H(f)V_1(f)$$

or

$$|V_o(f)|^2 = |H(f)|^2 |V_1(f)|^2 \tag{8-62b}$$

Consequently

$$R_{v0}(\tau) = R_h(\tau) \otimes R_{v1}(\tau) \tag{8-63}$$

which is analogous to

$$v_o(t) = h(t) \otimes v_1(t)$$

Similarly it can be shown that

$$R_{v0,v1}(\tau) = \int_{-\infty}^{\infty} h(t)R_{v1}(t + \tau)\, dt \tag{8-64}$$

Correlation of Bandpass Signals

The cross-correlation of two bandpass signals $m_1(t)$ and $m_2(t)$ of the same carrier frequency f_c is given by the relationship

$$R_{m1,m2}(\tau) = \int_{-\infty}^{\infty} m_1(t)m_2(t + \tau)\, dt \tag{8-65}$$

Using the relationships derived in a previous section for the product of bandpass signals, we obtain

$$R_{m1,m2}(\tau) = \text{Re}\left[e^{j2\pi f_c\tau}\int_{-\infty}^{\infty}\tfrac{1}{2}u_1(t)u_2(t+\tau)e^{j4\pi f_c t}\,dt\right] + \text{Re}\left[e^{j2\pi f_c\tau}\int_{-\infty}^{\infty}\tfrac{1}{2}u_1^*(t)u_2(t+\tau)\,dt\right] \tag{8-66}$$

But the first integral of Eq. (8-66) involving the factor $e^{j4\pi f_c t}$ will give zero, cycle by cycle, since the variations of $u_1(t)$ and $u_2(t)$ with time are much slower than this factor. Consequently,

$$R_{m1,m2}(\tau) = \text{Re}\left[e^{j2\pi f_c\tau}\int_{-\infty}^{\infty}\tfrac{1}{2}u_1^*(t)u_2(t+\tau)\,d\tau\right] \tag{8-67}$$

By defining

$$R_{u1,u2}(\tau) = \int_{-\infty}^{\infty}\tfrac{1}{2}u_1^*(t)u_2(t+\tau)\,d\tau \tag{8-68}$$

we obtain

$$R_{m1,m2}(\tau) = \text{Re}\,[R_{u1,u2}(\tau)e^{j2\pi f_c\tau}] \tag{8-69}$$

The complex autocorrelation of $m(t)$ is defined as

$$R_m(\tau) = \int_{-\infty}^{\infty} m(t)m(t+\tau)\,d\tau \tag{8-70}$$

By also defining

$$R_u(t) = \int_{-\infty}^{\infty}\tfrac{1}{2}u^*(t)u(t+\tau)\,d\tau \tag{8-71}$$

where $u(t)$ is the complex envelope of $m(t)$, we can obtain

$$R_m(\tau) = \text{Re}\,[R_u(\tau)e^{j2\pi f_c\tau}] \tag{8-72}$$

The energy density of the bandpass signal can be defined as

$$E(f) \leftrightarrows R_u(\tau) \tag{8-73}$$

and from Eq. (8-71) it can easily be shown that

$$E(f) = \tfrac{1}{2}|U(f)|^2$$

REFERENCES

Haykin, Simon: *Communication Systems,* John Wiley & Sons, New York, 1978.

Schwartz, M., W. R. Bennett, and S. Stein: *Communication Systems and Techniques,* McGraw-Hill, New York, 1966.

Stein, S., and J. J. Jones: *Modern Communication Principles with Application to Digital Signaling,* McGraw-Hill, New York, 1967.

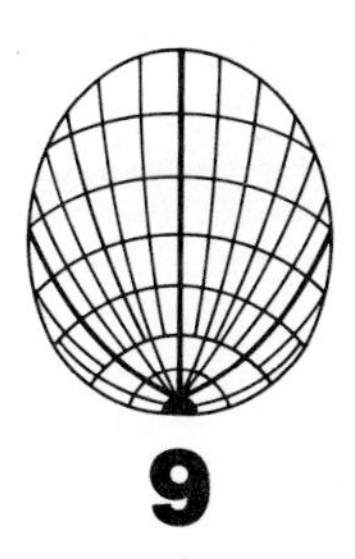

9
RANDOM PROCESSES AND NOISE

Very frequently, communications analysis has to deal with random signals such as voice, data, noise, and other nondeterministic signals. Whereas these signals are functions of time, they are not deterministic because, before they are transmitted, it is not possible to exactly describe the waveforms that will be evolving. As a result, the process is not dissimilar to the one resulting from an experiment in which the outcome for each set of trials is not deterministic. For example, in evaluating a solid-state amplifier, we may simultaneously test a large number of the solid-state devices under evaluation. The result will be a large number of time functions, presumably the output of each device for a test input signal. The ensemble comprising these functions of time is called a stochastic or random process.

This chapter will present only a brief review of this subject; for a more detailed review the reader should consult the references listed at the end of the chapter.

STOCHASTIC PROCESSES

Stationary Ergodic Processes

When dealing with random signals, such as voltage waves due to noise, we can describe these signals only in terms of their statistics.

Consider now a process consisting of a set of waveforms $x_i(t)$ that are functions of time, generally called an "ensemble of functions of time," depicted in Fig. 9-1. This process could be described in terms of the various statistics, for example, the mean $\overline{x}(t_1)$ of $x_1(t_1)$, $x_2(t_1)$, . . . , $x_n(t_1)$ for a given instant of time t_1, and the moments such as the variance

$$(x_1(t_1) - \overline{x}(t_1))^2$$

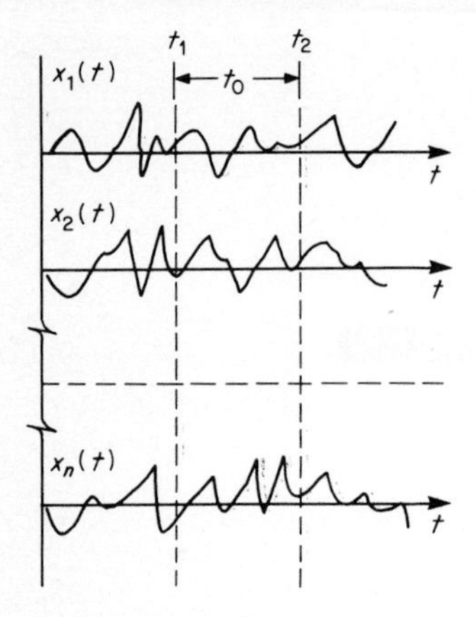

FIGURE 9-1 Ensemble $x_i(t)$.

for the same instant of time t_1. In addition, the joint statistics, such as the joint probability function $p(x_1, x_2, \ldots, x_n)$ could be considered at the same instant of time t_1. If all these derived statistics at time t_1 are identical with the statistics obtained at time t_2, that is, after we translate time by a fixed amount t_0, the process is said to be stationary in a strict sense. The process is stationary in a wide sense if the above statement is true for the mean and autocorrelation only.

Now let us consider the statistical averages taken over relatively long periods of time for a sample waveform $x(t)$. The time average of a function is denoted by an overhead bar; for example, the time average of $x(t)$ is denoted by

$$\overline{x(t)} = \lim_{T\to\infty} \frac{1}{2T} \int_{-T}^{T} x(t)\,dt \tag{9-1}$$

The variance σ^2 as a time average will be given by the relationship

$$\sigma^2 = \overline{[x(t) - \overline{x(t)}]^2} = \lim_{T\to\infty} \frac{1}{2T} \int_{-\infty}^{\infty} [x(t) - \overline{x(t)}]^2\,dt \tag{9-2}$$

Similarly, the autocorrelation function of the sample waveform $x(t)$ is

$$R_x(\tau) = \overline{x(t)x(t+\tau)} = \lim_{T\to\infty} \frac{1}{2T} \int_{-\infty}^{\infty} x(t)x(t+\tau)\,dt \tag{9-3}$$

In all three cases above, in taking the time average we considered a finite time segment from $-T$ to $+T$ and then we let this time segment increase from both ends to infinity.

The ensemble average, such as the expectation, is taken over the ensemble and is denoted as

$$E[x(t)] = \langle x(t) \rangle \tag{9-4}$$

where the angle brackets denote ensemble average. The variance and autocorrelation as ensemble averages will be written as

$$\sigma^2 = \langle [x(t) - \langle x(t) \rangle]^2 \rangle \tag{9-5}$$

and

$$R_x(\tau) = \langle x(t)x(t + \tau) \rangle \tag{9-6}$$

If the time averages of a process can be used to represent the ensemble averages taken at a fixed time, the process is called "ergodic." This is a simplification of a very involved and complicated subject. In the case of stationary ergodic processes, one can define the power spectrum density of the process and relate it through a Fourier transform to a covariance and autocorrelation function as in the case of deterministic signals. Since in the rest of this chapter we will deal only with stationary ergodic processes, ensemble statistics can be represented by time statistics.

In addition, if the process happens to be gaussian, as many processes are in the communications field, by application of the central limit theorem and a number of other principles we can define certain extremely useful concepts, such as the following:

- The joint probability density function (pdf) of a set of variates, which are the weighted sums of a large number of independent quantities, approaches a gaussian distribution. These independent quantities can have a variety of different distributions.
- The sum (linear sum) of gaussian variates is also gaussian. As a result, in a linear system when the input is a gaussian process, the output will also be a gaussian process.

The pdf of the continuous random variable X is defined as the probability of X being between the values of x and $x + dx$:

$$\text{Probability } (x < X < x + dx) = p(x)$$

The probability function $P(x)$ is defined as the probability that $X < x$, or

$$\text{Probability } (X < x) = P(x)$$

and it is clear that

$$p(x) = \frac{dP(x)}{dx} \tag{9-7}$$

or

$$P(x) = \int_{-\infty}^{x} p(x)\, dx \tag{9-8}$$

The mean is defined as

$$\langle x \rangle = \int_{-\infty}^{\infty} x p(x)\, dx \tag{9-9}$$

and the nth moment is

$$\langle x^n \rangle = \int_{-\infty}^{\infty} x^n p(x)\, dx \tag{9-10}$$

The central moment of nth power is defined as

$$\langle (x - \langle x \rangle)^n \rangle = \int_{-\infty}^{\infty} (x - \langle x \rangle)^n p(x)\, dx \tag{9-11}$$

and for $n = 2$ we have the definition of the variance $\sigma^2 = \langle (x - \langle x \rangle)^2 \rangle$, where σ is the standard deviation. For a distribution with zero mean, i.e., $\langle x \rangle = 0$, $\sigma^2 = \langle x^2 \rangle$.

If X is a random variable and Y is related to X through a functional relationship, namely,

$$y = g(x) \tag{9-12}$$

it is desirable to obtain the pdf $p_Y(y)$ when the pdf $p_X(x)$ is known. If the inverse of $g(x)$, namely $g^{-1}(y)$ is given by

$$x = w(y) = g^{-1}(y) \tag{9-13}$$

then the desired result (under certain conditions) is

$$p_Y(y) = p_X[w(y)] \left| \frac{dw}{dy} \right| \tag{9-14}$$

The proof of this transformation can be found in the reference by Papoulis (1965) listed at the end of this chapter.

Gaussian Distribution and Error Function erf *x*

In the case of a gaussian distribution with mean $\langle x \rangle = m$ and standard deviation σ, we have the pdf

$$p(x) = \frac{1}{\sigma\sqrt{2\pi}} e^{(x-m)2/2\sigma 2} \tag{9-15}$$

The gaussian distribution is shown in Fig. 9-2. The probability distribution function is obtained by integrating $p(x)\, dx$ from $-\infty$ to x:

$$P(x) = \frac{1}{2}\left(1 + \operatorname{erf} \frac{x - m}{\sigma\sqrt{2}}\right) \tag{9-16}$$

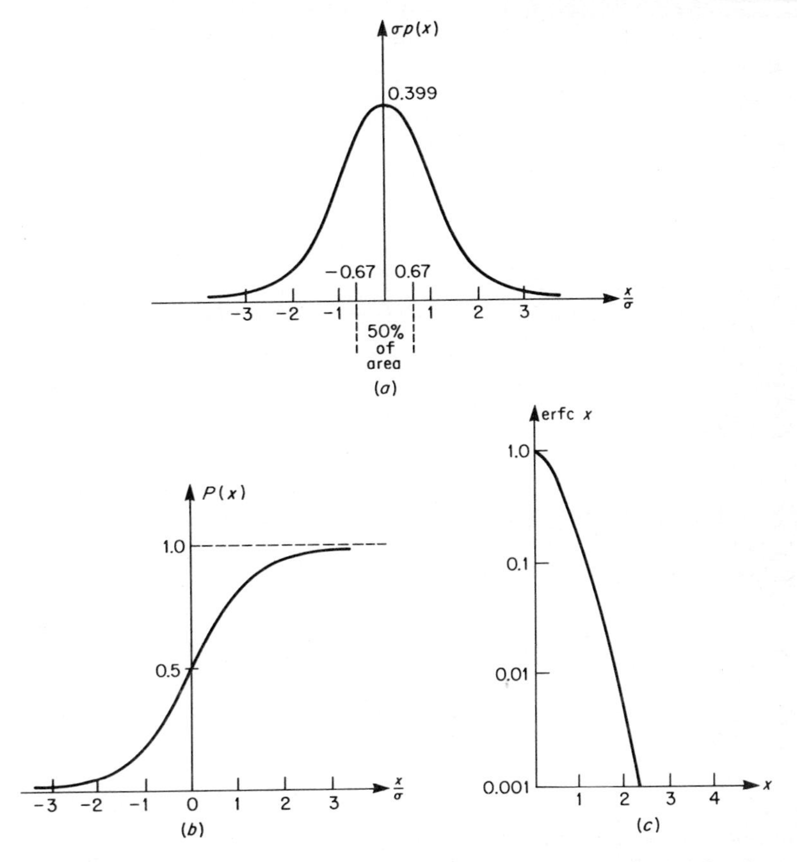

FIGURE 9-2 Characteristics of gaussian distribution. (*a*) Gaussian probability density function. (*b*) Gaussian probability distribution. (*c*) Complementary error function.

where

$$\operatorname{erf} x = \frac{2}{\sqrt{\pi}} \int_0^x e^{-y^2}\, dy \tag{9-17}$$

The complement of the error function, erfc, is defined as

$$\operatorname{erfc} x = 1 - \operatorname{erf} x = \frac{2}{\sqrt{\pi}} \int_x^\infty e^{-y^2}\, dy \tag{9-18}$$

Autocorrelation and Power Density

The autocorrelation function $R_x(\tau)$ was defined by Eq. (9-6) as

$$R_x(\tau) = \langle x(t)x(t + \tau)\rangle$$

We can compute $R_x(\tau)$ as a limit by averaging over time T as follows:

$$R_x(\tau) = \langle x(t)x(t+\tau)\rangle = \lim_{T\to\infty} \frac{1}{2T} \int_{-T}^{T} x(t)x(t+\tau)\,dt \tag{9-19}$$

For a stationary process in a wide sense, the autocorrelation is a function only of τ. The autocovariance is defined as

$$\begin{aligned}\mu(\tau) &= \langle [x(t) - \langle x(t)\rangle][x(t+\tau) - x(\langle t\rangle)]\rangle \\ &= \lim_{T\to\infty} \frac{1}{2T} \int_{-T}^{T} [x(t) - \langle x\rangle][x(t+\tau) - \langle x\rangle]\,dt\end{aligned} \tag{9-20}$$

and

$$\begin{aligned}\mu(\tau) &= R(\tau) - \langle x\rangle^2 \\ \mu(0) &= R(0) - \langle x\rangle^2 = \sigma^2\end{aligned} \tag{9-21}$$

The power density spectrum $S_T(f)$ is defined as the Fourier transform:

$$S_T(f) = \int_{-\infty}^{\infty} R_x(\tau)e^{-j2\pi f\tau}\,d\tau \tag{9-22}$$

$S_T(f)$ has not been related to the frequency characteristics of the original function $x(t)$.

In the deterministic case dealing with the energy density spectrum, we have

$$W(f) = |V(f)|^2 \tag{9-23}$$

where $W(f)$ is the Fourier transform of the autocorrelation function of the waveform $v(t)$ with Fourier transform $V(f)$.

In keeping with the concept of time averages in the stationary ergodic process, one intuitively feels that

$$S_T(f) = \lim_{T\to\infty} \frac{1}{2T} [X_T(f)]^2 \tag{9-24}$$

where $X_T(f)$ is the Fourier transform of a function that equals the original waveform $x(t)$ over the time segment from $-T$ to T, while it is considered to be zero outside of this time segment. This is not strictly correct. However, it can be proved that

$$S_T(f) = \lim_{T\to\infty} \frac{1}{2T} \langle [X_T(f)]^2\rangle \tag{9-25}$$

From Eq. (9-25), by exchanging power density spectra for energy density spectra, we can show that all the relationships derived for the deterministic case will have corresponding relationships for the case of the stationary ergodic process.

In general, a stationary random process has an infinite energy and therefore does not have a Fourier transform in an ordinary sense. However, time averages of a function $x(t)$ usually converge when time increases without limit. When the process is ergodic, these time averages equal the ensemble averages.

For a stationary ergodic process the autocorrelation $R_x(\tau)$ is a function of τ only, i.e., of the difference of the two time instants at which the observations were made. This property led to the definition of a stationary process in a wide sense.

The power density spectrum of $x(t)$ was defined as

$$S(f) = \int_{-\infty}^{\infty} R_x(\tau)e^{-j2\pi f\tau}\, d\tau$$

It is easy to show that $|S(f)|$ is an even function of f. The total power under $S(f)$ is given by the relationship

$$R_x(0) = \int_{-\infty}^{\infty} S(f)\, df \tag{9-26}$$

and from Eq. (9-19) we also have

$$R_x(0) = \langle [x(t)]^2 \rangle = \sigma^2 + \langle x \rangle^2 \tag{9-27}$$

From the definition of $R_x(\tau)$ it can easily be seen that

$$R_x(\tau) = R_x(-\tau) \tag{9-28}$$

It is also true that

$$R_x(\tau) \leq R_x(0) \tag{9-29}$$

White Noise

Consider the case of white gaussian noise, that is, noise with flat noise spectrum $n_0/2$ and zero mean.* The one-sided power spectrum will be n_0. In the case in which we consider both positive and negative frequencies, we have

$$S(f) = \frac{n_0}{2} \tag{9-30}$$

The autocorrelation function is

$$R_n(\tau) = \int_{-\infty}^{\infty} S(f)e^{j2\pi f\tau}\, df = \int_{-\infty}^{\infty} \frac{n_0}{2}\, e^{j2\pi f\tau}\, df = \frac{n_0}{2}\, \delta(\tau) \tag{9-31}$$

Figure 9-3 depicts the power density spectrum of white noise.

Since the noise voltage as a function of time is not time-limited, a Fourier transform of this voltage cannot be found. However, since in some of the communications systems we consider band-limited signals and since both signal and

*Occasionally N_0 may be used instead of n_0 in the term $n_0/2$.

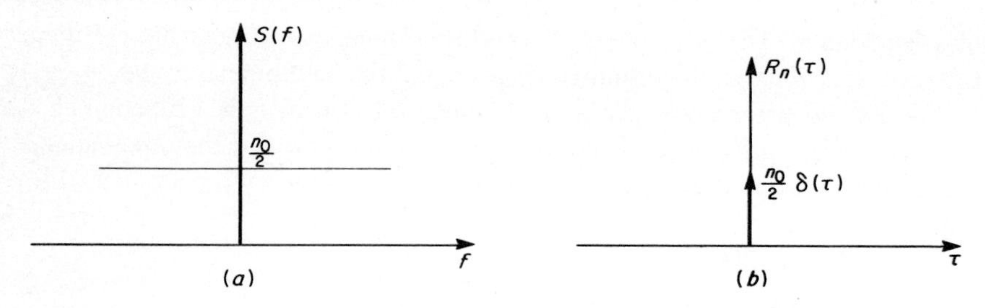

FIGURE 9-3 Characteristics of white noise. (*a*) Power spectrum. (*b*) Autocorrelation.

noise are processed through low-pass or bandpass filters, it is convenient to assume that the noise spectrum is spread uniformly over a certain range of frequencies (i.e., the band under consideration) and that it is zero outside this band of frequencies. The frequency spectrum $S(f)$ over this band assumes both positive and negative frequencies, and therefore

$$S(f) = \begin{cases} \dfrac{n_0}{2} & \text{for } f_1 < |f| < f_2 \\ 0 & \text{elsewhere} \end{cases} \tag{9-32}$$

Of course, this is an idealistic simplification since ideal low-pass filters do not exist.

As was discussed previously, the output of a linear filter will be a gaussian process when the input is a gaussian process, and if $H(f)$ is the transfer characteristic, we will have

$$S_0(f) = |H(f)|^2 S_1(f) \tag{9-33}$$

where $S_0(f)$ and $S_1(f)$ are the output and input power spectra functions.

EXAMPLE 9-1 NARROWBAND WHITE NOISE

Assume the white noise power density spectrum depicted in Figure 9-4*a*. The expression for this spectrum is

$$S(f) = \begin{cases} \dfrac{n_0}{2} & \text{for } f_1 < |f| < f_2 \\ 0 & \text{elsewhere} \end{cases} \tag{9-34}$$

The autocorrelation function will be

$$R_n(\tau) = \int_{-\infty}^{\infty} S(f) e^{j2\pi f\tau}\, df = \int_{-f_2}^{-f_1} \frac{n_0}{2} e^{j2\pi f\tau}\, df + \int_{f_1}^{f_2} \frac{n_0}{2} e^{j2\pi f\tau}\, df$$

and

$$R_n(\tau) = P[\text{sinc}\,(f_2 - f_1)\tau] \cos \pi(f_1 + f_2)\tau \tag{9-35}$$

where sinc $x = (\sin \pi x)/\pi x$ and $P = n_0(f_2 - f_1)$ is the total noise mean power. $R_n(\tau)$ is depicted in Fig. 9-4*b*.

EXAMPLE 9-2 AUTOCORRELATION OF SINE WAVE WITH RANDOM PHASE

Consider the probability distribution function $p(\phi)$ of the phase ϕ of a sinusoidal wave $x(t) = A \cos(2\pi f_c t + \phi)$ to be

$$p(\phi) = \begin{cases} \dfrac{1}{2\pi} & \text{for } 0 \le \phi \le 2\pi \\ 0 & \text{elsewhere} \end{cases} \tag{9-36}$$

Equation (9-33) expresses the fact that ϕ is uniformly distributed from 0 to 2π.

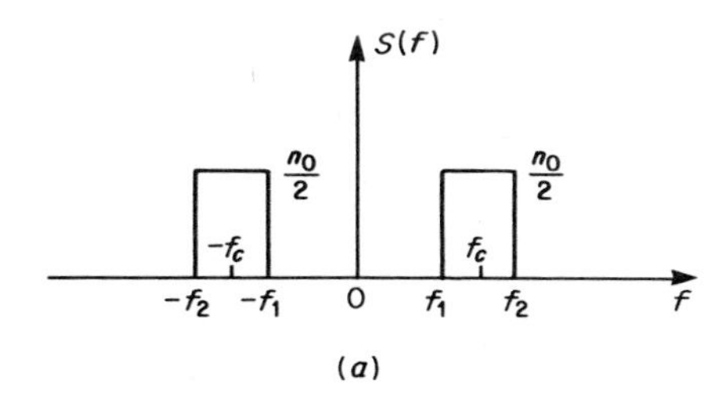

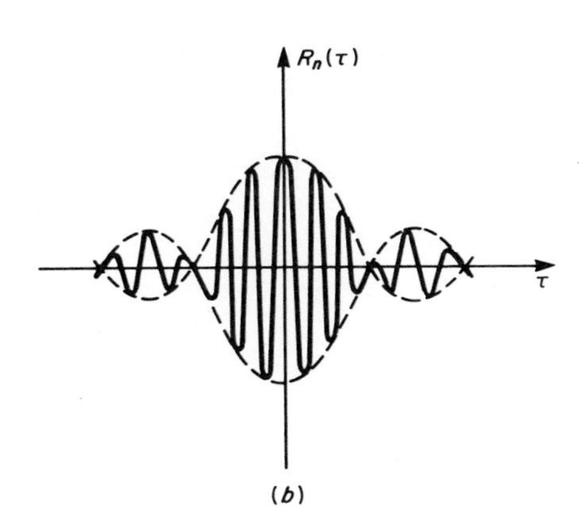

FIGURE 9-4 Characteristics of narrow-band white noise. (*a*) Band-limited power spectrum. (*b*) Autocorrelation.

Computing the autocorrelation, we have

$$\begin{aligned} R_x(\tau) &= \langle x(t+\tau)x(t)\rangle \\ &= \langle A\cos(2\pi f_c t + 2\pi f_c \tau + \phi)\, A\cos(2\pi f_c t + \phi)\rangle \\ &= \frac{A^2}{2}\langle \cos 2\pi f_c \tau + \cos(4\pi f_c t + 2\pi f_c \tau + 2\phi)\rangle \\ &= \frac{A^2}{2}\cos 2\pi f_c \tau + \frac{A^2}{2}\langle \cos(4\pi f_c t + 2\pi f_c r + 2\phi)\rangle \end{aligned} \tag{9-37}$$

But, because of the following relationship,

$$\langle \cos(4\pi f_c t + 2\pi f_c \tau + 2\phi)\rangle = \int_0^{2\pi} \frac{1}{2\pi} \cos(4\pi f_c t + 2\pi f_c \tau + 2\phi)\, d\phi = 0$$

we obtain

$$R_x(\tau) = \frac{A^2}{2}\cos 2\pi f_c \tau \tag{9-38}$$

NARROWBAND NOISE

Various types of noise produced by electronic devices were discussed in Chap. 5. In a previous section of this chapter, certain aspects of white noise were also presented. In this section we will cover certain analytical methods useful in the analysis of noise.

Analysis into Quadrature Components

Because of the selectivity of various filters in the communications channel, the transmission and the associated noise are limited to a certain band. Narrowband noise can be analyzed in a manner similar to that in which bandpass signals were analyzed in Chap. 8, with the aid of Hilbert transforms.

If we express the noise voltage in the same form as the bandpass signals, we have

$$n(t) = a(t)\cos[2\pi f_c t + \phi(t)] \tag{9-39}$$

where f_c is the carrier frequency.

The analytic signal, or preenvelope, is

$$g(t) = n(t) + j\hat{n}(t) = u(t)e^{j2\pi f_c t} \tag{9-40}$$

where $u(t)$ is the complex envelope.

If $x(t)$ and $y(t)$ are the defined by the relationship

$$u(t) = x(t) + jy(t) \tag{9-41}$$

then we can resolve the noise into two quadrature components,

$$n(t) = x(t) \cos 2\pi f_c t - y(t) \sin 2\pi f_c t \tag{9-42}$$

Similarly

$$\hat{n}(t) = x(t) \sin 2\pi f_c t + y(t) \sin 2\pi f_c t \tag{9-43}$$

The autocorrelation function $R_n(\tau)$ is defined as

$$R_n(\tau) = \langle n(t + \tau) n(t) \rangle \tag{9-44}$$

By inserting Eq. (9-42) in (9-44), we obtain

$$R_n(\tau) = \langle [x(t + \tau) \cos 2\pi f_c(t + \tau) - y(t + \tau) \sin 2\pi f_c(t + \tau)][x(t) \cos 2\pi f_c t - y(t) \sin 2\pi f_c t] \rangle$$

and by multiplying through and utilizing some common trigonometric identities, we get

$$\begin{aligned} R_n(\tau) = &\tfrac{1}{2}[\langle x(t + \tau)x(t) \rangle + \langle y(t + \tau)x(t) \rangle] \cos 2\pi f_c \tau \\ &+ \tfrac{1}{2}[\langle x(t + \tau)x(t) \rangle - \langle y(t + \tau)y(t) \rangle] \cos 2\pi f_c(2t + \tau) \\ &- \tfrac{1}{2}[\langle x(t + \tau)y(t) \rangle + \langle y(t + \tau)x(t) \rangle] \sin 2\pi f_c(2t + \tau) \\ &+ \tfrac{1}{2}[\langle x(t + \tau)y(t) \rangle - \langle y(t + \tau)x(t) \rangle] \sin 2\pi f_c \tau \end{aligned}$$

In order for $R_n(\tau)$ to be independent of t, if the ensemble of the noise waves $n(t)$ is to represent a stationary process,

$$\begin{aligned} R_x(\tau) &= \langle x(t + \tau)x(t) \rangle = \langle y(t + \tau)y(t) \rangle = R_y(\tau) \\ R_{xy}(\tau) &= \langle x(t + \tau)y(t) \rangle = -\langle y(t + \tau)x(t) \rangle = -R_{yx}(\tau) \end{aligned} \tag{9-45}$$

In addition, because of the general properties of the autocorrelation function, we have

$$\begin{aligned} R_x(\tau) &= R_x(-\tau) \\ R_{xy}(\tau) &= -R_{xy}(-\tau) \end{aligned} \tag{9-46}$$

The noise autocorrelation function now can be simplified to

$$R_n(\tau) = R_x(\tau) \cos 2\pi f_c \tau + R_{xy}(\tau) \sin 2\pi f_c \tau \tag{9-47}$$

In Chap. 8 the autocorrelation of the complex envelope $u(t)$ was defined as

$$R_u(\tau) = \tfrac{1}{2}\langle u^*(t) u(t + \tau) \rangle \tag{9-48}$$

If we substitute $u(t) = x(t) + jy(t)$, we obtain

$$R_u(\tau) = R_x(\tau) + jR_{xy}(\tau) \tag{9-49}$$

By making use of Eqs. (9-45) we can easily prove that

$$\langle u(t)u(t + \tau)\rangle = 0 \tag{9-50}$$

We can express the autocorrelation $R_n(\tau)$ in terms of $R_u(\tau)$ by utilizing the expression

$$n(t) = \tfrac{1}{2}[g(t) + g^*(t)] = \tfrac{1}{2}[u(t)e^{j2\pi f_c t} + u^*(t)e^{-j2\pi f_c t}]$$

By inserting this relationship into the expression for $R_n(\tau) = \langle n(t)n(t + \tau)\rangle$ and making use of Eq. (9-48), we obtain

$$\begin{aligned} R_n(\tau) &= \tfrac{1}{2}[R_u(\tau)e^{j2\pi f_c \tau} + R_u^*(\tau)e^{-j2\pi f_c \tau}] \\ &= \text{Re}\,[R_u(\tau)e^{j2\pi f_c \tau}] \end{aligned} \tag{9-51}$$

If $S_n(f)$ is the Fourier transform of $R_n(\tau)$, then $S_n(f)$ is the power density spectrum of $n(t)$, and

$$S_n(f) = \tfrac{1}{2}S_u(f - f_c) + \tfrac{1}{2}S_u^*(-f - f_c) \tag{9-52}$$

where $S_u(f)$ is the Fourier transform of $R_u(\tau)$. From Eqs. (9-49) and (9-46), it is easy to show that $S_u(f)$ is a real function, and therefore

$$S_u(f) = S_u^*(f) \tag{9-53}$$

and Eq. (9-52) becomes

$$S_n(f) = \tfrac{1}{2}S_u(f - f_c) + \tfrac{1}{2}S_u(-f - f_c) \tag{9-54}$$

We have shown before [in Eq. (9-46)] that the cross-correlation $R_{xy}(\tau) = \langle x(t)y(t + \tau)\rangle$ is an odd function of τ; as a result,

$$R_{xy}(0) = -R_{yx}(0) = \langle x(t)y(t)\rangle = 0 \tag{9-55}$$

This implies that $x(t)$ and $y(t)$ are statistically independent at any given instant of time. However, in general, for $\tau \neq 0$ they are not statistically independent. It is also easily seen that by using Eqs. (9-55) and (9-49) we can obtain

$$R_u(0) = R_x(0)$$

and consequently

$$\begin{aligned} \langle [n(t)]^2\rangle &= R_n(0) = \text{Re}\,[R_u(0)] \\ &= R_x(0) = R_y(0) = \langle [x(t)]^2\rangle = \langle [y(t)]^2\rangle \end{aligned} \tag{9-56}$$

That is, the mean square expectation of noise equals the mean square expectations of its quadrature components.

EXAMPLE 9-3 NOISE WITH GAUSSIAN DISTRIBUTION

Assume the noise $n(t)$ to have a gaussian distribution with zero mean and standard deviation σ^2. Then the quadrature components $x(t)$ and $y(t)$ will also have a gaussian distribution, each with zero mean, i.e., $\langle x(t)\rangle = \langle y(t)\rangle = 0$ and the same standard deviation $[x(t)]^2 = [y(t)]^2 = \sigma^2$. But, since $x(t)$ and $y(t)$ are statistically independent according to Eq. (9-55), the joint probability density function will be the product of the individual pdf's; that is,

$$p(x, y) = p(x)p(y) = \frac{1}{2\pi\sigma^2} e^{-(x^2+y^2)/2\sigma^2} \tag{9-57}$$

Narrowband Noise on a Carrier Envelope (Rayleigh and Rician Distributions)

The quadrature components $x(t)$ and $y(t)$ of the narrowband noise voltage are related to the envelope $r(t)$ and phase $\phi(t)$ by the equations

$$x(t) = r \cos \phi \tag{9-58}$$
$$y(t) = r \sin \phi$$

[For clarification, see Eq. (8-26) and set $r = a$.] By making a coordinate transformation from noise quadrature components to envelope and phase coordinates [polar coordinates $r(t)$ and $\phi(t)$], we obtain the following relationship between the joint pdf's:

$$\begin{aligned} p_{x,y}(x, y)\,dx\,dy &= p_{x,y}(r \cos \phi, r \sin \phi)\,dx\,dy \\ &= p_{r,\phi}(r, \phi) r\,dr\,d\phi \end{aligned} \tag{9-59}$$

[See Eq. (9-14) for clarification.]

Now if the noise $n(t)$ obeys a gaussian distribution with zero mean and standard deviation σ^2 according to Example 9-3, we have

$$p_{x,y}(x, y) = p_x(x)p_y(y) = \frac{1}{2\pi\sigma^2} e^{-(x^2+y^2)/2\sigma^2} \tag{9-60}$$

But because of Eq. (9-59) we obtain

$$p_{r,\phi}(r, \phi) = \frac{r}{2\pi\sigma^2} e^{-r^2/2\sigma^2} \tag{9-61}$$

The probability density function of the noise envelope could be found by obtaining the expectation over all phases; that is,

$$p_r(r) = \int_0^{2\pi} \frac{r}{2\pi\sigma^2} e^{-r^2/2\sigma^2} \, d\phi$$

and

$$p_r(r) = \frac{r}{\sigma^2} e^{-r^2/2\sigma^2} \tag{9-62}$$

The above distribution is called the Rayleigh distribution and represents the output of an envelope detector in the absence of a signal.

The phase probability could be obtained in a similar fashion by integrating the joint pdf over the radius r variations from 0 to ∞. The result is a rectangular distribution:

$$p(\phi) = \frac{1}{2\pi} \tag{9-63}$$

Figure 9-5 depicts a plot of the Rayleigh distribution.

Next we will examine the case in which an actual message signal is present. Given a sinusoidal signal

$$m(t) = A \cos 2\pi f_c t \tag{9-64}$$

and a noise signal as described by Eq. (9-42), the signal-plus-noise waveform is

$$v(t) = m(t) + n(t) = [A + x(t)] \cos 2\pi f_c t - y(t) \sin 2\pi f_c t \tag{9-65}$$

The waveform $v(t)$ can be written as

$$v(t) = r(t) \cos [2\pi f_c t + \phi(t)] \tag{9-66}$$

where

$$r(t) = \sqrt{[A + x(t)]^2 + [y(t)]^2} \tag{9-67}$$

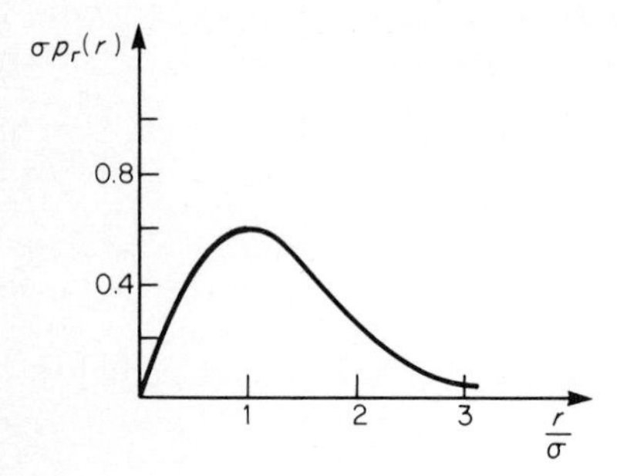

FIGURE 9-5 Rayleigh distribution.

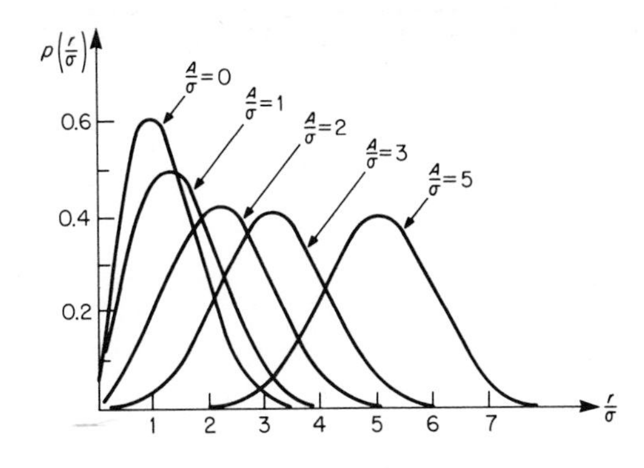

FIGURE 9-6 Rician distribution.

and

$$\phi(t) = \tan^{-1} \frac{-y(t)}{A + x(t)} \tag{9-68}$$

If $n(t)$ is gaussian with zero mean, then both $x(t)$ and $y(t)$ are gaussian with zero mean and variance $\sigma^2 = \langle x^2 \rangle = \langle y^2 \rangle$.

The joint probability density function of r and ϕ will be

$$p(r, \phi) = \frac{r}{2\pi\sigma^2} e^{-(r^2+A^2-2Ar\cos\phi)/2\sigma^2} \tag{9-69}$$

The integral $\int_0^{2\pi} p(r, \phi)\, d\phi$ gives the pdf of r as

$$p(r) = \frac{r}{\sigma^2} I_0\left(\frac{Ar}{\sigma^2}\right) e^{-(r^2+A^2)/2\sigma^2} \tag{9-70}$$

where $I_0(x)$ is a zero-order modified Bessel function of the first kind. This distribution is called a "Rician distribution" (see Fig. 9-6).

Similarly, the integral $\int_0^{\infty} p(r,\phi)\, dr$ gives the pdf of ϕ as

$$P(\phi) = \frac{1}{2\pi} e^{-A^2/2\sigma^2} + \frac{A\cos\phi}{2\sigma\sqrt{2\pi}}\left(1 + \operatorname{erf}\frac{A\cos\phi}{2\sigma^2}\right) e^{-(A^2\sin^2\phi)/2\sigma^2} \tag{9-71}$$

For $A = 0$, that is, only noise present, Eq. (9-70) becomes

$$p(r) = \frac{r}{\sigma^2} e^{-r^2/2\sigma^2} \tag{9-72}$$

which is a Rayleigh distribution.

REPRESENTATION OF NOISE

In the previous section, noise was analyzed into two quadrature components,

$$n(t) = x(t)\cos\omega t - y(t)\sin\omega t$$

where $x(t)$ and $y(t)$ are the real and imaginary parts of the complex envelope. In addition, it was shown that, for a stationary ergodic process, the two components $x(t)$ and $y(t)$ were statistically independent when considered at the same instant of time; namely, it was shown that

$$\langle x(t)y(t)\rangle = 0$$

The properties of the Hilbert transform were used in developing these analytical concepts. The functions sin ωt and cos ωt are orthogonal over the time interval 0 to T; that is,

$$\int_0^T \cos \omega t \sin \omega t \, dt = 0$$

where $T = 2\pi/\omega$. In fact, it can be shown that, in general, with a set of orthogonal functions, noise can be resolved into components that are statistically independent (uncorrelated) subject to certain conditions.

The expansion of noise into an orthogonal series with uncorrelated coefficients is known as the "Karhunen-Loève expansion."

General Concepts

A set of functions $\phi_i(t)$ with $i = 0, \pm 1, \pm 2, \ldots$, defined over the time interval 0 to T, is said to be "orthogonal" if it satisfies the relationship

$$\int_0^T \phi_k(t)\phi_e^*(t)\, dt = 0 \qquad \text{for } k \neq e \tag{9-73}$$

where $\phi_e^*(t)$ is the conjugate of $\phi_e(t)$ and k and e are various values of i.

In addition, if

$$\int_0^T \phi_k(t)\phi_e^*(t)\, dt = \delta_{ke} \tag{9-74}$$

where

$$\delta_{ke} = \begin{cases} 1 & \text{for } k = e \\ 0 & \text{for } k \neq e \end{cases} \tag{9-75}$$

the functions $\phi_i(t)$ are said to be "orthonormal." A function of time can be expanded in a series of such orthonormal functions if it satisfies a set of conditions similar to the Dirichlet conditions.

A noise-voltage function $n(t)$ with zero mean can be expressed in terms of a set of orthogonal (or orthonormal) functions over a desired domain 0 to T, so that

$$n(t) = \sum_{i=-\infty}^{\infty} n_i\phi_i(t) \tag{9-76}$$

where the coefficients n_i can be defined by the relationship

$$n_i = \int_0^T n(t)\phi_i^*(t)\,dt \tag{9-77}$$

Indeed, multiplying Eq. (9-76) by $\phi_k^*(t)$ and integrating from 0 to T, we get for the second side of the equation

$$\int_0^T \phi_k^*(t) \sum_{i=-\infty}^{\infty} n_i\phi_i(t)\,dt = n_k \int_0^T |\phi_k(t)|^2\,dt = n_k \tag{9-78}$$

Since, by exchanging the process of summation and integration, we make all the integrals vanish except the one for which $i = k$, the right-hand side of Eq. (9-78) proves the validity of Eq. (9-77).

Fourier Series Analysis

Orthogonality and Series Coefficients Expressing the function $n(t)$ in terms of a Fourier series expansion is one example of resolving $n(t)$ in a series of orthogonal functions. Specifically, over the interval 0 to T we can write

$$n(t) = \sum_{k=-\infty}^{\infty} n_k e^{jk\omega_0 t} \tag{9-79}$$

where $\omega_0 = 2\pi/T$. We can prove orthogonality by computing the integral

$$I = \frac{1}{T}\int_0^T n(t)e^{-je\omega_0 t}\,dt = \frac{1}{T}\int_0^T e^{-je\omega_0 t}\sum_{k=-\infty}^{\infty} n_k e^{jk\omega_0 t}\,dt$$

$$= \frac{1}{T}\sum_{k=-\infty}^{\infty} n_k \int_0^T e^{-je\omega_0 t}e^{jk\omega_0 t}\,dt$$

where

$$\frac{1}{T}\int_0^T e^{-je\omega_0 t}e^{jk\omega_0 t}\,dt = \frac{1}{T}\int_0^T e^{-j(e-k)\omega_0 t}\,dt$$

$$= \begin{cases} 1 & \text{for } e = k \\ 0 & \text{for } e \neq k \end{cases}$$

and therefore

$$I = \frac{1}{T}\int_0^T n(t)e^{-jk\omega_0 t}\,dt = n_k \tag{9-80}$$

The orthogonal functions $\phi_k(t) = e^{jk\omega_0 t}$ were used here for this purpose.

If $n(t)$ is gaussian noise, then the coefficients n_k will also have a gaussian distribution, since they are obtained through linear operations on $n(t)$. By defining $\omega_k = k\omega_0$ and $\omega_e = e\omega_0$, we also have

$$n_k^* n_e = \frac{1}{T}\int_0^T n(t_1)e^{j\omega_k t_1}\,dt_1 \frac{1}{T}\int_0^T n(t_2)e^{-j\omega_e t_2}\,dt_2$$

$$= \frac{1}{T^2}\int_0^T\int_0^T n(t_1)n(t_2)e^{j\omega_k t_1}e^{-j\omega_e t_2}\,dt_1\,dt_2$$

and, by a change of variable $t_1 = t_2 + \tau$,

$$n_k^* n_e = \frac{1}{T^2}\int_0^T\left[\int_{-t_2}^{T-t_2} n(t_2+\tau)n(t_2)e^{j\omega_k\tau}\,d\tau\right]e^{j(\omega_k-\omega_e)t_2}\,dt_2$$

Consequently, the ensemble average will be

$$\langle n_k^* n_e\rangle = \frac{1}{T^2}\int_0^T\left[\int_{-t_2}^{T-t_2} R_n(\tau)e^{j\omega_k\tau}\,d\tau\right]e^{j(\omega_k-\omega_e)t_2}\,dt_2 \tag{9-81}$$

since $R_n(\tau) = \langle n(t_2+\tau)n(t_2)\rangle$ is the autocorrelation function.

White Noise over a Limited Time Interval If we consider white noise, then

$$R_n(\tau) = \frac{n_0}{2}\,\delta(\tau) \tag{9-82}$$

where $n_0/2$ is the two-sided spectral density. For the case of white noise, Eq. (9-81) becomes

$$\langle n_k n_e^*\rangle = \frac{1}{T^2}\int_0^T\left[\int_{-t_2}^{T-t_2}\frac{n_0}{2}\,\delta(\tau)\,e^{-j\omega_k\tau}\,d\tau\right]e^{j(\omega_e-\omega_k)t_2}\,dt_2$$

and, since 0 is contained between $-t_2$ and $T - t_2$ (t_1 and t_2 were both between 0 and T),

$$\langle n_k n_e^*\rangle = \frac{1}{T^2}\int_0^T\frac{n_0}{2}\,e^{j(\omega_e-\omega_k)t_2}\,dt_2$$

or

$$\langle n_k n_e^*\rangle = \frac{1}{T^2}\int_0^T\frac{n_0}{2}\,e^{j(\omega_e-\omega_k)t}\,dt \tag{9-83}$$

Now, for $e \neq k$, the integral in Eq. (9-83) is zero, which means that n_k and n_e are uncorrelated.

Therefore the coefficients n_k are both gaussian and uncorrelated except, of course, when $e = k$, and then

$$T\langle|n_k|^2\rangle = \frac{n_0}{2} \tag{9-84}$$

Colored Noise over an Infinite Time Interval Now let us examine an infinite (timewise) strip of noise which is not necessarily white. The analysis can be facilitated by translating the time origin to the right by $T/2$. In this case we can derive Eq. (9-81) as

$$T\langle n_k n_e^*\rangle = \frac{1}{T}\int_{-T/2}^{T/2}\left[\int_{-(T/2)-t_2}^{(T/2)-t_2} R_n(\tau)e^{-j\omega_k\tau}\,d\tau\right]e^{j(\omega_e-\omega_k)t_2}\,dt_2 \qquad \text{(9-85)}$$

If we let $T \to \infty$, then $T/2 - t_2 \to T/2$ and $-T/2 - t_2 \to -T/2$. Therefore,

$$\lim_{T\to\infty} T\langle n_k n_e^*\rangle = \left[\lim_{T\to\infty}\int_{-T/2}^{T/2} R_n(\tau)e^{-j\omega_k\tau}\,d\tau\right]\left(\lim_{T\to\infty}\frac{1}{T}\int_{-T/2}^{T/2} e^{j(\omega_e-\omega_k)t_2}\,dt_2\right)$$

or, if we call $S_T(\omega)$ the power spectrum of the noise strip within $-T/2$ to $T/2$,

$$\lim_{T\to\infty} T\langle n_k n_e^*\rangle = \left[\lim_{T\to\infty} S_T(\omega_k)\right]\left(\lim_{T\to\infty}\frac{1}{T}\int_{-T/2}^{T/2} e^{j(\omega_e-\omega_k)t_2}\,dt_2\right)$$

But

$$\lim_{T\to\infty} S_T(\omega_k) = S_n(\omega_k)$$

where $S_n(\omega)$ is the power spectrum of $n(t)$, and

$$\lim_{T\to\infty}\frac{1}{T}\int_{-T/2}^{T/2} e^{j(\omega_e-\omega_k)t_2}\,dt_2 = \begin{cases}1 & \text{for } e = k\\ 0 & \text{for } e \neq k\end{cases}$$

Therefore

$$\lim_{T\to\infty} T\langle n_k n_e^*\rangle = \begin{cases}S_n(\omega_k) & \text{for } k = e\\ 0 & \text{for } k \neq e\end{cases} \qquad \text{(9-86)}$$

Thus we conclude that for an infinite strip of noise, not necessarily white, the coefficients n_k and n_e are uncorrelated. In addition,

$$\lim_{T\to\infty} T|n_k|^2 = S_n(\omega_k)$$

Band-Limited White Noise

In this case, we have the power spectrum

$$S_n(f) = \begin{cases}\dfrac{n_0}{2} & \text{for } -\dfrac{B}{2} < f < \dfrac{B}{2}\\ 0 & \text{elsewhere}\end{cases} \qquad \text{(9-87)}$$

The autocorrelation function, which is the Fourier transform of $S_n(f)$, will be

$$R_n(\tau) = n_0 B \frac{\sin 2\pi B\tau}{2\pi B\tau} \tag{9-88}$$

Samples of noise taken every $1/2B$ so that $2\pi B\tau = k\pi$, where $k = 0, 1, 2, \ldots$, will make the autocorrelation function zero. Therefore these samples will be uncorrelated.

In a strict sense, band-limited noise cannot be constrained over a limited time interval T in the time domain. However, we can still consider this case by making approximations if T is large enough so that $BT \gg 1$. The noise voltage can be expressed as

$$n(t) = \sum_{k=1}^{2BT} n\left(\frac{k}{2B}\right) \frac{\sin (2Bt - k)}{2Bt - k} \tag{9-89}$$

Here the samples of noise $n(k/2B)$ are taken so that they are uncorrelated. Equation (9-89) is derived in the same way that the time function is derived according to the sampling theorem, in terms of the samples at the Nyquist interval.

Equation (9-88) shows that $R_n(\tau) = 0$ for values of $\tau = k/2B$. Therefore the coefficients $n(k/2B)$ of the expansion, Eq. (9-89), are indeed uncorrelated. We have

$$\langle n(t)^2 \rangle = R_n(0) = n_0 B$$

and since

$$\langle n(t)^2 \rangle = \left\langle n^2\left(\frac{k}{2B}\right)\right\rangle$$

we also have

$$\left\langle n^2\left(\frac{k}{2B}\right)\right\rangle = n_0 B$$

Expansion to a Series with Uncorrelated Coefficients

Let us again consider the general case in which the noise function $n(t)$ is expanded in terms of a set of orthonormal functions $\phi_k(t)$, defined over the time interval 0 to T. We have

$$n(t) = \sum_{k=1}^{\infty} n_k \phi_k(t) \tag{9-90}$$

where

$$n_k = \int_0^T n(t)\phi_k^*(t)\,dt \tag{9-91}$$

If $n(t)$ is gaussian, then the n_k's will also have a gaussian distribution.

Now, if the functions $\phi_k(t)$ are such that the coefficients n_k are uncorrelated, then we have the so-called Karhunen-Loève expansion.

The product of two coefficients n_k and n_e will be

$$n_k n_e^* = \int_0^T n(t_1)\phi_k^*(t_1)\,dt_1 \int_0^T n(t_2)\phi_e(t_2)\,dt_2$$

By taking the ensemble average, we have

$$\langle n_k n_e^* \rangle = \int_0^T \int_0^T \langle n(t_1)n(t_2)\rangle \phi_k^*(t_1)\phi_e(t_2)\,dt_1\,dt_2$$

or

$$\langle n_k n_e^* \rangle = \int_0^T \left[\int_0^T R(t_2 - t_1)\phi_e(t_2)\,dt_2 \right] \phi_k^*(t_1)\,dt_1 \tag{9-92}$$

If we define the orthonormal functions in such a way as to satisfy the integral equation

$$\int_0^T R(t_2 - t_1)\phi_e(t_2)\,dt_2 = \sigma_e^2\phi_e(t_1) \tag{9-93}$$

then

$$\langle n_k n_e^* \rangle = \int_0^T \sigma_e^2\phi_e(t_1)\phi_k^*(t_1)\,dt_1$$

and, by making use of the orthogonality, we have

$$\langle n_k n_e^* \rangle = \begin{cases} \sigma_e^2 & \text{for } e = k \\ 0 & \text{for } e \neq k \end{cases}$$

Therefore, when Eq. (9-93) is satisfied, the coefficients of the expansion are uncorrelated.

In Fourier series expansion for white noise, we have seen that the coefficients of the expansion are uncorrelated, and therefore Eq. (9-93) must be satisfied.

The same was true in the case of band-limited white noise; therefore the functions

$$\phi_k(t) = \frac{\sin \pi(2Bt - k)}{\pi(2Bt - k)}$$

are orthogonal and satisfy the Karhunen-Loève integral equation, Eq. (9-93). The expansion of band-limited gaussian stationary noise

$$n(t) = x(t) \cos \omega_c t - y(t) \sin \omega_c t$$

by use of the Hilbert transform and the analytic signal $g(t) = u(t)e^{j\omega ct}$ so that

$$n(t) = \text{Re}\,[u(t)e^{j\omega ct}]$$

where

$$u(t) = x(t) + jy(t)$$

results in the relationship

$$\langle x(t)y(t) \rangle = 0 \tag{9-94}$$

This again shows that $x(t)$ and $y(t)$ are uncorrelated when considered at the same instant of time.

REFERENCES

Haykin, Simon: *Communication Systems*, John Wiley & Sons, New York, 1978.

Members of the Technical Staff of Bell Telephone Laboratories: *Transmission Systems for Communications*, rev. 4th ed., Bell Telephone Laboratories, Winston Salem, N.C., 1971.

Papoulis, Athanasios: *Probability Random Variables and Stochastic Processes*, McGraw-Hill, New York, 1965.

Parzen, Emanuel: *Modern Probability Theory and Its Applications*, John Wiley & Sons, New York, 1960.

Schwartz, M., W. R. Bennett, and S. Stein: *Communication Systems and Techniques*, McGraw-Hill, New York, 1966.

Stein, S., and J. J. Jones: *Modern Communication Principles with Application to Digital Signaling*, McGraw-Hill, New York, 1967.

10
ANALOG MODULATION

DEFINITIONS

The objective of a communications channel is to transmit signals carrying information. This transmission takes place in the presence of noise. The message to be transmitted is a band-limited signal with a frequency content much lower than the radio-transmission frequencies. (Video signals have a bandwidth of several megahertz.) This message signal is called the "baseband signal." Radio transmission, such as satellite transmission, takes place at frequencies much higher than the baseband signal.

Satellite communications systems utilize carriers at the gigahertz range for the carriage of baseband signals (see Chap. 5). The baseband signal modulates the carrier, utilizing for this purpose a particular characteristic of the carrier, such as phase. Of course, the carrier characteristic that is used for carrying the message will vary with time, since the message usually is time-varying.

The analysis of a modulated wave in general terms can be very complicated, if not impossible. However, we can understand most of the behavior of this wave if we assume that the baseband signal consists of a sum of sinusoidal components and conduct the analysis for one of these components. This is justifiable since most baseband signals can be expressed through either a Fourier series or a Fourier transform.

A general expression for the modulated carrier $m(t)$ is given by the relationship

$$m(t) = a(t) \cos [2\pi f_c t + \phi(t)]$$

where the amplitude $a(t)$ and the angle $\theta(t) = 2\pi f_c t + \phi(t)$ are functions of time. The message information is carried either in $a(t)$ or in $\theta(t)$ as follows:

1. If it is carried in $a(t)$, we have amplitude modulation (AM).
2. If it is carried in $\theta(t) = 2\pi f_c t + \phi(t)$, we have angle modulation.
   a. If it is carried in $\phi(t)$, we have phase modulation (PM).
   b. If it is carried in $d\phi(t)/dt = \phi'(t)$, we have frequency modulation (FM).

We define

$$\theta(t) = 2\pi f_c t + \phi(t) = \text{instantaneous phase}$$

$$\theta'(t) = 2\pi f_c + \phi'(t) = \text{instantaneous frequency}$$

where f_c = carrier frequency.

AMPLITUDE MODULATION

General Description

The general expression for the AM wave is

$$m(t) = [v_0 + v(t)] \cos (2\pi f_c t + \phi) \tag{10-1}$$

where v_0 is a bias to prevent the amplitude from going negative. Most often, the modulated signal is expressed as

$$m(t) = A_c[1 + mv(t)] \cos 2\pi f_c t \tag{10-2}$$

where m is the modulation index and $v(t)$, the baseband signal, is normalized so that $|v(t)| \leq 1$. In addition, the modulation index m is less than 1. For $m = 1$ we have 100 percent modulation, and for $m = 0.5$ we have 50 percent modulation.

Spectral Representation of an AM Wave

If the modulating signal is a single sinusoid $v(t) = a \cos 2\pi ft$, then

$$m(t) = A_c[1 + ma \cos 2\pi ft] \cos 2\pi f_c t \tag{10-3}$$

or

$$m(t) = A_c \cos 2\pi f_c t + \tfrac{1}{2} maA_c \cos [2\pi(f_c + f)t] + \tfrac{1}{2} maA_c \cos [2\pi(f_c - f)t] \tag{10-4}$$

The single-sinusoid frequency spectrum is shown in Fig. 10-1*a*. The more general case is depicted in Fig. 10-1*b*.

By taking the Fourier transform of Eq. (10-2), we obtain

$$M(f) = \tfrac{1}{2}A_c[\delta(f - f_c) + \delta(f + f_c)] + \tfrac{1}{2}mA_c[V(f - f_c) + V(f + f_c)] \quad (10\text{-}5)$$

Figure 10-2 depicts the spectra of the baseband and the AM wave. For most of the practical applications, $f_c \gg f_m$, and therefore there is no intermixing of the spectra. Then one can filter out all other frequencies and transmit only the modulated wave by transmitting the carrier and the two sidebands, namely, the upper sideband and the lower sideband.

If $f_c \gg f_m$, the modulated wave in the time domain looks like the one depicted in Fig. 10-3. The envelope closely duplicates the information signal $v(t)$. In order to detect the signal $v(t)$ at the receiving end, all one has to do is detect the envelope.

Modulators

Several types of amplitude modulators have been developed, among them the following.

Curvature Modulators The most common modulator of this type is the square-law modulator. A nonlinear device—such as a semiconductor diode, properly biased—can be used to provide an output-input characteristic of the form

$$v_0 = k_1 v_1 + k_2 v_1^2 \quad (10\text{-}6)$$

The input voltage v_1 is the sum of the carrier and the modulating signal. By proper filtering of the output, the modulated carrier can be obtained.

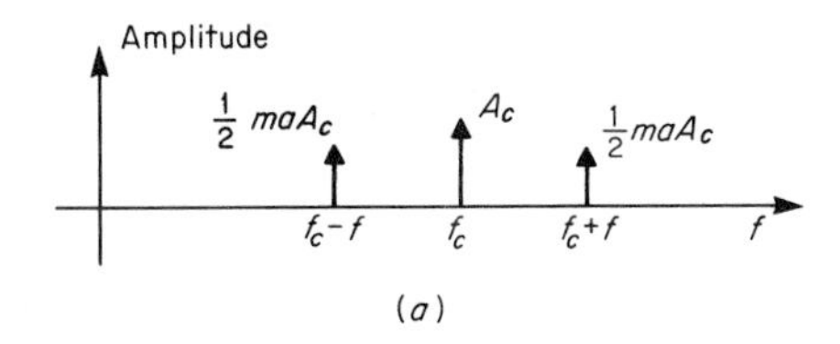

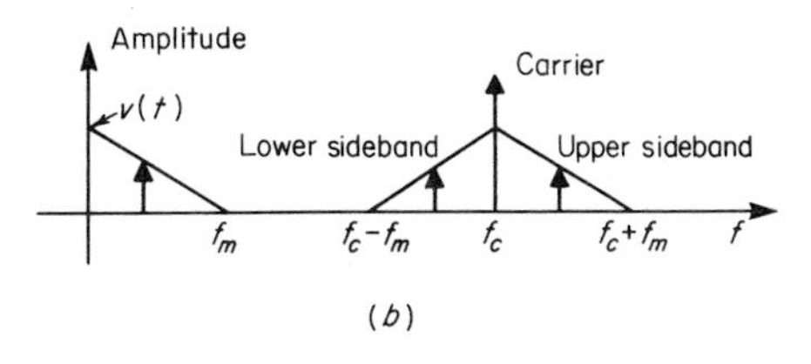

FIGURE 10-1 One-sided spectra of AM wave. (*a*) Single-sinusoid frequency spectrum. (*b*) Triangular-baseband spectrum.

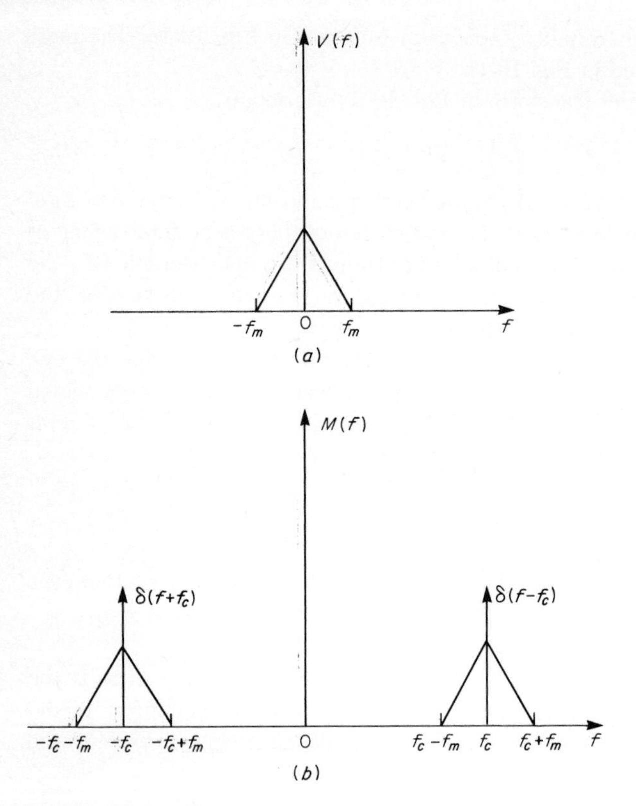

FIGURE 10-2 **AM spectra. (*a*) Spectrum of baseband signal. (*b*) Spectrum of the AM wave.**

Switching Modulators One of the most useful modulators of this type is the ring modulator. Figure 10-4 shows a typical ring modulator.

It is easy to show that the output of this modulator consists only of the desired products. For this reason it is also known as a product modulator.

Demodulators and Detectors

Several types of demodulators are being used, among them the following.

Envelope Detector Figure 10-5 shows the schematic of the most frequently used envelope detector. The high frequency passes through a capacitor C, charging the capacitor to its maximum value, and the low-frequency envelope passes through R, becoming the detector output.

Square-Law Detector A square-law modulator can also be used for demodulation. The output-input characteristic is the same as the one indicated in Eq.

(10-6). If we assume v_1 to be expressed by Eq. (10-3), it is easy to see that the output, after filtering, will provide the message signal.

Synchronous Detection By having a reference carrier available at the receiving end, with accurate frequency and phase, one can reproduce the baseband by multiplying the received modulated wave by the reference carrier and filtering out the high frequencies (the homodyne principle). The proof is obtained quite simply with the use of trigonometric identities. If the receiver carrier has a phase error $\Delta\phi$, the output of the coherent detector will be multiplied by $\cos \Delta\phi$. As a result, a signal loss will be experienced.

Amplitude-Modulation Schemes

The full amplitude-modulated signal includes both sidebands and the carrier. The transmission of such a signal is inefficient in bandwidth use, since one sideband alone contains all the necessary information for the reproduction of the baseband signal; it also makes inefficient use of power, since the carrier power

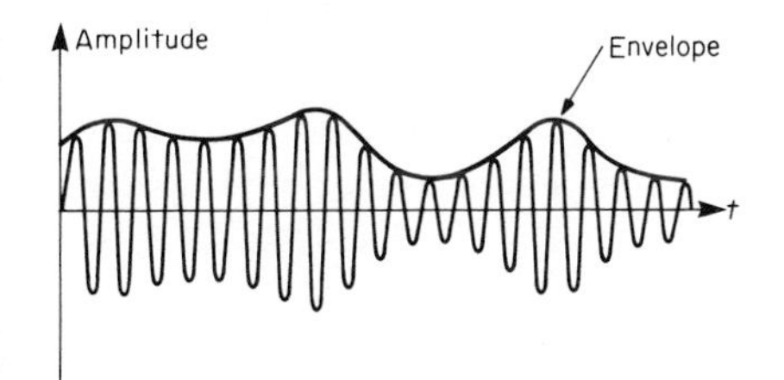

FIGURE 10-3 Modulated wave.

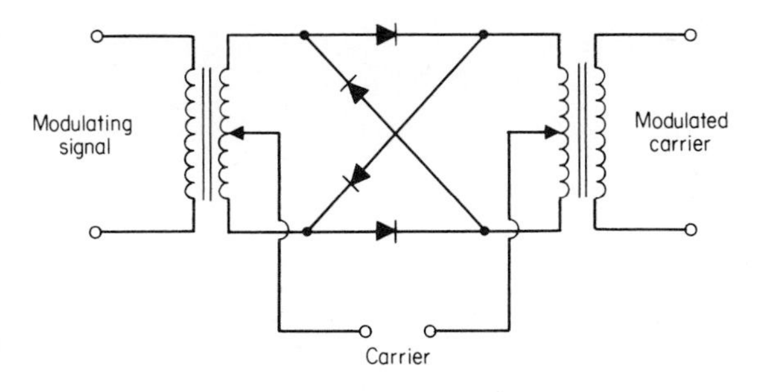

FIGURE 10-4 Double-balanced ring modulator.

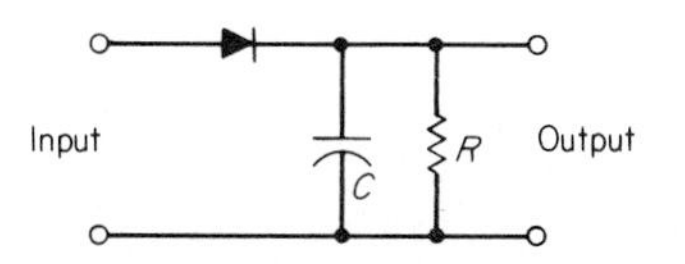

FIGURE 10-5 Envelope detector.

and the second-sideband power are not necessary. On the other hand, eliminating one sideband or the carrier, or both, complicates the detection.

A number of concepts with improved bandwidth or power efficiency, or both, have been developed and have varying degrees of complication at the detector end. Some examples follow.

Double-Sideband Suppressed Carrier Only the carrier is eliminated in double-sideband suppressed-carrier (DSBSC) transmission. Carrier recovery at the detector is essential for the demodulation process. The carrier can be recovered either by detecting a small carrier transmitted separately or by operating on the sidebands. It is important also to recover accurately the phase ϕ of the carrier. A phase error will serve to reduce the size of the recovered signal. As has been previously mentioned, the magnitude of the recovered signal will be proportional to the cos $\Delta\phi$, where $\Delta\phi$ is the phase error in the recovered carrier.

Single Sideband Single-sideband (SSB) transmission eliminates one of the sidebands and the carrier. Thus both a power advantage and a bandwidth advantage are obtained. However, the detection becomes more complicated. A modification of this approach is to transmit a small portion of the second sideband and carrier while reducing equally the corresponding portion of the transmitted sideband. At the detection end, by folding around zero the transmitted portion of the second sideband, the total spectrum of the baseband is reproduced. This transmission is called vestigial-sideband transmission.

Postdetection Signal-to-Noise Ratio

Envelope Detection (Full-AM Transmission) In the analysis of narrowband noise $n(t)$ appearing at the envelope of a modulated carrier, it was pointed out that the probability density function for the envelope radius r is

$$p(r) = \frac{r}{N_1} e^{-r^2/2N_1} e^{-\gamma} I_0 \left(\frac{rA_c}{N_1} \right) \tag{10-7}$$

where

N_1 = input mean square of the gaussian noise with zero mean

γ = carrier-to-noise ratio

$= \dfrac{A_c^2}{2N_1} = \dfrac{C}{N_1}$

A_c = amplitude of the carrier

I_0 = modified Bessel function of the first kind and zero order

r = envelope radius

The envelope radius is expressed as

$$|r|^2 = |x + A_c|^2 + y^2$$

where x and y are the quadrature components of the noise:

$$n(t) = x \cos 2\pi f_c t - y \sin 2\pi f_c t$$

The noise components $x(t)$ and $y(t)$ are gaussian with zero mean and are uncorrelated; that is, $\langle x(t)y(t)\rangle = 0$.

We also have for the correlation functions

$$R_x(\tau) = \langle x(t)x(t + \tau)\rangle = \langle y(t)y(t + \tau)\rangle = R_y(\tau)$$

$$R_{xy}(\tau) = \langle x(t)y(t + \tau)\rangle = -\langle y(t)x(t + \tau)\rangle = -R_{yx}(\tau)$$

In addition we have

$$R_x(\tau) = R_y(\tau) = R_n(\tau) \cos 2\pi f_c\tau - \hat{R}_n(\tau) \sin 2\pi f_c\tau$$

and

$$R_x(\tau) \leftrightarrows S_x(f) = \begin{cases} S_n(f + f_c) + S_n(f - f_c) & \text{for } -f_c < f < f_c \\ 0 & \text{elsewhere} \end{cases}$$

where $S_n(f)$ is the power spectral density of the noise $n(t)$, $R_n(\tau)$ the autocorrelation of $n(t)$, and $\hat{R}_n(\tau)$ is the Hilbert transform of $R_n(\tau)$.

The definition of the output signal-to-noise ratio is not always consistent. We will use the definition in which the output signal is considered to be equal to the output in the absence of noise.

In the case in which the modulating signal is a single sinusoid,

$$v_1(t) = m(t) + n(t) = A_c(1 + m \cos 2\pi ft) \cos 2\pi f_c t + n(t) \qquad (10\text{-}7)$$

where $n(t)$ is noise. (We assume $a = 1$.) If $\langle n(t)^2\rangle = N$, we have for the input carrier-to-noise ratio

$$\frac{C}{N} = \frac{A_c^2}{2N} \qquad (10\text{-}8)$$

and for the input signal-to-noise ratio

$$\left(\frac{S}{N}\right)_1 = m^2 \frac{A_c^2}{4N} \quad \text{or} \quad \left(\frac{S}{N}\right)_1 = \left(1 + \frac{m^2}{2}\right)\frac{A_c^2}{2N} \qquad (10\text{-}9)$$

Equation (10-9) is derived by computing $\langle v_1^2\rangle$ as

$$\langle v_1^2\rangle = \frac{1}{T}\int_0^T [m(t)]^2\, dt + \langle n^2\rangle = \frac{A_c^2}{2}\left(1 + \frac{m^2}{2}\right) + N \qquad (10\text{-}10)$$

Depending on whether or not we take the carrier power $A_c^2/2$ into account, we obtain the first or the second expression in Eq. (10-9).

The noise $n(t)$ can be expressed in polar coordinates as

$$n(t) = r_n(t) \cos [2\pi f_c t + \phi_n(t)] \tag{10-11}$$

The square of the output of the linear envelope detector will be

$$\begin{aligned} v_0^2 = A_c^2[1 + m \cos 2\pi ft]^2 + [r_n(t)]^2 \\ + 2A_c[1 + m \cos 2\, \pi ft]r(t) \cos \phi_n(t) \end{aligned} \tag{10-12}$$

The analysis of this envelope is complicated, but we can examine two specific cases and make simplifying assumptions. First, consider the case in which

$$\frac{C}{N} = \frac{A_c^2}{2N} \gg 1$$

Then the output mean square will be

$$\langle v_0(t)^2 \rangle = A_c^2 \left(1 + \frac{m^2}{2}\right) + N \tag{10-13}$$

and the output signal-to-noise ratio will be

$$\left(\frac{S}{N}\right)_0 = m^2 \frac{A_c^2}{2N} \tag{10-14}$$

Therefore

$$\left(\frac{S}{N}\right)_0 = m^2 \frac{C}{N} = 2\left(\frac{S}{N}\right)_1 \quad \text{or} \quad \left(\frac{S}{N}\right)_0 = \frac{m^2}{1 + m^2/2}\left(\frac{S}{N}\right)_1 \tag{10-15}$$

The alternative expression in Eq. (10-15) applies if we include the carrier power in the input signal.

Second, consider the case in which C/N is small (threshold approximately 7 dB). The resulting signal at the output may be lost below a threshold value of C/N. The analytic expression for S/N in this case is complicated and is usually obtained by making certain simplifying assumptions (see references).

Synchronous Detection Consider a DSBSC wave in which noise has been added:

$$v_1(t) = A_c \cos 2\pi ft \cos 2\pi f_c t + n(t) \tag{10-16}$$

We will have

$$\langle v_1^2(t) \rangle = \frac{A_c^2}{4} + N \tag{10-17}$$

and

$$\left(\frac{S}{N}\right)_1 = \frac{A_c^2}{4N} \tag{10-18}$$

Since, in coherent detection, the receiver multiplies the input by a carrier synchronized in phase and frequency, it is useful to express the noise in terms of its quadrature components; that is,

$$n(t) = x \cos 2\pi f_c t - y \sin 2\pi f_c t$$

where

$$\langle n^2(t) \rangle = \langle x^2 \rangle = \langle y^2 \rangle = N$$

The output of the product detector will be

$$v_0(t) = v_1(t) \cos 2\pi f_c t$$

or

$$v_0(t) = A_c \cos 2\pi f t \cos^2 2\pi f_c t + n(t) \cos 2\pi f_c t$$

After low-pass filtering

$$v_0(t) = \frac{A_c}{2} \cos 2\pi f t + \tfrac{1}{2}x(t)$$

and

$$\langle v_0^2(t) \rangle = \frac{A_c^2}{8} + \frac{N}{4} \tag{10-19}$$

Consequently,

$$\left(\frac{S}{N}\right)_0 = \frac{A_c^2}{2N} = \frac{C}{N} \tag{10-20}$$

and

$$\left(\frac{S}{N}\right)_0 = 2\left(\frac{S}{N}\right)_1 \tag{10-21}$$

In the SSB case, the analysis is similar to that of the DSBSC case. The relationship for $(S/N)_0$ is

$$\left(\frac{S}{N}\right)_0 = \left(\frac{S}{N}\right)_1 \tag{10-22}$$

However, the noise bandwidth is half that of the noise bandwidth of the DSBSC case; therefore the signal-to-noise ratio of SSB equals that of DSBSC in spite of the fact that there is no improvement in the detection similar to the improvement obtained in the DSBSC case.

In conclusion:

1. In full-AM transmission with high $(S/N)_1$ and $m = 1$, DSBSC and SSB provide postdetection $(S/N)_0$ equal to the carrier-to-noise ratio C/N.
2. Full-AM transmission with poor $(S/N)_1$ is characterized by a threshold limit. Below this threshold, the noise takes over. A typical value for this threshold is 7 dB.
3. In obtaining the full-AM $(S/N)_0$, we did not take into account the carrier power as a signal. However, if we take the total transmitted power into account, the DSBSC has a significant advantage:

$$\frac{(S/N)_{0,\mathrm{DSBSC}}}{(S/N)_{0,\mathrm{AM}}} = 4.8 \text{ dB}$$

ANGLE MODULATION

Definitions and Sinusoidal Analysis

The expression for a sinusoidal carrier is

$$m(t) = A_c \cos [\omega_c t + \phi(t)]$$

where

$$\begin{aligned} A_c &= \text{peak amplitude of the carrier} \\ \omega_c t + \phi(t) &= \text{instantaneous phase, rad} \\ \omega_c &= 2\pi f_c \\ \phi(t) &= \text{instantaneous phase deviation, rad} \end{aligned}$$

The instantaneous frequency is

$$\frac{d}{dt}[\omega_c t + \phi(t)] = \omega_c + \phi'(t) \qquad \text{rad/s}$$

where $\phi'(t)$ = instantaneous frequency deviation, rad/s. Given a modulating signal $v(t)$, we have phase modulation when $\phi(t) = k_1 v(t)$ and frequency modulation when $\phi'(t) = k_2 v(t)$ or $\phi(t) = k_2 \int v(t)\, dt$.

For a sinusoidal wave $v(t) = A_m \cos \omega_m t$, where $\omega_m = 2\pi f_m$, we have the PM case when

$$m(t) = A_c \cos (\omega_c t + k_1 A_m \cos \omega_m t) \tag{10-23}$$

We define the peak phase deviation, or index of modulation, as

$$D_f = k_1 A_m \qquad \text{rad} \tag{10-24}$$

We have the FM case when

$$m(t) = A_c \cos\left(\omega_c t + \frac{k_2 A_m}{\omega_m} \sin \omega_m t\right) \tag{10-25}$$

We define the peak frequency deviation* as

$$\Delta F = \frac{k_2 A_m}{2\pi} \quad \text{Hz} \tag{10-26}$$

and the index of modulation as

$$\beta = \frac{k_2 A_m}{2\pi f_m} \tag{10-27}$$

We also have

$$\beta = \frac{\Delta F}{f_m} \tag{10-28}$$

Equations (10-23) and (10-25) represent sinusoidal functions of an argument which is a sinusoidal function itself. These types of functions can be analyzed by using the definition of the Bessel function $J_n(X)$. A Bessel function of the first kind and of order n is defined as

$$J_n(X) = \sum_{m=0}^{\infty} \frac{(-1)^m (X/2)^{2m+n}}{m!(m+n)!} \tag{10-29}$$

Using this definition, we can prove that

$$\begin{aligned}
\cos(a + X \cos b) &= \sum_{n=-\infty}^{\infty} J_n(X) \cos\left(a + nb + \frac{n\pi}{2}\right) \\
\cos(a + X \sin b) &= \sum_{n=-\infty}^{\infty} J_n(X) \cos(a + nb) \\
\sin(a + X \sin b) &= \sum_{n=-\infty}^{\infty} J_n(X) \sin(a + nb) \\
\sin(a + X \cos b) &= \sum_{n=-\infty}^{\infty} J_n(X) \sin\left(a + nb + \frac{n\pi}{2}\right)
\end{aligned} \tag{10-30}$$

We can now simplify Eq. (10-23) by equating $X = D_f = k_1 A_m$ and applying—for the PM case—the first of the Eq. (10-30) identities:

$$m(t) = A_c \sum_{n=-\infty}^{\infty} J_n(X) \cos\left(\omega_c t + n\omega_m t + \frac{n\pi}{2}\right) \tag{10-31}$$

*ΔF could also be defined as $k_2 A_m$ to actually correspond to angular frequency.

Since

$$J_{-n}(X) = (-1)^n J_n(X)$$

we have, by developing the series in Eq. (10-31) for the various values of n,

$$\begin{aligned} m(t) = {} & A_c J_0(X) \cos \omega_c t \\ & + A_c J_1(X) \left\{ \cos \left[(\omega_c + \omega_m)t + \frac{\pi}{2} \right] + \cos \left[(\omega_c - \omega_m)t + \frac{\pi}{2} \right] \right\} \\ & - A_c J_2(X) [\cos (\omega_c + 2\omega_m)t + \cos (\omega_c - 2\omega_m)t] + \cdots \end{aligned} \qquad (10\text{-}32)$$

which is the expression for a phase-modulated wave with a modulating signal

$$v(t) = A_m \cos \omega_m t$$

The amplitude spectrum of Eq. (10-32) is depicted in Fig. 10-6.

This spectrum resembles the spectrum of an AM signal. In the PM case, however, we have all the multiples $f_c + nf_m$ and $f_c - nf_m$.

For small values of X and $X < 1$, the function $J_n(X)$ decreases rapidly with n. For example, for $X = \frac{1}{2}$, $J_0(X) = 0.938$, $J_1(X) = 0.242$, $J_2(X) = 0.031$, $J_3(X) = 0.003$, etc. Consequently, all values of n above $n = 1$ could be neglected, and we would have a spectrum like the AM modulation spectrum.

For $X > 1$, $J_n(X)$ decreases rapidly as soon as n becomes larger than X.

Similar analysis could be carried out by applying two sinusoidal signals simultaneously to an angle modulator. It can be shown that, in the case of two sinusoids,

$$m(t) = A_c \sum_{n=-\infty}^{\infty} \sum_{p=-\infty}^{\infty} J_n(X_1) J_p(X_2) \cos \left[(\omega_c + n\omega_1 + p\omega_2)t + \frac{(n + p)\pi}{2} \right] \qquad (10\text{-}33)$$

This relation shows that the frequency spectrum will be fairly complicated.

In the FM case, we can analyze the signal $m(t)$ of Eq. (10-25) by equating $\beta = k_2 A_m / \omega_m$. We have

$$m(t) = A_c \sum_{n=-\infty}^{\infty} J_n(\beta) \cos [(\omega_c + n\omega_m)t] \qquad (10\text{-}34)$$

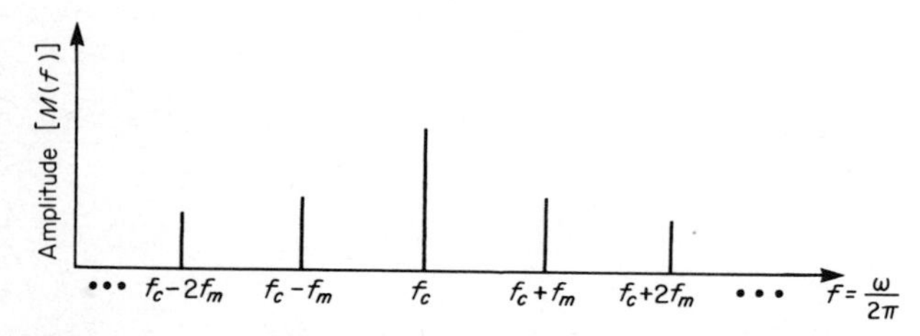

FIGURE 10-6 Amplitude spectrum of an angle-modulated signal.

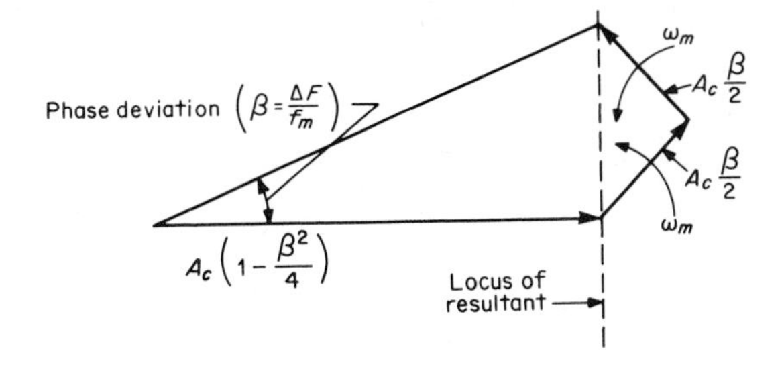

FIGURE 10-7 Low-index FM phasor representation.

The amplitude spectrum will be similar to the one in Fig. 10-6.

The phasor representation of the frequency-modulated signal $m(t)$ with a low index of modulation β (where $\beta \ll 1$) is given in Fig. 10-7, where the approximate expression for the instantaneous value of the signal is

$$m(t) = A_c J_0(\beta) \cos \omega_c t + A_c J_1(\beta) \cos [(\omega_c + \omega_m)t] - A_c J_1(\beta) \cos [(\omega_c - \omega_m)t] \tag{10-35}$$

The locus of the resultant is a line at right angles to the direction of the carrier phasor, since the horizontal components of the two sideband frequencies cancel each other. One can see that, although the frequency spectrum resembles that of an amplitude-modulated signal, the phases are such that there is no change of the amplitude of the resultant for small phase deviations. In fact, the small change of amplitude is only a result of the approximation used in Eq. (10-35).

Figure 10-8 depicts the phasor representation for a large modulation index β where two frequencies are used, namely, ω_1 and $2\omega_1$. The locus of the resultant is a circle with radius equal to the amplitude of the unmodulated carrier.

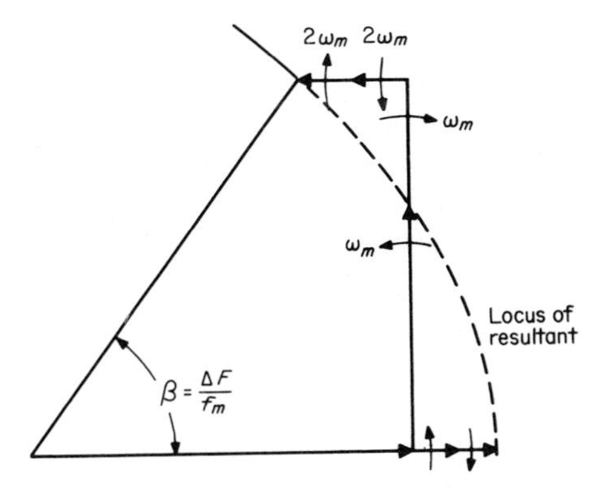

FIGURE 10-8 High-index FM phasor representation.

Occasionally, because of system nonlinearities, an AM component is added to the carrier. For example, this is the case for TWT amplifiers such as the ones used in satellite repeaters. It can be seen that phase nonlinearities will result in a change of the phase deviation β and consequently will cause a PM interference. This is known as an AM-PM conversion. The AM component can be minimized by using a limiter which maintains the amplitude of the modulated carrier constant. Such a limiter will also decrease the amplitude of the sidebands and, as a result, will also decrease the signal power. In a frequency-modulated wave, the power of the carrier is not increased by the modulation, but is all contained in the carrier and dispensed to the sidebands. One can see this by computing the power into a 1-Ω resistive load of the angle-modulated wave $m(t) = A_c \cos[\omega_c t + \phi(t)]$:

$$W(t) = m^2(t) = A_c^2 \cos^2[\omega_c + \phi(t)]$$

$$= \frac{A_c^2}{2}\{1 - 2\sin^2[\omega_c t + \phi(t)]\}$$

The average power is

$$\overline{W}(t) = \frac{A_c^2}{2} \tag{10-36}$$

which is the average power of the carrier.

The required RF bandwidth B_R for transmitting a frequency-modulated carrier can be computed easily in the case of the low modulation index. From Eq. (10-35), one can see that enough bandwidth must be allowed to transmit the two sidebands $f_c + f_m$ and $f_c - f_m$; therefore,

$$B_R > 2f_m$$

For the general case of a frequency-modulated carrier with peak frequency deviation ΔF, the required RF bandwidth is given by Carson's rule:

$$B_R \geq 2(f_m + \Delta F) \tag{10-37}$$

where f_m is the highest significant frequency in the modulating signal.

Figure 10-9*a* represents a typical FM receiver. The superheterodyne receiver converts, with the aid of a mixer, the radio frequency, say 4 GHz, to an intermediate frequency, say 70 MHz, without affecting the modulation. The IF filter is a bandpass filter of a bandwidth B_R such that it allows the whole RF band to pass through with minimum distortion. The limiter takes out any amplitude variations and the bandpass filter eliminates the harmonics. The discriminator consists of a "slope network," or "differentiator," and an envelope detector. The differentiator output is the first derivative of the phase of the input signal, i.e., the instantaneous frequency, and the envelope detector recovers the envelope, i.e., the message signal. However, below a certain input carrier-to-

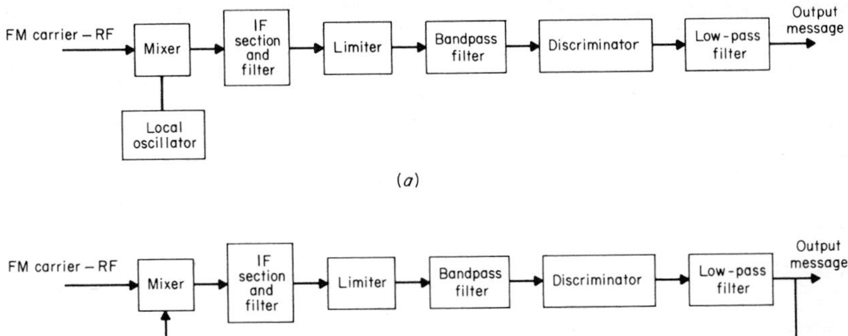

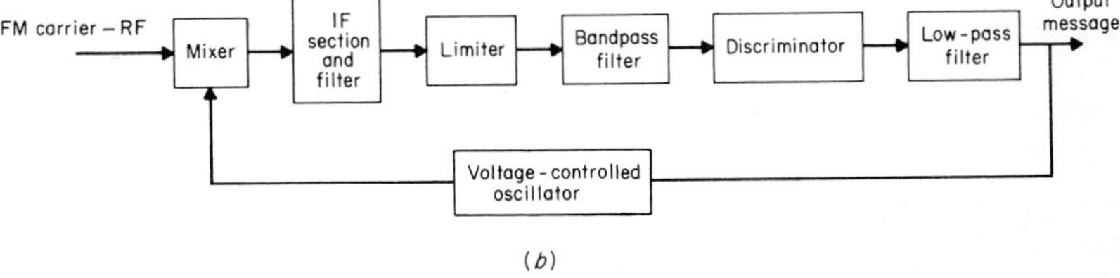

FIGURE 10-9 **(*a*) Typical FM receiver. (*b*) FM receiver with threshold extension.**

noise level, the receiver reaches a "breaking point," and noise takes over, producing a high number of "clicks." In cases in which the carrier signal-to-noise level is not very strong, a "threshold extension" receiver is used. Space systems widely utilize this type of receiver. The local oscillator is replaced by a voltage-controlled oscillator (VCO), and this allows an extension of the receiver breaking point due to noise. A 3- to 7-dB improvement is achievable. (See Fig. 10-9*b*.)

Narrowband FM

First consider an amplitude-modulated wave,

$$m_{AM}(t) = A_c[1 + mv(t)] \cos 2\pi f_c t \qquad (10\text{-}38)$$

where m is the AM index of modulation. The frequency spectrum of this wave is given by the relationship

$$M_{AM}(f) = \tfrac{1}{2}A_c [\delta(f - f_c) + \delta(f + f_c)] + \frac{mA_c}{2}[V(f - f_c) + V(f + f_c)] \qquad (10\text{-}39)$$

The same signal $v(t)$ in the FM case will give

$$m_{FM}(t) = A_c \cos [2\pi f_c t + \beta g(t)] \qquad (10\text{-}40)$$

where β is the index of modulation and

$$\beta g'(t) = \beta \frac{dg(t)}{dt} = K_2 v(t) \qquad (10\text{-}41)$$

Equation (10-40) can be written as

$$m_{FM}(t) = A_c \cos 2\pi f_c t \cos \beta g(t) - A_c \sin 2\pi f_c t \sin \beta g(t) \qquad (10\text{-}42)$$

Now, if $\beta \ll 1$, that is, for narrowband FM,*

$$\cos \beta g(t) \approx 1 \qquad \sin \beta g(t) \approx \beta g(t)$$

and therefore

$$m_{FM}(t) = A_c[\cos 2\pi f_c t - \beta g(t) \sin 2\pi f_c t] \qquad (10\text{-}43)$$

Equation (10-43) resembles Eq. (10-38) in form except that the former contains a minus sign followed by a sine function, instead of the cosine function in Eq. (10-38). The spectrum of $g(t)$ can be given in terms of the spectrum of $v(t)$ if the latter spectrum is divided by $j2\pi f$ (integration). Equation (10-43) indicates that

$$M_{FM}(f) = \tfrac{1}{2}A_c[\delta(f - f_c) + \delta(f + f_c)] + \frac{A_c k_2}{4\pi}\left[\frac{V(f - f_c)}{f - f_c} - \frac{V(f + f_c)}{f + f_c}\right] \qquad (10\text{-}44)$$

Comparing Eqs. (10-39) and (10-44), we can see that the lower sidebands are at a 180° phase difference and that a distortion factor $1/f$ has been introduced in Eq. (10-44).

EXAMPLE 10-1 COMPARISON OF NARROWBAND FM WITH AM

Assume a sinusoidal modulating signal $\cos 2\pi f_m t$. From Eq. (10-38), we obtain

$$m_{AM} = A_c(1 + m \cos 2\pi f_m t) \cos 2\pi f_c t$$

or, if we assume $m = 1$ for simplicity,

$$m_{AM} = A_c \cos 2\pi f_c t + \frac{A_c}{2} \cos 2\pi(f_c - f_m)t + \frac{A_c}{2} \cos 2\pi(f_c + f_m)t \qquad (10\text{-}45)$$

Similarly, for $\beta = 1$, Eq. (10-43) becomes

$$m_{FM} = A_c \cos 2\pi f_c t - \frac{A_c}{2\pi f_m} \sin 2\pi f_m t \sin 2\pi f_c t \qquad (10\text{-}46)$$

and

$$m_{FM} = A_c \cos 2\pi f_c t + \frac{A_c}{4\pi f_m}[\cos 2\pi(f_m - f_c)t - \cos 2\pi(f_m + f_c)t] \qquad (10\text{-}47)$$

*Usually when $\beta < 0.5$ we define FM as narrowband.

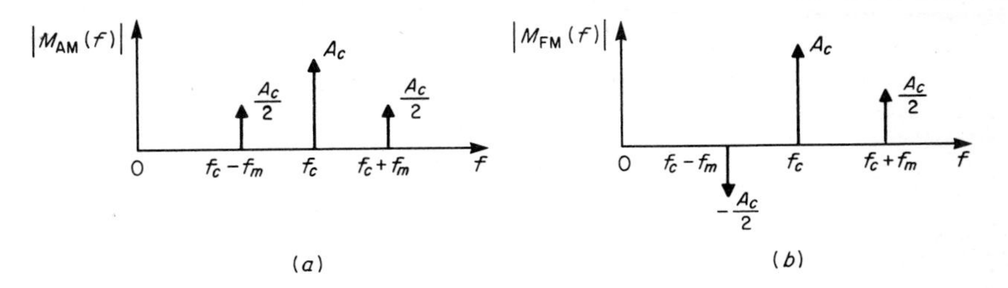

FIGURE 10-10 Comparison of spectra: narrowband FM vs. AM. (*a*) AM representation. (*b*) FM representation.

In Fig. 10-10, which shows amplitude vs. frequency for Eqs. (10-45) and (10-47), it is easy to see that in the FM case the lower sideband has a 180° phase shift.

Wideband FM

For a general type of signal $v(t)$, the frequency-modulated wave is given by the relationship

$$m(t) = A_c \cos\left[2\pi f_c t + \beta g(t)\right] \tag{10-48}$$

where

$$\beta g'(t) = \beta \frac{dg(t)}{dt} = K_2 v(t) \tag{10-49}$$

and β is the modulation index or deviation ratio. The function $g(t)$ is normalized with respect to its own maximum so that

$$|g(t)| \leq 1 \tag{10-50}$$

The instantaneous frequency is

$$\frac{1}{2\pi}\frac{d}{dt}\left[2\pi f_c t + \beta g(t)\right] = f_c + \frac{\beta}{2\pi} g'(t) \tag{10-51a}$$

and the peak frequency deviation is

$$\Delta F = \frac{\beta}{2\pi}\,|\max g'(t)| \tag{10-51b}$$

Now, assuming a sinusoidal modulating signal

$$v(t) = A_m \cos 2\pi f_m t \tag{10-52}$$

we have

$$\beta g(t) = \int k_2 A_m \cos 2\pi f_m t \; dt = k_2 \frac{A_m}{2\pi f_m} \sin 2\pi f_m t + C$$

Assuming the constant of integration $C = 0$, we find

$$\beta g(t) = k_2 \frac{A_m}{2\pi f_m} \sin 2\pi f_m t \tag{10-53}$$

Consequently, the peak frequency deviation is

$$\Delta F = \frac{k_2 A_m}{2\pi} \tag{10-54}$$

and the modulation index is

$$\beta = \frac{\Delta F}{f_m} = \frac{k_2 A_m}{2\pi f_m} \tag{10-55}$$

Equation (10-48) then becomes

$$m(f) = A_c \cos [2\pi f_c t + \beta \sin 2\pi f_m t] \tag{10-56}$$

By applying the Bessel function relationship, Eq. (10-34), we obtain

$$m(t) = A_c \sum_{n=-\infty}^{\infty} J_n(\beta) \cos 2\pi(f_c + nf_m)t \tag{10-57}$$

where

$$J_{-n}(\beta) = (-1)^n J_n(\beta) \tag{10-58}$$

For odd values of n, the $f_c - |n| f_m$ components will be 180° out of phase with the $f_c + |n| f_m$ components. We have seen that this is true in the narrowband case, in which we considered only the $n = \pm 1$ components.

The amplitude of these sinusoidal components is shown in Fig. 10-11.

The bandwidth requirements for transmitting all these components is much greater than in the AM case, which has only the $f_c + f_m$ and $f_c - f_m$ components. For this exchange of bandwidth, one obtains a significant postdetection signal-to-noise ratio improvement, assuming the signal level is above a certain threshold.

As previously indicated, the amplitude of $J_n(\beta)$ decreases very rapidly for

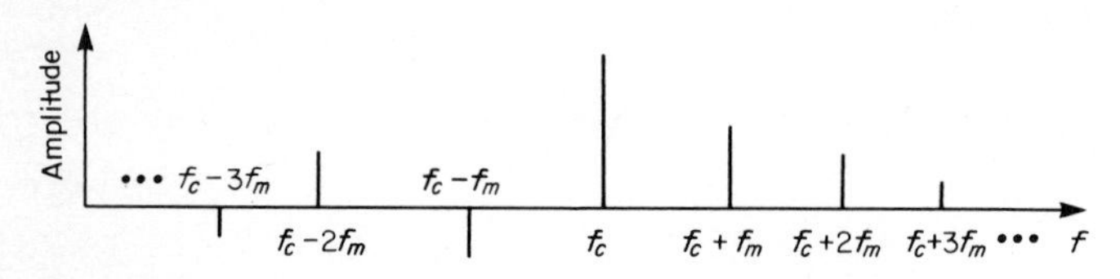

FIGURE 10-11 Components of an FM carrier modulated by a cosine wave.

values of n higher than β. Therefore the minimum required bandwidth corresponds to the value of $n = \beta$. Then, taking into account the positive and negative components, we find that this bandwidth must be at least twice ΔF, since $2\,\Delta F = 2nf_m = 2\beta f_m$. For a signal $v(t)$ with a maximum frequency f_m, one must also include twice the baseband width f_m. Consequently,

$$B_R = 2(\Delta F + f_m) = 2(\beta + 1)f_m \tag{10-59}$$

Equation (10-59) is known as Carson's rule for FM.

The received frequency-modulated wave usually first passes through a bandpass limiter, which eliminates amplitude fluctuations. Since the message is carried in the frequency part of the wave, the limiting process does not affect the ability to detect the message.

The frequency discriminator, which follows the limiter, detects the message by differentiating the phase. Both the bandwidth of the limiter and that of the discriminator must accommodate the total RF bandwidth $2(\Delta F + f_m)$. The lowpass filter that follows restricts the bandwidth to the maximum message frequency f_m.

The received signal plus noise will be

$$s(t) = m(t) + n(t)$$

or

$$s(t) = A_c \cos\,[2\pi f_c t + \beta g(t)] + n(t) \tag{10-60}$$

But the bandpass noise $n(t)$ will be

$$n(t) = \text{Re}\,[u(t)e^{j2\pi f_c t}] \tag{10-61}$$

The complex envelope $u(t)$ can be written as

$$u(t) = re^{j\phi} = x + jy \tag{10-62}$$

where r, ϕ, x, and y are functions of time. When $n(t)$ is gaussian noise, the envelope $r(t)$ will have a Rayleigh distribution and the phase $\phi(t)$ will have a uniform distribution (see Chap. 9). In terms of r and ϕ, $n(t)$ is expressed as

$$n(t) = \text{Re}\,(re^{j\phi}e^{j2\pi f_c t}) = r\cos\,(2\pi f_c t + \phi) \tag{10-63}$$

As shown in Chap. 9, if N is the mean square noise power,

$$N = \langle n^2 \rangle = \langle x^2 \rangle = \langle y^2 \rangle = \tfrac{1}{2}|\langle x^2 + y^2 \rangle| = \tfrac{1}{2}\langle r^2 \rangle$$

Equation (10-60) can be written as

$$s(t) = A_c \cos\,[2\pi f_c t + \beta g(t)] + r\cos\,(2\pi f_c t + \phi) \tag{10-64}$$

We can now examine two extreme cases that allow a simplified analysis.

Case 1 For the case in which $A_c \gg r$, the receiver carrier-to-noise ratio can be described as follows:

$$\frac{C}{N} = \frac{A_c^2}{2N} \gg 1$$

In this case, we can rewrite Eq. (10-64) as

$$s(t) = A_c h(t) \cos [2\pi f_c t + \theta(t)] \tag{10-65}$$

where $h(t)$ and $\theta(t)$ are functions of $\beta g(t)$, r/A_c, and ϕ. The amplitude $h(t)$ fluctuates with noise, but these fluctuations can be eliminated by a limiter. The angle $\theta(t)$ containing the message will be

$$\theta(t) = \beta g(t) + \tan^{-1} \frac{r \sin [\phi - \beta g(t)]}{A_c + r \cos [\phi - \beta g(t)]}$$

But again for $A_c/r \gg 1$ or $r/A_c \ll 1$, we get

$$\theta(t) \approx \beta g(t) + \frac{r}{A_c} \sin [\phi - \beta g(t)] \tag{10-66}$$

The FM discriminator will differentiate this angle before passing the demodulated message through a low-pass filter. The discriminator output after differentiation will consist of two components,

$$v_0(t) = k\beta \frac{dg(t)}{dt} \tag{10-67}$$

and

$$n_0(t) = k \frac{d}{dt} \left\{ \frac{r}{A_c} \sin [\phi - \beta g(t)] \right\} \tag{10-68a}$$

where k is a discriminator constant.

Equation (10-67) provides the output signal and Eq. (10-68a) provides the output noise. Under certain simplifying assumptions, for rms noise power computations utilizing Eq. (10-68a), the deterministic function $\beta g(t)$ can be neglected. Then, if $y(t)$ is the quadrature component of noise,

$$n_0(t) = \frac{k}{A_c} \frac{d}{dt} [r(t) \sin \phi] = \frac{k}{A_c} \frac{dy(t)}{dt} \tag{10-68b}$$

where $y(t) = r(t) \sin \phi$. (See definition of quadrature component of noise in Chap. 9.)

Differentiation in the time domain corresponds to multiplication by $j2\pi f$ in the frequency domain; therefore, Eq. (10-68) suggests that the noise is passed through a linear filter with transfer characteristic $H(f) = (k/A_c)j2\pi f$. Consequently, if $S_n(f)$ is the power spectrum of the noise after the discriminator and if $N_y(f)$ is the power spectrum of $y(t)$, we will have

$$S_n(f) = |H(f)|^2 N_y(f)$$

or

$$S_n(f) = \frac{4\pi^2 k^2}{A_c^2} f^2 N_y(f) \tag{10-69a}$$

The discriminator constant k is usually adjusted to be

$$k = \frac{1}{2\pi}$$

For white noise with $N_y(f) = n_0$, the spectral density relationship in Eq. (10-69a) becomes

$$S_n(f) = \begin{cases} \dfrac{n_0}{A_c^2} f^2 & \text{for } |f| \leq \dfrac{B_R}{2} \\ 0 & \text{elsewhere} \end{cases} \tag{10-69b}$$

where B_R is the IF bandwidth.

The predetection carrier-to-noise ratio C/N is given by the relationship $C/N = A_c^2/2N = A_c^2/2n_0B_R$. By substituting into Eq. (10-69b), we obtain

$$S_n(f) = \begin{cases} \left(\dfrac{C}{N}\right)^{-1} \dfrac{f^2}{2B_R} & \text{for } |f| \leq \dfrac{B_R}{2} \\ 0 & \text{elsewhere} \end{cases} \tag{10-69c}$$

If the ideal low-pass filter after the discriminator has a bandwidth $2W$, we will have $2W < B_R$ and often $2W = 2f_m$, where f_m is the largest frequency in the baseband signal. The plot of the output noise power spectrum over this bandwidth is shown in Fig. 10-12.

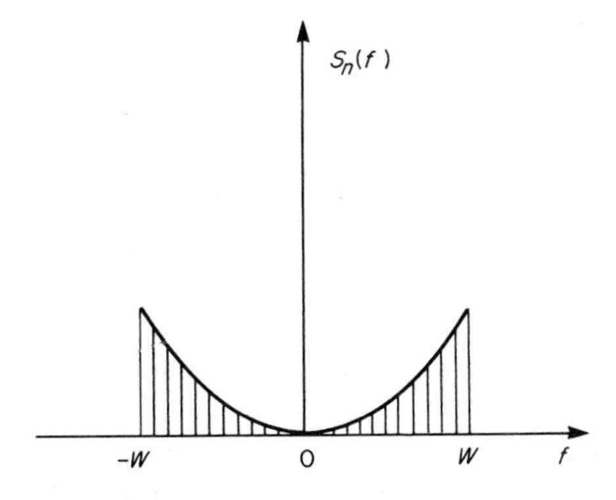

FIGURE 10-12 Noise spectrum for post-detection FM.

The total output noise power will be

$$(N)_o = \int_{-W}^{W} S_n(f)\, df$$

and, because of Eq. (10-69*c*),

$$(N)_o = \frac{W^3}{3B_R}\left(\frac{C}{N}\right)^{-1} \tag{10-69d}$$

Equation (10-51) gives the output of the discriminator as $f_c + (\beta/2\pi)g'(t)$. The peak signal at the output of the low-pass filter equals the peak frequency deviation. This peak frequency deviation is given by Eq. (10-51*b*); that is,

$$\Delta F = \frac{1}{2\pi}\beta|\max g'(t)|$$

The signal power will be $S_0 = (\Delta F_{\text{rms}})^2$, and, because of Eq. (10-69*d*),

$$\left(\frac{S}{N}\right)_0 = 3\left(\frac{\Delta F_{\text{rms}}}{W}\right)^2 \frac{B_R}{W}\frac{C}{N} \tag{10-70a}$$

where C/N is the predetection carrier-to-noise ratio. For single-tone modulation, in which $W = f_m$, Carson's bandwidth $B_R = 2(\Delta F + f_m)$, and $\beta = \Delta F/f_m$ [see Eqs. (10-54) and (10-55)],

$$\left(\frac{S}{N}\right)_0 = 3\beta^2(\beta + 1)\frac{C}{N} \tag{10-70b}$$

Case 2 If $A_c \ll r$, the relationship giving the instantaneous phase angle $\theta(t)$ in Eq. (10-65) should be written as

$$\theta(t) = \phi + \frac{A_c}{r}\sin\left[\beta g(t) - \phi\right] \tag{10-71}$$

In this case $\theta(t)$ is dominated by the noise parameter $\phi(t)$ and the message will be lost. Equations (10-66) and (10-71) demonstrate the "capture" characteristic of an FM signal, in which the stronger signal will capture the receiver. The threshold carrier-to-noise ratio, below which the output signal-to-noise ratio deteriorates significantly, is approximately 12 to 15 dB. Most often,

$$\left(\frac{C}{N}\right)_{\text{threshold}} \approx 13 \text{ dB}$$

This can be improved and extended by using the threshold extension receiver depicted in Fig. 10-9*b*. A threshold improvement in the range of 3 to 7 dB can be achieved. This represents a significant improvement and justifies in many cases the added implementation complexity of a voltage-controlled oscillator.

Preemphasis and Deemphasis

Additional improvement can be obtained by taking advantage of the signal- and noise-frequency characteristics with the well-known preemphasis-deemphasis technique, in which the message signal, before being introduced to the modulator, passes through a filter which emphasizes (boosts) the high frequencies, when the signal has usually low-amplitude components. At the receiver, after demodulation, the signal passes through an opposite filter that reduces the high frequencies, including noise. Thus the signal is restored to its original balance with the noise high frequencies highly suppressed. Because of the noise characteristic, as depicted in Fig. 10-12, we will obtain a substantial improvement in the postdetection signal-to-noise ratio. In effect, we boosted the input high-frequency signal components, which are normally low, and we decreased the output high-frequency noise components, which are high. By using properly matched preemphasis and deemphasis filters, an improvement in $(S/N)_0$ in the range of 7 to 13 dB can be obtained.

EXAMPLE 10-2 FM SPACE LINK

Consider the case described in Example 7-2. For a satellite downlink frequency $f_c = 4$ GHz, the carrier-to-noise ratio C/N for the described space link was computed to be $C/N = 96.1$ dB in front of the receiver. Let us further assume that the downlink carrier was operating as an FM carrier with a peak frequency deviation $\Delta F = 2$ MHz and a maximum baseband message frequency $f_m = 4$ MHz. Then the radio frequency bandwidth should be

$$B = 2(f_m + \Delta F) = 2(4 \text{ MHz} + 2 \text{ MHz}) = 12 \text{ MHz}$$

The modulation index is

$$\beta = \frac{\Delta F}{f_m} = \frac{1}{2}$$

and the Bessel function coefficients are

$$J_0(\beta) = J_0(\tfrac{1}{2}) = 0.938$$

$$J_1(\beta) = J_1(\tfrac{1}{2}) = 0.242$$

$$J_2(\beta) = J_2(\tfrac{1}{2}) = 0.031$$

$$J_3(\beta) = J_3(\tfrac{1}{2}) = 0.003$$

If the baseband signal is a sinusoid with frequency f_m, then the expression for the modulated carrier will be

$$m(t) = A_c[0.938 \cos 2\pi f_m t + 0.242 \cos 2\pi(f_c + f_m)t - 0.242 \cos 2\pi(f_c - f_m)t]$$

For $n \geq 2$, the coefficients $J_n(\beta)$ are negligible.

The postdetection signal-to-noise ratio will be

$$\left(\frac{S}{N}\right)_0 = 3\beta^2(1 + \beta)\frac{C}{N} = 3(\tfrac{1}{2})^2(1 + \tfrac{1}{2})\frac{C}{N} = 1.125\frac{C}{N}$$

and

$$\left(\frac{S}{N}\right)_0 = 96.6 \text{ dB}$$

FM System Performance

In Example 10-2, a modulation index of 0.5 was employed. This is considered the dividing line between narrowband and wideband FM.

If the modulation index β were set equal to 3, then an FM improvement of 20.3 dB would have resulted. Expressed mathematically,

$$\text{FM improvement} = 3\beta^2(1 + \beta) = 108 \doteq 20.3 \text{ dB}$$

This improvement is relative to the input carrier-to-noise ratio. As we have seen in the AM case, $(S/N)_0 = C/N$. Therefore, FM can greatly improve over AM.

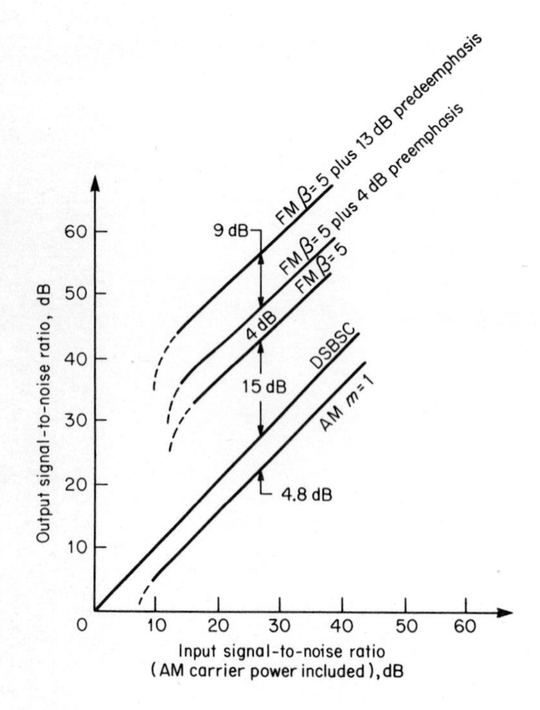

FIGURE 10-13 Performance comparison for analog modulation.

However, the RF bandwidth requirement for $\beta = 3$ will be

$$B_{RF} = 2(\Delta F + f_m) = 2(\beta + 1)f_m$$

and

$$B_{RF} = 8f_m$$

where f_m is the maximum baseband frequency. In the AM case, we required $2f_m$ and in the SSB case we required only the baseband width. A substantial signal-to-noise improvement is achieved at the expense of bandwidth. This is a key characteristic of a wideband FM system. However, FM signals with low C/N ratio exhibit a breaking characteristic when the output is extremely noisy. The threshold is approximately $(C/N)_T = 13$ dB. A substantial threshold improvement could be obtained by employing a threshold extension circuit. A feedback circuit with a voltage-controlled oscillator could provide a threshold reduction of about 3 to 7 dB.

Figure 10-13 shows a comparison of performance of the various analog modulation schemes.

REFERENCES

Haykin, Simon: *Communication Systems,* John Wiley & Sons, New York, 1978.

Members of the Technical Staff of Bell Telephone Laboratories: *Transmission Systems for Communications,* rev. 4th ed., Bell Telephone Laboratories, Winston Salem, N.C., 1971.

Schwartz, M., W. R. Bennett, and S. Stein: *Communication Systems and Techniques,* McGraw-Hill, New York, 1966.

Stein, S., and J. J. Jones: *Modern Communication Principles with Application to Digital Signaling,* McGraw-Hill, New York, 1967.

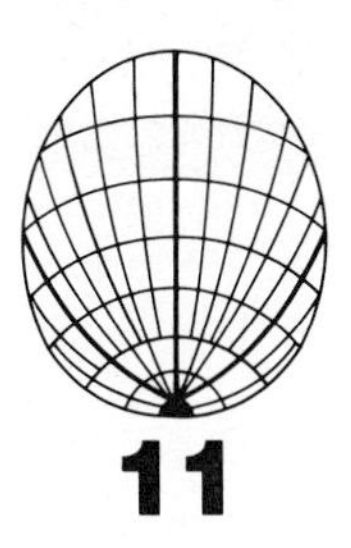

11

DIGITAL COMMUNICATIONS

The use of digital transmission systems has been greatly facilitated by the advances in large-scale integration (LSI) technology. The ability to implement complicated digital circuits at low cost, volume, and power, as well as the ability to provide high reliability and accuracy, has been substantially enhanced with the application of LSI technology. At the same time, the need for high-speed data transmission, distributed computer capacity, and remote data bases and all the other needs that stem from the wide application of digital signals have resulted in heavy requirements for digital communications systems.

Satellite communications, because of its broadband capability and low bit error rate, has been extensively used for high-speed data transmission.

In this chapter we will examine only the fundamentals that are required for understanding digital communications via satellite. More detailed discussions can be found in the suggested references.

SAMPLING PRINCIPLES

Sampling Theorem

Most physical signals possess a frequency spectrum that does not extend to infinite frequencies and, in fact, is band-limited. The sampling theorem refers to band-limited signals and can be stated as follows:

> If a signal in the time domain is sampled at regular intervals and at a rate at least twice as high as the highest frequency contained in its frequency spectrum, then the samples contain all the information of the original signal.

Figure 11-1*a* and *b* depicts the time function $v(t)$ as a band-limited signal with Fourier transform $V(f)$ vanishing above the maximum frequency f_m. If we sample the signal $v(t)$ with pulses at regular intervals T where $T \leq 1/2f_m$, then this train of pulses contains all the necessary information to reconstruct the original signal. The Fourier transform $V_s(f)$ of the pulse train is a periodic function in the frequency domain. Figure 11-1*c* is drawn for a sampling rate $T = 1/2f_m$. The original double spectrum $V(f)$ is repeated with a period $2f_m$. Now, if we pass this train of pulses through an ideal low-pass filter with a passband from 0 to f_m, we will recover the original signal $v(t)$. The fact that $T \leq 1/2f_m$ ensures that the original shape of the spectrum is preserved by providing adequate separation from period to period in the spectrum of Fig. 11-1*c*. The sampling interval $T = 1/2f_m$ is called the "Nyquist interval."

In practice, the signal is sampled with pulses of small but finite duration, and, in effect, we obtain a pulse-amplitude-modulated (PAM) signal. This case is depicted in Fig. 11-2.

Unfortunately, it is not feasible to construct an ideal low-pass filter. In practice, the filter will have a certain slope at the cutoff frequency f_m as indicated in Fig. 11-3.

As a result of all these implementation imperfections, the reconstructed signal will not be the exact replica of the original. The effect will be similar to that resulting from the introduction of interfering noise.

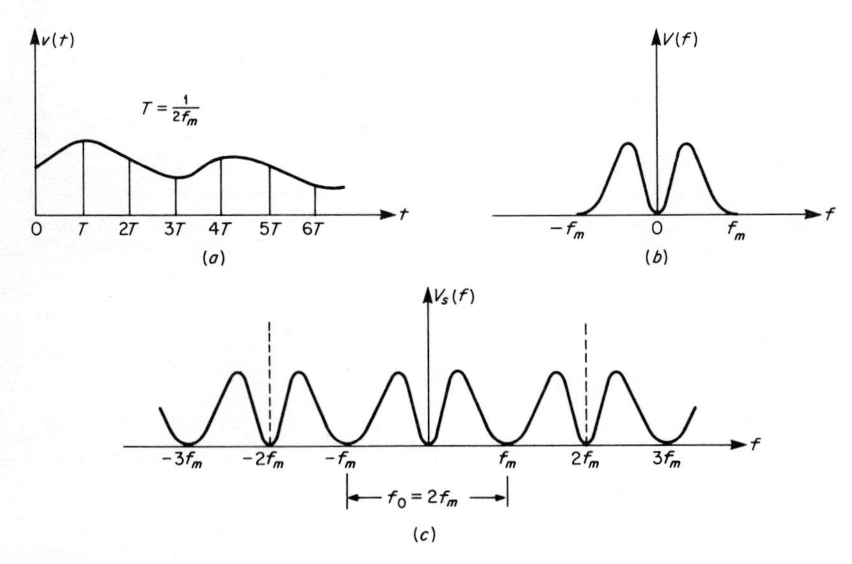

FIGURE 11-1 Signal sampling. (*a*) Signal $v(t)$ being sampled. (*b*) Spectrum of $v(t)$. (*c*) Spectrum of sample pulses; period $f_0 = 2f_m$.

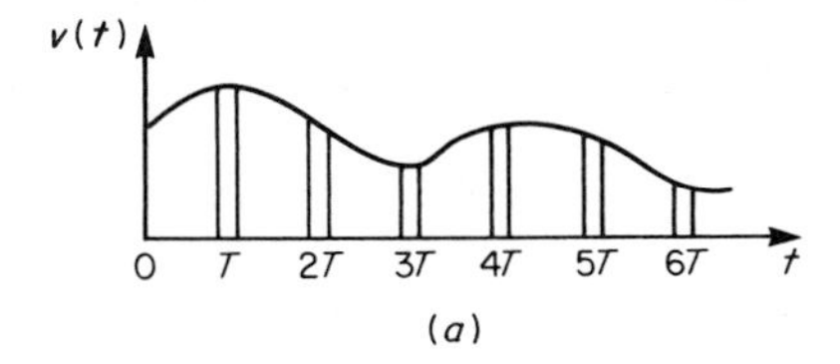

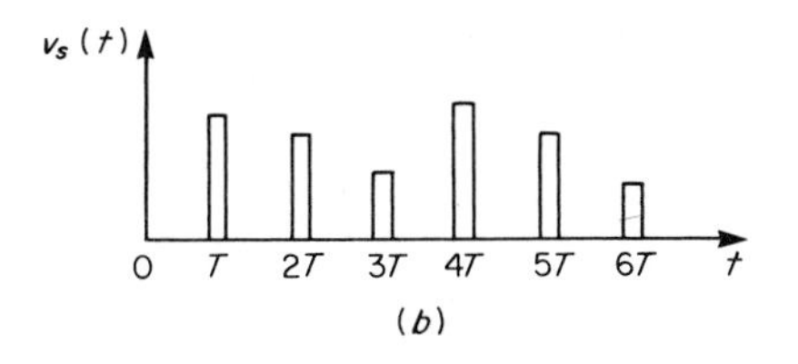

FIGURE 11-2 Sampling with pulses of finite width. (*a*) Pulses. (*b*) Finite samples.

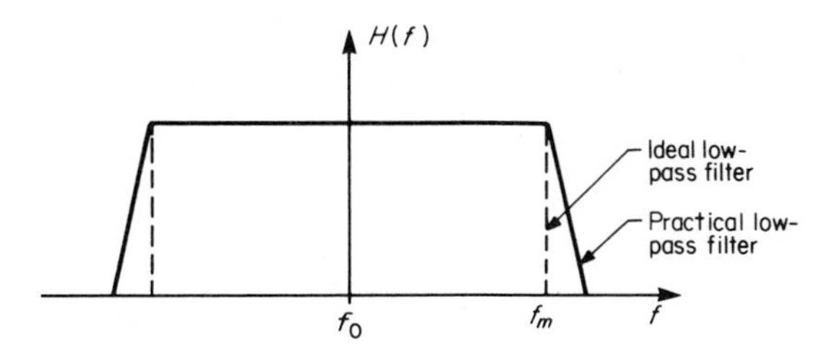

FIGURE 11-3 Low-pass-filter characteristics.

Sampling with a Train of Unit-Impulse Functions

A unit-impulse function or delta function $\delta(t - t_0)$ has previously been described as having zero amplitude everywhere except at $t = t_0$, where it is infinitely tall but its waveform includes an area of magnitude 1. A unit-impulse function can also be defined from its integral properties; that is,

$$\int_A^B f(t)\,\delta(t - t_0)\,dt = \begin{cases} f(t_0) & \text{for } A < t_0 < B \\ 0 & \text{for } t_0 < A,\ t_0 > B \end{cases}$$

Consider a train of impulse functions $S(t - mT)$, where $m = 0, \pm 1, \pm 2, \ldots$. This periodic function can be written as

$$s(t) = \sum_{m=-\infty}^{\infty} \delta(t - mT) \tag{11-1}$$

The Fourier series representing $s(t)$ will be

$$s(t) = \frac{1}{T} + \frac{2}{T}\sum_{k=1}^{\infty} \cos 2\pi k f_0 t \tag{11-2}$$

where $f_0 = 1/T$. The Fourier transform of this train can be written as

$$S(f) = \frac{1}{T} \sum_{k=-\infty}^{\infty} \delta(f - kf_0) \tag{11-3}$$

Now consider the signal function $v(t)$ of Fig. 11-1 sampled with the impulse train $s(t)$ (Fig. 11-4), by multiplication of $s(t)$ by $v(t)$.

The multiplication of $v(t)$ by a unit-impulse function $\delta(t - t_0)$ changes the impulse area to the value of $v(t)$ at the time t_0.

The sampled signal $v_s(t)$ will be

$$v_s(t) = v(t)s(t) = \sum_{m=-\infty}^{\infty} v(mT)\, \delta(t - mT) \tag{11-4}$$

The Fourier transform $V_s(f)$ of $v_s(t)$ will be the convolution of the Fourier transforms of $v(t)$ and $s(t)$:

$$V_s(f) = V(f) \otimes S(f) \tag{11-5}$$

From Eq. (11-3), we obtain

$$V_s(f) = \frac{1}{T} \sum_{k=-\infty}^{\infty} V\left(f - \frac{k}{T}\right)$$

But $1/T = 2f_m$, and

$$V_s(f) = \frac{1}{T} \sum_{k=-\infty}^{\infty} V(f - k2f_m) \tag{11-6}$$

Equation (11-6) represents a periodic spectrum with period T such as the one depicted in Fig. 11-1c. One can see from Fig. 11-1c that, as long as $1/2T \geq f_m$

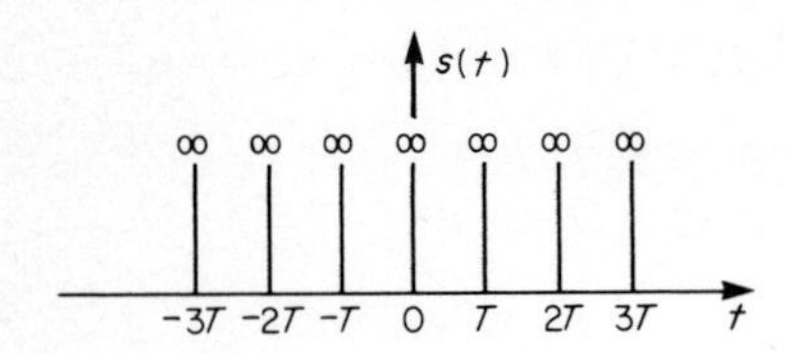

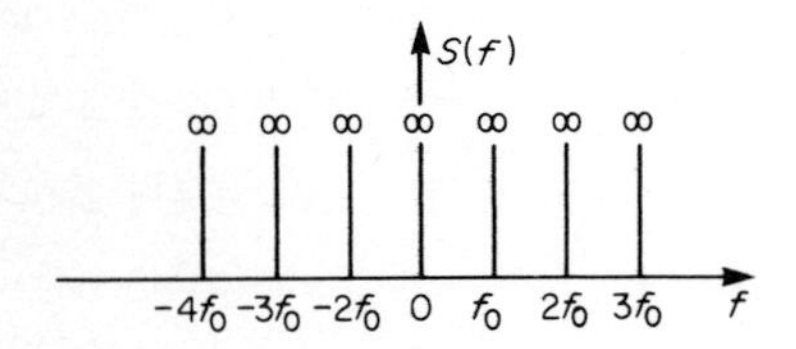

FIGURE 11-4 A periodic impulse train and its frequency spectrum.

or $T \leq 1/2f_m$, that is, as long as the sampling rate is higher than twice the maximum frequency of $v(t)$, we preserve the spectrum and we can reconstruct $v(t)$ by using an ideal low-pass filter. The spectrum resembles the one of an amplitude-modulated wave except for the repetitive element. If $T > 1/2f_m$, then the inverted spectrum will overlap the original one. Consequently, for $T > 1/2f_m$, the original spectrum cannot be recovered and the signal $v(t)$ cannot be reconstructed.

Since $V(f)$ is band-limited, if one assumes that $V(f)$ is periodic with period $1/2f_m$, the original frequency spectrum up to frequency $\pm f_m$ is not affected. The new periodic spectrum $V_s(f)$ then can be represented in terms of a Fourier series. It is easy, therefore, to find the inverse transform and determine the original $v(t)$. This process will provide $v(t)$ as

$$v(t) = \sum_{k=-\infty}^{\infty} V\left(\frac{k}{2f_m}\right) \frac{\sin 2\pi f_m t_k}{2\pi f_m t_k} \tag{11-7}$$

where

$$t_k = t - \frac{k}{2f_m} \tag{11-8}$$

Indeed, if $V_s(f)$ is the periodic spectrum with period $f_0 = 1/2f_m$, its Fourier series expansion will be

$$V_s(f) = \sum_{k=-\infty}^{\infty} a_k e^{j2\pi(k/2f_m)f} \tag{11-9}$$

where

$$a_k = \frac{1}{2f_m} \int_{-f_m}^{f_m} V_s(f) e^{-j2\pi(k/2f_m)f}\, df \tag{11-10}$$

But in Eq. (11-10), the limits of integration can be changed to $-\infty$ and ∞ instead of $-f_m$ and f_m if $V_s(f)$ is replaced by $V(f)$. [The spectrum of $V(f)$ is zero beyond the limits $-f_m$ and $+f_m$.]

Consequently,

$$a_k = \frac{1}{2f_m} \int_{-\infty}^{\infty} V(f) e^{j2\pi(-k/2f_m)f}\, df = \frac{1}{2f_m} v\left(-\frac{k}{2f_m}\right) \tag{11-11}$$

because $V(f)$ was defined as the Fourier transform of $v(t)$. In addition, we will have

$$v(t) = \int_{-\infty}^{\infty} V(f) e^{j2\pi ft}\, df = \int_{-f_m}^{f_m} V_s(f) e^{j2\pi ft}\, df$$

$$= \int_{-f_m}^{f_m} \sum_{k=-\infty}^{\infty} a_k e^{j2\pi(k/2f_m)f} e^{j2\pi ft}\, df$$

$$= \sum_{k=-\infty}^{\infty} a_k \int_{-f_m}^{f_m} e^{j2\pi f(t+k/2f_m)}\, df$$

$$= \sum_{k=-\infty}^{\infty} a_k \frac{\sin 2\pi f_m(t + k/2f_m)}{\pi(t + k/2f_m)} \tag{11-12}$$

By substituting Eq. (11-11) into Eq. (11-12) and replacing k by $-k$, we obtain

$$v(t) \sum_{k=-\infty}^{\infty} = v\left(\frac{k}{2f_m}\right) \frac{\sin 2\pi f_m(t - k/2f_m)}{2\pi f_m(t - k/2f_m)}$$

which is Eq. (11-7). Equation (11-7) indicates that $v(t)$ is determined by its samples $v(k/2f_m)$, confirming the statement of the sampling theorem.

PULSE-AMPLITUDE MODULATION

In Fig. 11-2, the signal $v(t)$ is sampled with narrowband pulses so that the resulting pulse train carries the information on the height (amplitude) of the pulses. The Fourier transform of a single pulse with amplitude v and width T_p is

$$s(f) = \int_{-T_p/2}^{T_p/2} ve^{-j2\pi ft}\, dt = vT_p \frac{\sin \pi f T_p}{\pi f T_p} \tag{11-13}$$

Figure 11-5 depicts this pulse and its Fourier transform.

A train of unit-amplitude pulses $P_T(t)$, with pulse width T_p and repetition rate T, can be represented as the Fourier series

$$p_T(t) = \sum_{k=-\infty}^{\infty} a(k) e^{j2\pi kt/T} \tag{11-14}$$

where

$$a(k) = \frac{1}{T} \int_{-T/2}^{T/2} P_T(t) e^{-j2\pi(k/T)t}\, dt$$

$$= \frac{1}{T} \int_{-T_p/2}^{T_p/2} ve^{-j2\pi kt/T}\, dt$$

$$= \frac{T_p}{T} \frac{\sin (k\pi T_p/T)}{k\pi T_p/T}$$

Now, if the signal $v(t)$ of Fig. 11-2 is multiplied by this train of pulses, we will get the PAM signal

$$v_{\text{PAM}}(t) = v(t)P_T(t)$$

and the Fourier transform of $V_{\text{PAM}}(t)$ will be the convolution of the Fourier transforms of $v(t)$ and $p_T(t)$.

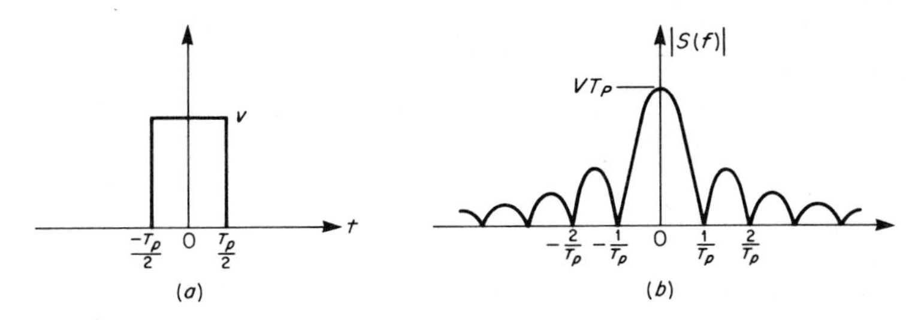

FIGURE 11-5 Fourier transform of a single pulse.

The Fourier transform of the train of pulses can be found easily from the series expression of $P_T(t)$ given by Eq. (11-14):

$$P_T(f) = \frac{T_p}{T} \sum_{k=-\infty}^{\infty} \frac{\sin (k\pi T_p/T)}{k\pi T_p/T} \delta(f - k/T) \tag{11-15}$$

The spectrum $V_{PAM}(f)$ of $v_{PAM}(t)$ is depicted in Fig. 11-6.

There is an attenuation factor involved, but otherwise the spectrum shape is preserved, and it can be recovered by a low-pass filter.

In the time domain, the tops of the pulses are shaped by the signal, and, as a result, they are not flat. In practice, because of the noise introduced during transmission, this shaping of the sample pulses cannot be maintained. Usually this shaping is eliminated, and a flat pulse is transmitted. As a result, a certain minor distortion is introduced which will appear at the output of the low-pass filter.

PULSE-CODE MODULATION

Definition and Bandwidth Requirement

In a PAM system, it is usually difficult to accurately preserve the height of the sampling pulses over long and complex transmission systems. In order to avoid

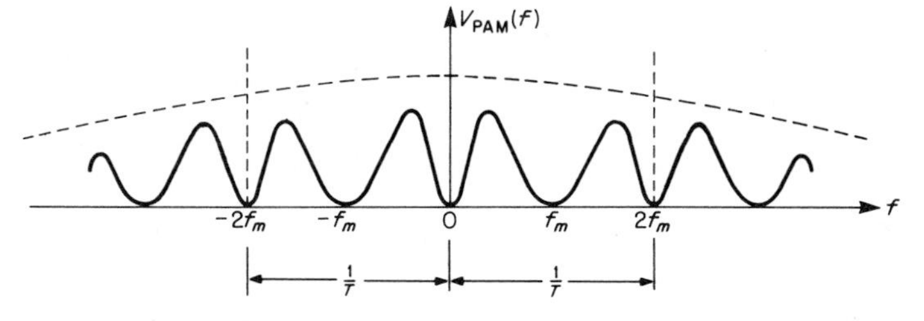

FIGURE 11-6 Spectrum of a signal sampled with pulses having a finite width.

the consequences of this difficulty, the height of the pulse is read before it is transmitted, and it is coded with a series of constant-amplitude pulses. As long as the existence of these new pulses can be detected at the receiving end, the code can be read and the PAM pulses can be accurately reproduced. The noise introduced during transmission will not influence the reconstruction of the original signal unless the new pulses do not have adequate height. This process of height coding is known as a pulse-code modulation (PCM).

At the transmitting end, the PAM pulses are quantized in a finite number of steps. Since most of the time the pulse height will not fall exactly at the end of one of these steps, quantization noise is introduced. The step closest to the actual signal will be selected in this case to represent the signal. Of course, this "round-off" error can be reduced by increasing the number of quantization levels. This increase in the number of quantization levels will result in an increased bandwidth requirement.

In PCM, quite often 3 binary digits are used to encode the 8 possible levels that result from the 2^3 combinations. The following list tabulates this process:

Level	Code
0	000
1	001
2	010
3	011
4	100
5	101
6	110
7	111

If an 8-digit word is to be used, then we can have $2^8 = 256$ levels and a significant reduction of the quantization noise.

In theory, one can make the transmission free of all transmission noise effects as long as the receiver detector can detect the presence or absence of a pulse. In addition, one can also reduce the quantization noise at will by increasing the bandwidth.

If f_m is the maximum frequency of the signal $v(t)$, then one needs at least a $2f_m$ pulse rate with PAM to transmit the information, but if each PAM pulse is coded to n pulses, then at least a $2nf_m$ pulse rate is required. Since a bandwidth B corresponds to a $2B$ pulse rate, we have $2B = 2nf_m$, and

$$B = nf_m \tag{11-16}$$

This relation indicates that we decrease the effect of quantization noise by increasing the code length n and simultaneously increasing the transmission bandwidth in proportion to n.

Quantization Noise

In conversion of analog signals into digital signals, quantization noise is generated because there is a finite number of quantization levels. If the level of the signal does not correspond exactly with one of the quantization levels, an error will be generated. For a linear quantizer, these errors are equally likely and vary in magnitude from minus one-half step to plus one-half step. The result is noise with rectangular density function. Figure 11-7 represents this density function with V_q being the quantization step.

By increasing the number of steps, i.e., by increasing the code word length, we can decrease this noise. Of course, this will increase the required bit rate and, consequently, the required bandwidth. Each added binary bit improves the signal-to-noise ratio by 6 dB.* The table below lists S/N versus the number of quantization levels.

Number of quantization levels	Number of binary digits	S/N, dB
8	3	20
16	4	26
32	5	32
64	6	38
128	7	44
256	8	50

Error Probability in PCM Systems

Now assume that the amplitude of the PCM pulses is constant at V_0. These pulses are received at the receiving end and sampled. The sampler is synchronized to sample at the time when the maximum pulse amplitude should be present if a pulse is transmitted, or when no amplitude at all exists if a pulse is not transmitted. As a result, the pulses do not necessarily have to be flat. Superimposed noise will also be detected if it exceeds the threshold level. As a result, a pulse may be detected in error because of noise spikes, and, conversely, an

*Signal-to-noise ratio is defined as the ratio of the mean signal-wave power to the mean quantization-noise power.

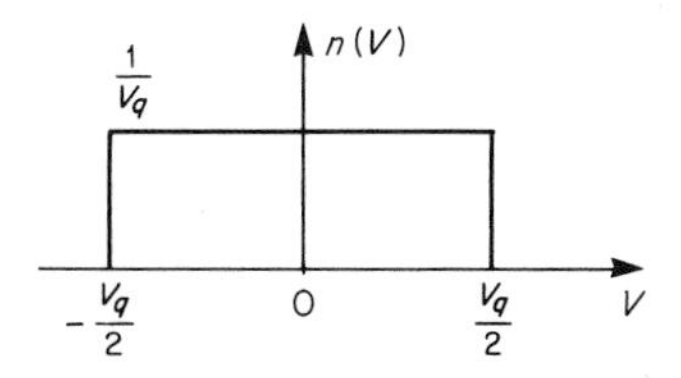

FIGURE 11-7 Quantizing noise density function.

absence of pulse may be detected when negative noise spikes at the sampling time reduce the pulse amplitude below threshold. Threshold is usually set at $\frac{1}{2}V_0$.

If the noise has a gaussian distribution with zero mean and σ_n standard deviation, the probability that an error will occur is

$$P_e = \frac{1}{\sigma_n\sqrt{2\pi}} \int_{V_0/2}^{\infty} e^{-v_n^2/2\sigma_n^2}\, dv_n = \text{prob}\left(v_n > \frac{V_0}{2}\right) \tag{11-17}$$

By substituting

$$x = \frac{v_n}{\sigma_n\sqrt{2}}$$

we get

$$P_e = \frac{1}{\sqrt{\pi}} \int_{V_0/\sigma_n 2\sqrt{2}}^{\infty} - e^{-x^2}\, dx$$

or

$$P_e = \tfrac{1}{2}\,\text{erfc}\, \frac{V_0}{2\sqrt{2\sigma_n^2}} \tag{11-18}$$

By defining $S = V_0^2$ (the peak-signal-power voltage), $N = \sigma_n^2$ (the rms noise power), and $\gamma = S/2N$, we obtain

$$P_e = \tfrac{1}{2}\,\text{erfc}\, \frac{1}{2}\sqrt{\frac{S}{2N}} = \tfrac{1}{2}\,\text{erfc}\, \frac{\sqrt{\gamma}}{2} \tag{11-19}$$

This error probability is plotted in Fig. 11-8. It is important to note that γ was defined by using one-half the peak power of the signal.

Suppose that a single pulse is used to transmit 1 bit of information, and let us call the energy of this pulse E_b. Then, if T_p is the duration of the pulse, the energy E_b will be given by the relationship

$$E_b = V_0^2 T_p = S T_p \tag{11-20}$$

If $n_0/2$ is the two-sided noise density for a spectrum with a bandwidth extending from $-B$ to $+B$, then the noise power N will be

$$N = (2B)\left(\frac{n_0}{2}\right) = B n_0 \tag{11-21}$$

The rms signal-to-noise ratio will be

$$\frac{S}{N} = \frac{E_b}{n_0}\left(\frac{1}{BT_p}\right) \tag{11-22a}$$

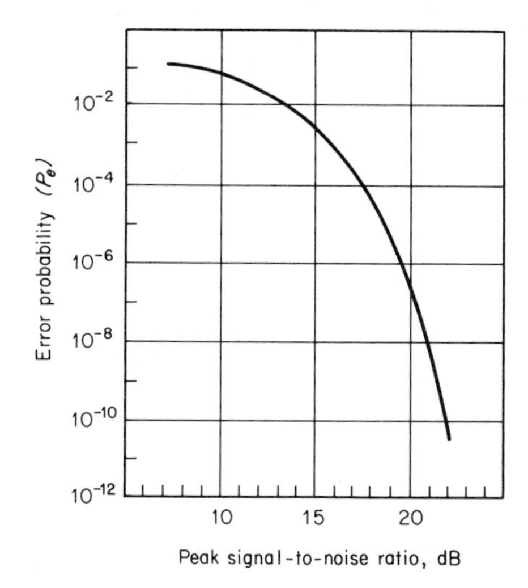

FIGURE 11-8 Error probability in PCM due to transmission noise.

Consequently, the error probability will depend on both the ratio E_b/n_0 and the BT_p product. The bandwidth-time product BT_p is an important parameter in the design of a digital system. Under certain conditions, this product is set to be close to 1. Then

$$\frac{S}{N} = \frac{E_b}{n_0} \qquad \text{for } BT_p = 1 \tag{11-22b}$$

Equation (11-22) can be written as

$$\frac{C}{N} = \frac{E_b}{n_0}\frac{B_p}{B} \tag{11-23}$$

where $B_p = 1/T_p$.

It is clear from Fig. 11-8 that when S/N is larger than 20 dB, the probability of error P_e rapidly diminishes. This is easily achievable, and usually PCM systems can be made to be immune to this kind of noise. On the other hand, the quantization noise remains with the signal and it will be detected.

Consider in Fig. 11-9 the input signal $v(t)$ quantized to $V_q(t)$ by equal steps of length b. The values of $V_q(t)$ correspond to the middle of each step. The expected mean square error can be computed if the amplitude distribution of the input samples is known. If v is the actual value of $v(t)$ at the ith sample interval and V_{qi} is the corresponding value of $V_q(t)$, the error e will have a mean square value

$$\overline{e^2} = \sum_{i=1}^{N} \int_{V_{qi}}^{V_{q,i+1}} (V - \overline{V}_{qi})^2 P(v)\, dv$$

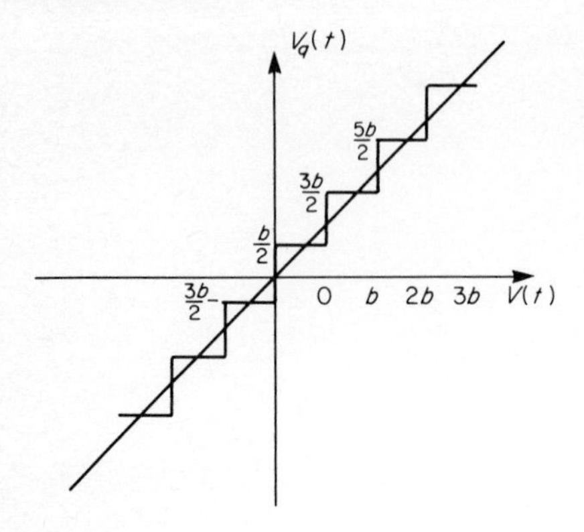

FIGURE 11-9 Quantized signal.

For a uniform distribution of the error with each step of width b, the above integral will be evaluated as

$$\overline{e^2} = \frac{b^2}{12} \tag{11-24}$$

If the rms value of the baseband signal power at the quantizer input is S, then, assuming zero error rate at reception ($P_e = 0$), we find that the signal-to-noise ratio after the signal reconstruction is

$$\left(\frac{S}{N}\right)_0 = \frac{S}{b^2/12} \tag{11-25}$$

If we assume the use of a binary code with 2^n quantizing levels and therefore a code word length of n, then

$$b = \frac{V_0}{2^n} \tag{11-26}$$

and Eq. (11-25) becomes

$$\left(\frac{S}{N}\right)_0 = \left(\frac{S}{V_0^2}\right)\left(\frac{12}{2^{2n}}\right) \tag{11-27}$$

The ratio S/V_0^2 represents the square of the form factor $\tilde{V}_0$, that is, the rms-to-peak ratio.

EXAMPLE 11-1 DIGITIZATION OF VOICE

The baseband of a voice signal contains significant frequencies up to 4 kHz. As a result, in digitizing voice, a sampling rate of $2 \times 4000 = 8000$ samples per second is required. If an 8-bit PCM system is used, then a transmission rate of $8 \times 8000 = 64$ kb/s is required. Each of the 256 quantization levels will be represented by 8 pulses. The signal-to-quantization noise ratio will be 50 dB.

PULSE-TIME MODULATION

A different type of pulse modulation system is one in which the timing between pulses is used to encode the message. In the simplest form, the amplitude of the PAM pulse could be encoded as the time duration of another pulse. This is the pulse-duration-modulation (PDM) system, or pulse-width-modulation system. The main disadvantage of this system is the fact that long-duration pulses may require substantial power. This disadvantage may be cured by coding the signal as the time duration between the leading edges of two standard-size pulses. Thus we derive the pulse-position-modulation (PPM) system.

DELTA MODULATION

When a signal is sampled at a rate higher than the minimum rate required to satisfy the relationship $T = 1/2B$ (Nyquist rate), the transmitted samples contain more information than that required to reconstruct the signal. Successive samples will be correlated, and the knowledge of the past samples can provide a good prediction of the next forthcoming sample. The differential pulse-code-modulation (DPCM) technique utilizes this property to correct an erroneously detected sample. A feedback loop compares the predicted sample value with the actual detected sample value and provides a correction factor. Delta-modulation (DM) schemes transmit only the difference between samples, quantizing the original signal in steps by oversampling at a constant sampling rate, higher than the Nyquist rate. In the 1-bit version of DPCM scheme, if the prediction falls below the detected value, a correction $+\delta$ is applied. However, if the prediction falls above the detected value, a correction $-\delta$ is applied. As a result, the error always stays within $\pm\delta$ of the original signal. Several variations of this basic technique have been developed.

DIGITAL SIGNALING

Amplitude-Shift Keying

The amplitude-shift keying (ASK) technique is a special form of amplitude modulation and sometimes is referred to as binary on-off keying (OOK). In satellite communications, ASK is one of the least-used digital-signaling techniques.

The basic waveform is a sinusoidal one, although other types could be used. The presence of a pulse, or the binary symbol 1, is indicated by the presence of the sinusoidal signal, whereas the absence of a pulse, or the binary symbol 0, is indicated by the absence of the sinusoidal signal.

Figure 11-10 depicts such a sequence of waveforms, each one extended over a time interval T which is in effect the pulse duration. The amplitude is depicted as constant, but that is not a necessary condition. The waveforms of an ASK signal can be expressed as

$$V_T(t) = \begin{cases} U_T(t) \cos (2\pi f_c t + \phi_T) & \text{for 1} \\ 0 & \text{for 0} \end{cases} \tag{11-28}$$

For the waveform in the idealized case of Fig. 11-10, we have $\phi_T = 0$ and $U_T(t) =$ constant. One can see that neither of these two conditions is significant for this type of signaling.

Binary Frequency-Shift Keying

In binary frequency-shift keying (FSK), we utilize two carrier frequencies, f_1 and f_2, in order to transmit a 0 and a 1, respectively. The transmitted waveforms are given by the relationship

$$V_T(t) = \begin{cases} u_T \cos 2\pi f_1 t & \text{for 1} \\ u_T \cos 2\pi f_2 t & \text{for 0} \end{cases} \tag{11-29}$$

It is obvious that this signaling is analogous to frequency modulation. Again, it is not necessary to use constant-amplitude carriers; the selection depends on the particular requirements of the system. Figure 11-11 depicts the waveforms of an FSK signal.

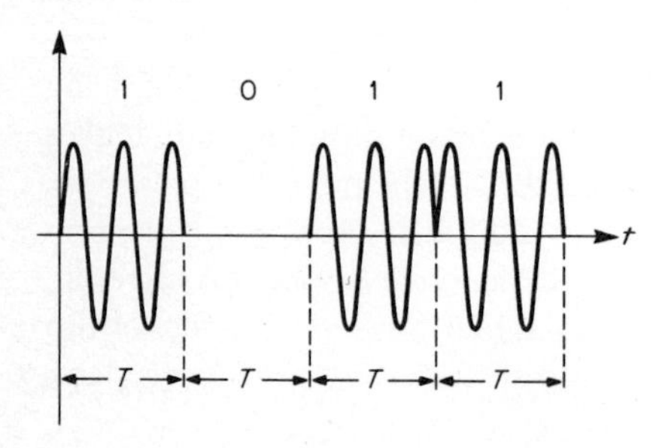

FIGURE 11-10 ASK or OOK signal.

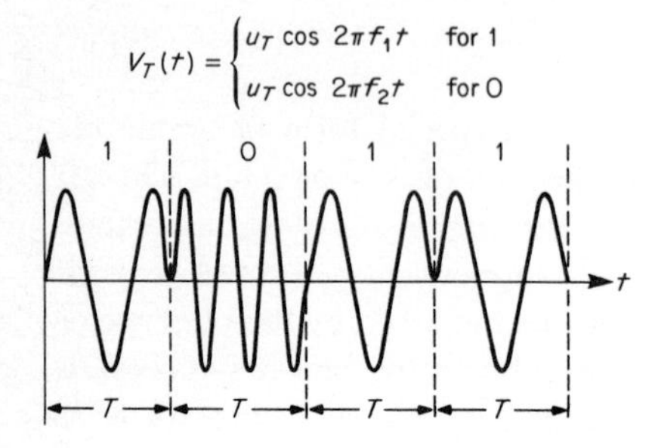

FIGURE 11-11 FSK signals.

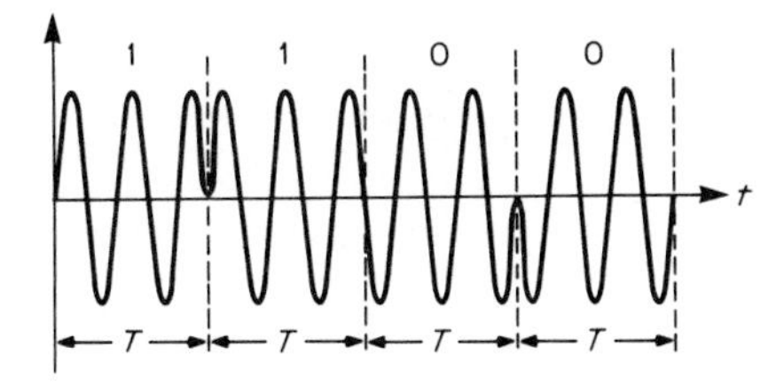

FIGURE 11-12 Two-phase, or binary, PSK (BPSK).

Binary and Differential Phase-Shift Keying

Binary phase-shift keying (BPSK), usually called two-phase PSK, is equivalent to angle modulation. A constant-amplitude carrier is utilized over the pulse interval T. However, the phase of the carrier changes by 180° to indicate a 0 instead of a 1. For example, all pulses designating 1 have 0° phase, and all pulses designating 0 have 180° phase:

$$V_T(t) = \begin{cases} u_T \cos 2\pi f_c t & \text{for } 1 \\ u_T \cos (2\pi f_c t + \pi) & \text{for } 0 \end{cases} \tag{11-30}$$

It is possible to use other waveforms besides a sinusoid for each pulse. The transmitted information is encoded in the phase of the waveform; i.e., the waveform is inverted, its algebraic sign changing, to indicate a change from 1 to 0.

A variation of the BPSK technique is differential PSK (DPSK), in which the change of phase by 180° designates a 0 following the previous pulse, while maintaining the same phase indicates a 1 following the previous pulse. Of course, one has to start with an arbitrary pulse, say 1, well-known in advance at the decoding end. Figure 11-12 depicts the word 1100 in a BPSK system.

FOUR-PHASE PHASE-SHIFT KEYING

In four-phase, or quadriphase, PSK (QPSK), a constant-amplitude carrier is utilized, but four distinct phases are considered, each one displaced by 90° from the previous one.

One could consider this technique as utilizing two orthogonal carriers, one $\cos 2\pi f_c t$ and the other $\sin 2\pi f_c t$, each one being used to transmit a binary PSK signal. Since the two carriers are 90° apart, no cross talk exists, and, as a result, one can double the rate of transmission of information. For a constant-amplitude carrier, the transmitted signal will be

$$v_T(t) = u_c \cos\left(2\pi f_c t + \frac{\pi}{4}\right) + u_s \sin\left(2\pi f_c t + \frac{\pi}{4}\right) \tag{11-31}$$

where $u_c = \pm 1$ and $u_s = \pm 1$, and the whole signal is transmitted at time T. Here each symbol transmits 2 bits of information, each bit independent of the other. The phasor diagram of this technique is represented in Fig. 11-13. Again, one can use differential encoding for each pair of bits, allowing the detector to

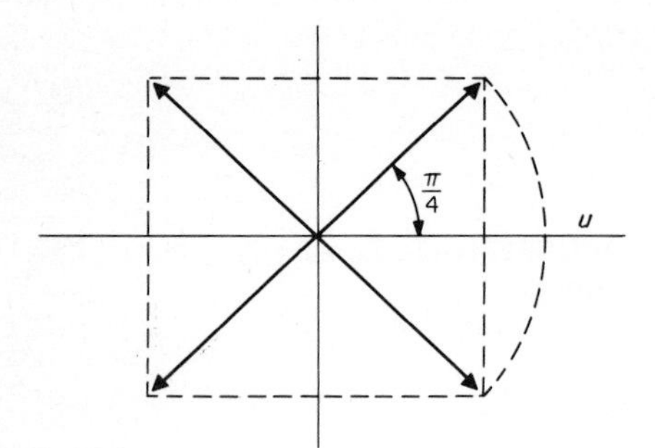

FIGURE 11-13 Phasor diagram for four-phase, or quadriphase, PSK (QPSK).

select the phase change rather than the absolute phase to determine 0 or 1. The ambiguities can be eliminated by using a unique synchronization word or a specific structure in the coding scheme.

In addition, one can delay the transmission of the in-phase bit stream by $T/2$ seconds, thus staggering it with the transmission of the quadrature channel. In this case, no envelope nulls will occur, and the performance with bandpass filters can be improved. This signaling technique is known as sequential PSK.

M-ary PSK

One can expand the concept of QPSK to eight-phase PSK and, in general, to M-ary PSK, where $M = 2^k$ (k = integer). But, of course, in this case, since certain of the signals will not be in quadrature with each other, the ability to discriminate against noise will be decreased. If R is the rate of transmission in bits per second, each symbol transmitted at time T will have k bits, and therefore $R = k/T$ or $k = RT$, where $k = \log_2 M$.

DETECTION OF DIGITAL SIGNALS

Idealized Digital Receiver

An idealized digital receiver is depicted in Fig. 11-14. At the sampling network, a timing pulse is available. This pulse is synchronized with the signal so that sampling takes place at the optimum time for a decision.

The detector will be either an envelope detector, in the case of noncoherent detection, or a product detector, when the carrier frequency and phase are available for synchronous or coherent detection.

If the input signal is

$$s(t) = a(t) \cos 2\pi f_c t$$

and the noise is

$$n(t) = x \cos 2\pi f_c t - y \sin 2\pi f_c t$$

the signal presented for decision will be either

$$r = \sqrt{(a + x)^2 + y^2}$$

in the case of envelope detection or

$$v = a + x$$

in the case of coherent detection, where r and v are functions of time.

Predetection Optimization of Signal-to-Noise Ratio

When a signal embedded in noise is received, the decision to be made, in digital transmission, is whether a pulse is present or absent. A variation of this case is the decision between a positive polarity pulse vs. a negative polarity pulse. In either case, the actual shape of the pulse is not as important as the presence or absence of the pulse.

It is therefore important, before a decision is made and before the time t_0 of sampling, to maximize the signal-to-noise ratio, expressed as $A^2/2N$, where A is the peak amplitude of the signal and N is the mean square noise power.

Prefiltering the received signal with a bandpass filter or a low-pass filter will have a significant effect on this ratio. If the filter bandwidth is too broad, then the output mean square noise value will be large. In the case of a bandpass filter with flat transfer characteristic $|H(f)| = H_0$ over the bandwidth B and white input noise, the output mean square noise power will be directly proportional to the bandwidth. On the other hand, if the bandwidth of the filter is too narrow, the signal amplitude of the received rectangular pulse will be suppressed well below its maximum value. As an example, let us consider the specific case of a signal represented by a rectangular pulse and the filter by an ideal low-pass filter. The response of such a filter to a square pulse is covered in Chap. 8, in particular in Example 8-1 and Fig. 8-3. If the signal-pulse duration is T and the bandwidth of the filter is $B = 1/T$, then the output will reach approximately the maximum height. The maximum output amplitude is a monotonically increasing function of B for values of B up to $1/T$. For $B > 1/T$, the height of the response will not exceed the height obtained for $B = 1/T$ (see Fig. 8-3). As a result, in this case it is obvious that the output signal-to-noise ratio will improve as B increases from very low values. However, it is obvious

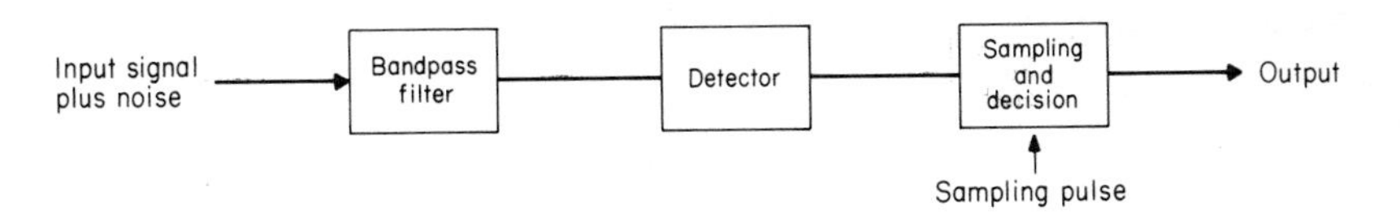

FIGURE 11-14 Idealized digital receiver.

that, for $B > 1/T$, this ratio deteriorates as B increases. Consequently an optimum value exists for $B \leq 1/T$. In the general case in which the received signal $g_1(t)$ is the input to the optimum filter, i.e., the filter that optimizes the signal-to-noise ratio, the filter transfer function must be related to the signal $g_1(t)$. This relationship will be derived in this section.

Assume a filter with unit-impulse response $h(t)$ and transfer characteristic $H(f)$. If the input signal is $g_1(t)$, the output will be

$$g_0(t) = \int_{-\infty}^{\infty} g_1(\tau)h(t - \tau)\, dt \qquad (11\text{-}32)$$

If, in addition, the input noise is white with a two-sided spectral density $n_0/2$, the output noise power will be

$$N = \int_{-\infty}^{\infty} \frac{n_0}{2}\, |H(f)|^2\, df = \frac{n_0}{2} \int_{-\infty}^{\infty} [h(\tau)]^2\, d\tau \qquad (11\text{-}33)$$

The energy of the input-pulse signal $g_1(t)$ is

$$E = \int_{-\infty}^{\infty} [g_1(\tau)]^2\, d\tau \qquad (11\text{-}34)$$

and it is assumed to be constant.

The peak signal-to-noise ratio at the output will be

$$\frac{S}{N} = \frac{g_0^2(t)}{N} = \frac{\left[\int_{-\infty}^{\infty} g_1(\tau)h(t - \tau)\, dt\right]^2}{(n_0/2)\int_{-\infty}^{\infty} [h(\tau)]^2\, d\tau} \qquad (11\text{-}35)$$

The normalized signal-to-noise ratio with respect to the energy E of the input pulse will be

$$\left(\frac{S}{N}\right)_0 = \frac{S/N}{E} = \frac{\left[\int_{-\infty}^{\infty} g_1(\tau)h(t - \tau)\, d\tau\right]^2}{(n_0/2)\int_{-\infty}^{\infty} [g_1(\tau)]^2\, d\tau \int_{-\infty}^{\infty} [h(\tau)]^2\, d\tau} \qquad (11\text{-}36)$$

But Schwartz's inequality states that

$$\left[\int_{-\infty}^{\infty} g_1(\tau)h(t - \tau)\, d\tau\right]^2 \leq \int_{-\infty}^{\infty} [g_1(\tau)]^2\, d\tau \int_{-\infty}^{\infty} [h(\tau)]^2\, d\tau \qquad (11\text{-}37)$$

In order to maximize $(S/N)_0$ at the sampling time t_0, we must satisfy the condition for which Eq. (11-37) becomes an equality:

$$g_1(\tau) = h(t_0 - \tau) \qquad (11\text{-}38)$$

Denoting $t = t_0 - \tau$, we find

$$h(t) = g_1(t_0 - t) \tag{11-39}$$

Equation (11-39) defines the "matched-filter" transfer characteristic, which, in the frequency domain, becomes

$$H(f) = G_1^*(f)e^{-j2\pi f t_0} \tag{11-40}$$

By using the equality in Eq. (11-37), we change Eq. (11-35) into

$$\left(\frac{S}{N}\right)_{\max} = \frac{\int_{-\infty}^{\infty} [g_1(\tau)]^2 \, d\tau}{n_0/2}$$

or

$$\left(\frac{S}{N}\right)_{\max} = \frac{E}{n_0/2} \tag{11-41}$$

Equation (11-41) gives S/N in terms of the peak signal. For sinusoidal signals, in which the mean square must be considered, E becomes $E/2$, and

$$\left(\frac{S}{N}\right)_{\max} = \gamma_{\max} = \frac{E}{n_0} \tag{11-42}$$

The maximum value of S/N at the output of the filter does not depend at all on the shape of the pulse, but only on the total energy E of the input pulse and the noise spectral density.

The matched-filter characteristic does not always represent a physically realizable network, but usually a good approximation is feasible with a network having the proper passband.

EXAMPLE 11-2 MATCHED FILTER FOR A SQUARE INPUT PULSE

Let us examine the case of an input signal represented by the rectangular pulse depicted in Fig. 11-15*a*. The expression for this pulse is

$$g_1(t) = \begin{cases} A & \text{for } 0 \le t \le T \\ 0 & \text{elsewhere} \end{cases}$$

The Fourier transform of this pulse can be found by multiplying Eq. (B-22) by $e^{-j2\pi fT/2}$ to indicate the time shift of the origin from the middle to the beginning of the pulse. Consequently,

$$G_1(f) = AT\,\frac{\sin \pi fT}{fT}\,e^{-j\pi fT}$$

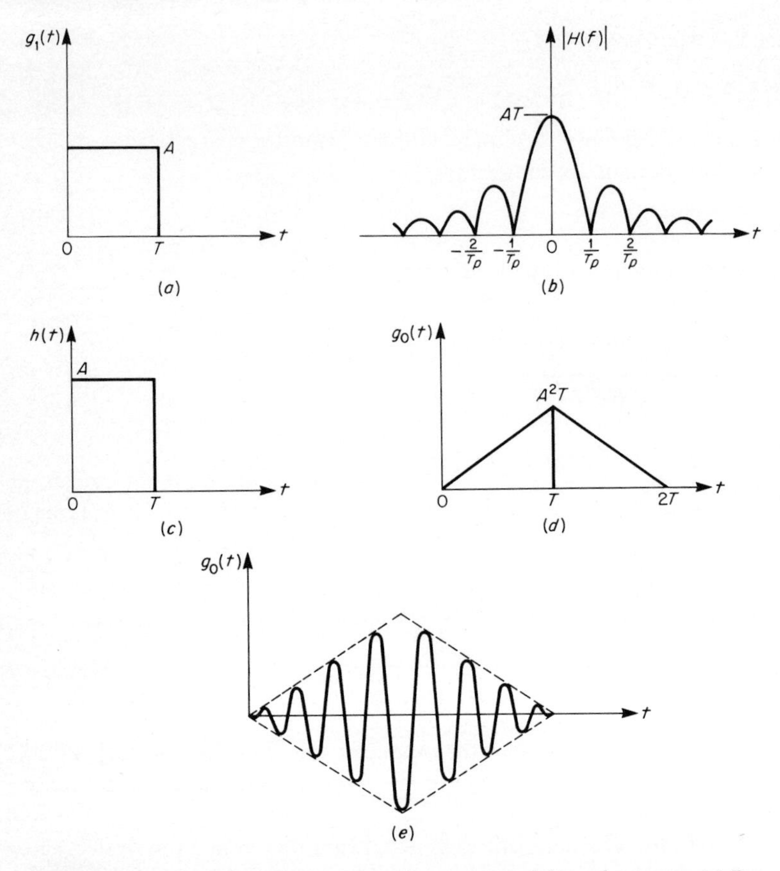

FIGURE 11-15 **Matched filter response for square and RF pulse inputs. (*a*) Rectangular pulse. (*b*) Matched filter transfer function $|H(f)|$ for a rectangular pulse. (*c*) Matched filter $h(t)$ for a rectangular pulse. (*d*) Output of matched filter for a rectangular pulse. (*e*) Output of matched filter for an RF pulse.**

The Fourier transform $H(f)$ of the matched filter will be

$$H(f) = G_1^*(f)e^{-j2\pi f t_0}$$

if we want to maximize the output signal-to-noise ratio for $t = t_0$. Let us set $t_0 = T$ for simplicity. Then we obtain

$$H(f) = AT \frac{\sin \pi f T}{\pi f T} e^{-j\pi f T}$$

The amplitude of $H(f)$ is the function $|H(f)| = AT \operatorname{sinc} fT$, depicted in Fig. 11-15*b*. [Again, the function $\operatorname{sinc} x = (\sin \pi x)/\pi x$.]

The unit-impulse response of $h(t)$ will be the Fourier pair of $AT(\sin \pi fT)/\pi fT$, translated in time by $T/2$ (see "Time Shift" in Appendix B). This is a rectangular pulse of amplitude A and duration T with origin at the left extreme of the leading edge, expressed as

$$h(t) = \begin{cases} A & \text{for } 0 \leq t \leq T \\ 0 & \text{elsewhere} \end{cases}$$

Of course, this was expected since the matched-filter unit-impulse response should have the same shape as the input pulse (Fig. 11-15*c*).

The output of the matched filter will have an output transfer function

$$G_0(f) = G_1(f)H(f) = |G_1(f)|^2 e^{-j2\pi f t_0}$$

Therefore the output signal will be the autocorrelation function of the input shifted by t_0. Since the last relationship is general and does not depend on the parameters selected for this example, the preceding conclusion relating the output signal with the autocorrelation of the input signal is also general. That is,

$$g_0(t) = R_1(t - t_0) \tag{11-43}$$

In the present example, because we selected $t_0 = T$, we have

$$g_0(t) = R_1(t - T)$$

But it is easy to prove that the autocorrelation $R_1(t)$ is a triangular pulse, as depicted in Fig. 11-15*d*, expressed as

$$g_0(t) = \begin{cases} A^2 t & \text{for } 0 \leq t \leq T \\ A^2(2T - t) & \text{for } T \leq t \leq 2T \\ 0 & \text{elsewhere} \end{cases}$$

The energy of the pulse is $E = A^2T$. If the noise power density is n_0, then the output signal-to-noise ratio will be

$$\left(\frac{S}{N}\right)_0 = \frac{E}{n_0/2} = \frac{A^2T}{n_0/2}$$

The matched filter represented by $H(f)$ and $h(t)$ of this example could be physically approximated very accurately by a high-Q integrating filter. However, because a new pulse will be transmitted immediately after time $2T$ and before the output has completely decayed, intersymbol interference will be experienced. With the aid of a switch, the trailing edges could be eliminated and symbol interference avoided. A filter using such a switch is known as an "integrate-and-dump filter."

EXAMPLE 11-3 MATCHED FILTER FOR AN RF PULSE

Assume that the RF pulse is

$$m(t) = \begin{cases} A \cos 2\pi f_c t & \text{for } 0 \le t \le T \\ 0 & \text{elsewhere} \end{cases}$$

where $f_c \gg 1/T$.

The Fourier transform of this RF pulse is, from the general equation given in Appendix B under "Frequency Shift,"

$$M(f) = \frac{AT}{2} [\text{sinc } (f - f_c)T + \text{sinc } (f + f_c)T] e^{-j\pi fT}$$

where the factor $e^{-j\pi fT}$ is necessary in this example because the pulse is translated in time by $T/2$.

For $f_c \gg 1/T$, the above relationship could be approximated as

$$M(f)e^{j\pi fT} = \begin{cases} \dfrac{AT}{2} \text{ sinc } (f - f_c)T & \text{for } f > 0 \\ \dfrac{AT}{2} \text{ sinc } (f + f_c)T & \text{for } f < 0 \end{cases}$$

The Fourier transform $H(f)$ of the matched filter will be

$$H(f) = M^*(f)e^{-j2\pi fT}$$

or

$$H(f)e^{j\pi fT} = \begin{cases} \dfrac{AT}{2} \text{ sinc } (f - f_c)T & \text{for } f > 0 \\ \dfrac{AT}{2} \text{ sinc } (f + f_c)T & \text{for } f < 0 \end{cases}$$

The matched-filter output will be

$$g_0(t) = \begin{cases} A^2 t \cos 2\pi f_c(t - T) & \text{for } 0 \le t \le T \\ A^2(2T - t) \cos 2\pi f_c(t - T) & \text{for } T \le t \le 2T \end{cases}$$

This output is depicted in Fig. 11-15*e*.

STATISTICAL DECISION IN THE DETECTION PROCESS

Likelihood and Bayes' Rule

Given a train of pulses, let us designate with 1 (mark) the presence of a pulse and 0 (space) the absence of a pulse. In addition, let us assume that P_1 is the probability (frequency of occurrence) of a pulse (mark or 1) within the train of pulses and that P_0 is the probability of no pulse (space or 0). If this train of pulses is transmitted in the presence of noise, let $P_1(v)$ be the probability that a pulse will result in a signal $v(t_0) = v$ in front of the sampling network at the appropriate sampling time t_0, and let $P_0(v)$ be the probability that the absence of a pulse will result in a signal $v(t_0) = v$ at the same sampling time. Assume that the receiver is the same as the one depicted in Fig. 11-14.

The detector, sensing a voltage v, must decide if a mark or a space was transmitted. A way to arrive at such a decision is to examine the a posteriori probabilities $P(1/v)$ and $P(0/v)$, where $P(1/v)$ is the probability that a 1 was sent if an amplitude v was detected and $P(0/v)$ is the probability that a 0 was sent if an amplitude v was detected.

But it is clear that

$$P_1 P_1(v) = P(1/v)p(v) = P(1, v) \tag{11-44a}$$

where $p(v)$ is the probability of v being received and $P(1, v)$ is the probability of 1 and v occurring jointly.

Similarly

$$P_0 P_0(v) = P(0/v)p(v) = P(0, v) \tag{11-44b}$$

By comparing Eqs. (11-44*a*) and (11-44*b*), we obtain

$$\frac{P(1/v)}{P(0/v)} = \frac{P_1 P_1(v)}{P_0 P_0(v)} \tag{11-45}$$

If $P(1/v)/P(0/v) > 1$, then it is logical that the decision should be made in favor of a pulse having been transmitted. But the ratio of the a posteriori probabilities is also expressed by the right side of Eq. (11-45), and therefore the inequality can be written as

$$\frac{P_1 P_1(v)}{P_0 P_0(v)} > 1$$

or

$$L = \frac{P_1(v)}{P_0(v)} > \frac{P_0}{P_1} \tag{11-46}$$

where L is the likelihood ratio. Equation (11-46) is known as Bayes' decision rule.

When we have more than one sample of the signal at the decision point, Eq. (11-46) can be generalized to read

$$L = \frac{P_1(v_1, v_2, \ldots, v_n)}{P_0(v_1, v_2, \ldots, v_n)} > \frac{P_0}{P_1} \tag{11-47}$$

where $P_1(v_1, v_2, \ldots, v_n)$ and $P_0(v_1, v_2, \ldots, v_n)$ are the joint probabilities of the sampled values $v_1, v_2, \ldots, v_n$; that is, $P_1(v_1, v_2, \ldots, v_n)$ is the probability that, when 1 is transmitted, the amplitudes $v_1, v_2, \ldots, v_n$ will be observed at the receiving end if the same transmitted signal is repeated n times in order to increase the likelihood of correct detection.

If the random variables are statistically independent, then the joint probabilities in Eq. (11-47) become the product of the individual probabilities.

As an example, consider the case in which gaussian noise is superimposed on the binary transmission with probabilities $P_1 = P_0 = \frac{1}{2}$. If the gaussian noise has 0 mean and σ^2 variance and if the pulses have amplitude A at the time of sampling, we will have

$$P_0(v) = \frac{1}{\sqrt{2\pi\sigma^2}} e^{-v^2/2\sigma^2} \tag{11-48}$$

$$P_1(v) = \frac{1}{\sqrt{2\pi\sigma^2}} e^{-(v-A)^2/2\sigma^2} \tag{11-49}$$

For a decision favoring mark, we must have

$$L = \frac{P_1(v)}{P_0(v)} > \frac{P_0}{P_1} = 1$$

The threshold is $v = A/2$. For $v > A/2$, a 1 will be decided, and for $v < A/2$, a 0 will be decided. As can be seen from the graphical representation of Fig. 11-16, for $v = A/2$ the two probabilities $P_0(A/2)$ and $P_1(A/2)$ become equal.

Neyman-Pearson Criterion

When the a priori probabilities are not known, the likelihood ratio cannot be used as a criterion for a threshold decision.

In this case, the Neyman-Pearson process of decision is widely used, especially in radar applications. An acceptable error level for wrongly rejecting the presence of a space (0) is first established, and the error probability for accepting a space (0) while a mark (1) is transmitted is minimized.

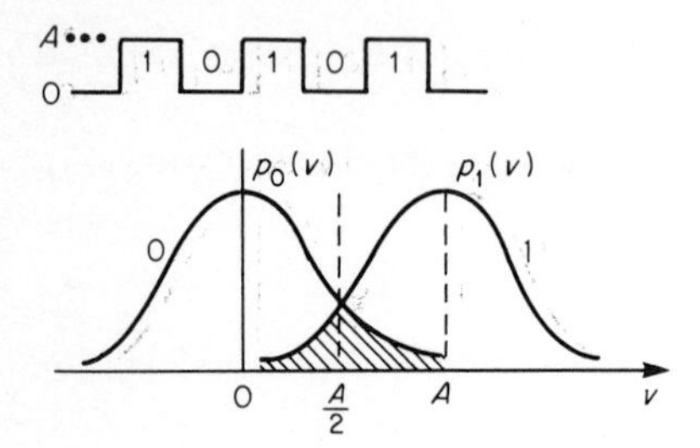

FIGURE 11-16 Maximum-likelihood threshold: coherent detection.

The probability of rejecting wrongly the presence of a space is

$$P_e'(0) = \int_{v_1} P_0(v)\, dv \tag{11-50}$$

where v_1 is the region in which a mark decision ought to be made.

The probability of accepting a space wrongly is

$$P_e(0) = \int_{v_0} P_1(v)\, dv \tag{11-51}$$

where v_0 is the region in which a space decision ought to be made.

By defining an acceptable level for $P_e'(0)$, we determine the region v_1 so that $P_e(0)$ is minimized.

Threshold Detection

Gaussian Noise and Binary Transmission Assume that the transmitted signal consists of a train of pulses in which 1s and 0s occur with equal probability. Assume also that superimposed on the signal is gaussian noise $n(t)$ with probability density function

$$p(n) = \frac{1}{\sqrt{2\pi N}} e^{-n^2/2N} \tag{11-52}$$

where $N = \sigma^2$ (the variance of the distribution) represents the mean square of the noise power with zero mean. The probability density functions of mark and space are given by Eqs. (11-49) and (11-48), respectively. Let us designate $A/2$ as the threshold above which the receiver decides it has received a pulse 1, and below which it decides it has received a 0. The probability of making an error when 1 is transmitted is equal to the probability of making an error when 0 is transmitted. This probability, depicted as the shaded area under the graphs of the distribution in Fig. 11-16, can be computed as

$$P_e = \int_{A/2}^{\infty} p_0(v)\, dv = \tfrac{1}{2}\left(1 - \operatorname{erf}\frac{\sqrt{\gamma}}{2}\right) = \tfrac{1}{2}\operatorname{erfc}\frac{\sqrt{\gamma}}{2} \tag{11-53}$$

where

$$\gamma = \frac{A^2}{2N} \tag{11-54}$$

The plot of P_e versus γ is depicted in Fig. 11-17.

Carrier System and Envelope Detection Consider the noise at the output of a narrow-bandpass linear network with center frequency f_c. If the input noise

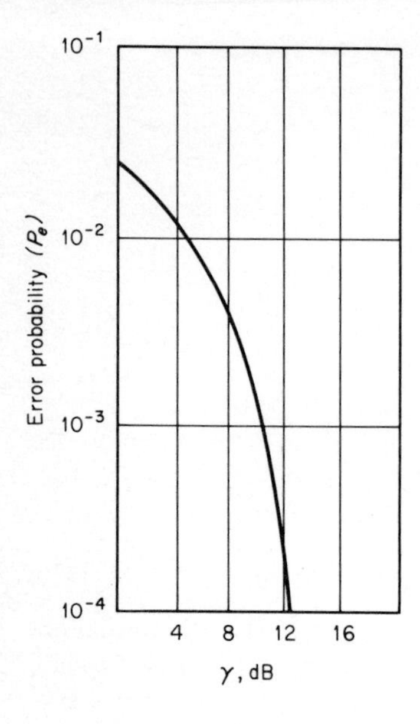

FIGURE 11-17 Error probability.

is gaussian only, the output will be a narrowband noise wave centered around f_c.

When no other signal is present, this noise wave will look like a carrier with frequency f_c and a slowly varying envelope $n(t)$:

$$n(t) = re^{j\phi} \tag{11-55}$$

where both r and ϕ are random functions of time. The probability density function of the envelope r will be the Rayleigh distribution:

$$P_0(r) = \frac{r}{N} e^{-r^2/2N} \tag{11-56}$$

The shape of this function is depicted in Fig. 11-18*a*. Now, if a signal represented by RF pulse $A \cos 2\pi f_c t$ is present, the envelope of the modulated signal $v(t)$ will be

$$v(t) = re^{j\phi} \tag{11-57}$$

where r and ϕ are random variables. The distribution of r will be a Rician distribution with probability density function

$$P_1(r) = \frac{r}{N} e^{-r^2/2N} I_0\left(\frac{rA}{N}\right) e^{-A^2/2N} \tag{11-58}$$

where $I_0(x)$ = modified Bessel function of the first kind and zero order. For a large signal-to-noise ratio $\gamma \gg 1$, this distribution resembles a gaussian distribution; it is depicted in Fig. 11-18*b*.

Now consider the same binary transmission of 0s and 1s as in the first example. When we have 0 (space), i.e., no signal, the Rayleigh distribution will be applicable, and when we have 1 (mark), i.e., a signal, the Rician distribution will be applicable. If we try to establish a threshold a above which the receiver assumes mark and below which it assumes space, it is not obvious that the optimum selection is $a = A/2$.

If we wanted the probability P_e of an error when 0 is transmitted to be equal to the probability of error when 1 is transmitted, we would have

$$P_e = \int_a^{\infty} \frac{r}{N} e^{-r^2/2N}\, dr = \int_0^a \frac{r}{N} e^{-r^2/2N} I_0\left(\frac{rA}{N}\right) e^{-A^2/2N}\, dr \qquad (11\text{-}59)$$

By selecting a desirable P_e from the first relationship, we can determine a. Then from the second relationship, we can determine the required signal-to-noise ratio, namely, $\gamma = S/N = A^2/2N$. There is an optimum threshold a_0 for a given $S/N = A^2/2N$ which minimizes the probability of an error P_e. A good approximation of this optimum threshold a_0, for which $\partial P_e/\partial a = 0$, is

$$a_0 = \sqrt{2 + \frac{\gamma}{2}} \qquad (11\text{-}60)$$

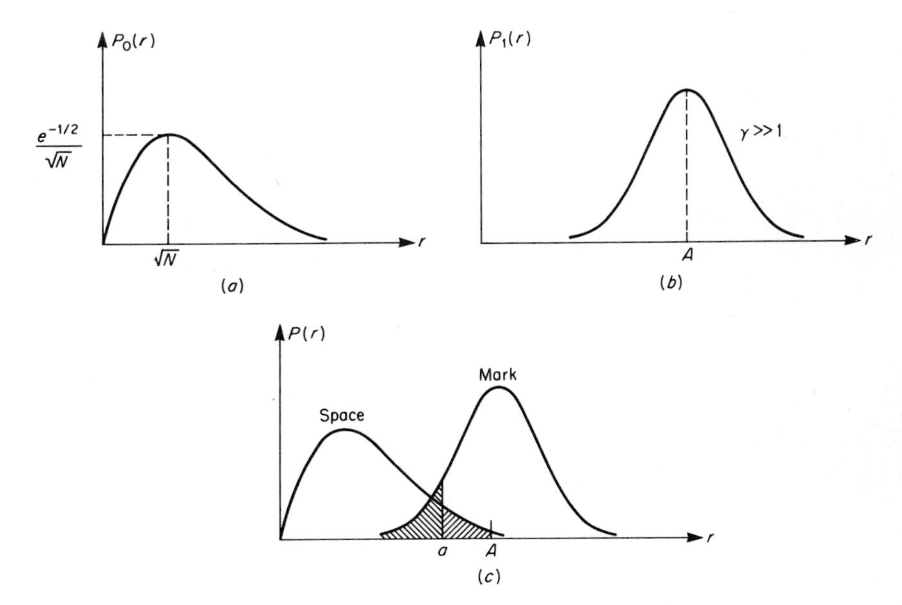

FIGURE 11-18 Envelope detection. (*a*) Rayleigh distribution. (*b*) Rician distribution. (*c*) Threshold determination.

For this optimum threshold,

$$P_e = \tfrac{1}{4}\,\text{erfc}\,\frac{\sqrt{\gamma}}{2} + \tfrac{1}{2}e^{-\gamma/4} \tag{11-61a}$$

The error probability when $\gamma \gg 1$ [Eq. (11-61*a*)] can be approximated as

$$P_e = \tfrac{1}{2}e^{-\gamma/4} \tag{11-61b}$$

COHERENT DETECTION IN DIGITAL SIGNALING

General Concepts

In coherent detection, there is an exact replica of the transmitted signal, with respect to frequency and phase, at the receiver. Of course, this is an idealized situation, and, in practice, both frequency and phase errors will be present. When the signal is received, it undergoes predetection filtering for smoothing purposes and rejection of those components that are not necessary for the detection process. The output of the filter will be the smoothed signal $m(t)$ plus the noise $n(t)$. The RF pulse $m(t)$ can be expressed as

$$m(t) = \begin{cases} A \cos 2\pi f_c t & \text{for } 0 \le t \le T \\ 0 & \text{elsewhere} \end{cases}$$

The energy of this pulse will be

$$E = \int_0^T (A \cos 2\pi f_c t)^2 \, dt = \frac{A^2 T}{2}$$

where $S = A^2/2$ the rms signal power.

Consequently, the ratio E/n_0 for white noise with spectral density $n_0/2$ will be

$$\frac{E}{n_0} = \frac{A^2 T}{2n_0}$$

As was shown in the section on idealized digital receivers, the signal-to-noise ratio γ is maximized at the output of a matched filter and takes the value

$$\gamma = \frac{E}{n_0}$$

[see Eq. (11-42)]. If each pulse transmits a bit of information and E_b is the energy per bit, then $E = E_b$ and

$$\gamma_b = \frac{E_b}{n_0} = \frac{A^2 T}{2n_0} \tag{11-62}$$

The last section of the low-pass filter (before the detector) is designed to approximate a matched filter, with maximum predetection S/N. As a result, the signal-to-noise ratio of the waveform entering the sign detector will be given by

$$\frac{S}{N} = \frac{E_b}{n_0}$$

A generalized block diagram of an ideal coherent or synchronous detector is depicted in Fig. 11-19. The synchronized detector will sample the incoming pulse at the appropriate time, provided that the receiver knows in advance the timing of the incoming pulses (waveforms). The detectors used in this process are known as "zero-memory" or "memoryless" detectors.

By resolving the noise $n(t)$ into two quadrature components, we obtain

$$n(t) = x(t) \cos 2\pi f_c t - y(t) \sin 2\pi f_c t \tag{11-63}$$

If $n(t)$ is gaussian noise, then $x(t)$ and $y(t)$ are gaussian processes. Now assume that the message signal $m(t)$ is

$$m(t) = a(t) \cos 2\pi f_c t \tag{11-64}$$

The signal $s(t)$ entering the detector will be

$$s(t) = m(t) + n(t) = [a(t) + x(t)] \cos 2\pi f_c t - y(t) \sin 2\pi f_c t \tag{11-65}$$

This signal is multiplied by $\frac{1}{2} \cos 2\pi f_c t$ and is then filtered so that the double frequency components are rejected while the low-frequency term is passed without distortion. The output of the low-pass filter will be

$$v(t) = a(t) + x(t)$$

The function $x(t)$ is gaussian with zero mean $[\langle x(t) \rangle = 0]$ and average power $\langle x(t)^2 \rangle = \sigma^2$. At the sampling instant when $a(t) = a$ and $v(t) = v$, the probability density function $p(v)$ will be

$$p(v) = \frac{1}{\sigma\sqrt{2\pi}} e^{-(v-a)^2/2\sigma^2} \tag{11-66}$$

Equation (11-66) is the fundamental relationship for computing the error probability at the output of the receiver. We will apply this relationship in the various digital signaling schemes in order to obtain this error probability.

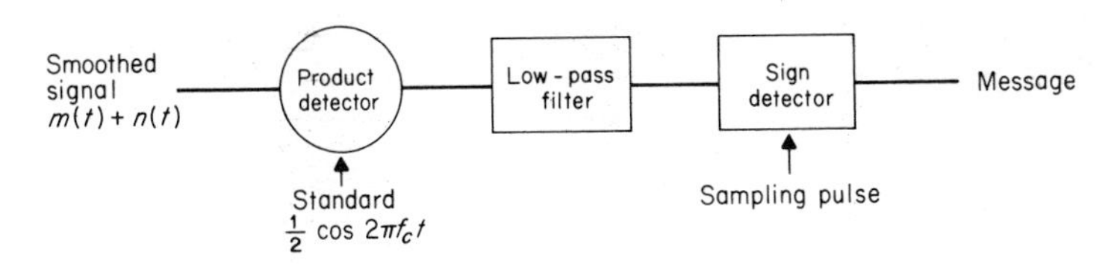

FIGURE 11-19 Coherent detector.

Binary PSK

In the case of BPSK we have 1 (mark) for $a(t) = a$ (positive) and 0 (space) for $a(t) = -a$ (negative). The probability of error is equal to the probability of $v(t)$ being negative when $a(t)$ is positive, that is,

$$P_e = \text{prob}\,(v < 0) \qquad \text{for } a(t) > 0$$

$$P_e = \int_{-\infty}^{0} \frac{1}{\sigma\sqrt{2\pi}}\, e^{-(v-a)^2/2\sigma^2}\, dv = \tfrac{1}{2}\,\text{erfc}\,\sqrt{\frac{a^2}{2\sigma^2}} \tag{11-67}$$

It is obvious that we have the same probability of error for $v(t)$ being positive when $a(t)$ is negative, that is, $P_e(v > 0)$. Since $a^2/2 = S =$ signal power and $\sigma^2 = N =$ average noise power, Eq. (11-65) becomes

$$P_e = \tfrac{1}{2}\,\text{erfc}\,\sqrt{\frac{S}{N}} = \tfrac{1}{2}\,\text{erfc}\,\sqrt{\frac{E_b}{n_0}} \tag{11-68}$$

(For the case in which the filtering is done by a matched filter maximizing S/N, we have $S/N = E_b/n_0$, where E_b is the energy per bit and n_0 is the noise power density of the one-sided spectrum.)

Four-Phase PSK

By a similar analysis, we can find that the probability of symbol* error rate in QPSK is

$$P_e = \left(\text{erfc}\sqrt{\frac{1}{2}\frac{S}{N}}\right)\left(1 - \tfrac{1}{4}\,\text{erfc}\,\sqrt{\frac{1}{2}\frac{S}{N}}\right) \tag{11-69a}$$

For large S/N, that is, when $S/N \gg 1$, Eq. (11-69a) becomes

$$P_e = \text{erfc}\,\sqrt{\frac{1}{2}\frac{S}{N}} = \text{erfc}\,\sqrt{\frac{1}{2}\frac{E}{n_0}} \tag{11-69b}$$

Equation (11-69b) provides the average error probability per symbol. But since we have 2 bits per symbol, $E = 2E_b$, where E_b is the energy per bit. As a result, Eq. (11-69b) becomes

$$P_e = \text{erfc}\,\sqrt{\frac{E_b}{n_0}} \tag{11-70}$$

In effect, a QPSK signal can be considered to consist of two orthogonal BPSK signals that are uncorrelated.

*Symbol is defined to include both quadrature components.

Binary Frequency-Shift Keying

In order to isolate the 0 and 1 pulses, we have two bandpass filters before the detection, one at f_1 and the second at f_2 (Fig. 11-20). When the f_1 frequency pulse is transmitted, the output of the bandpass filter at f_1 will have a signal,

$$m_1(t) + n_1(t) = [a(t) + x_1(t)] \cos 2\pi f_1(t) - y_1(t) \sin 2\pi f_1 t \qquad (11\text{-}71)$$

and the output of the filter at f_2 will have only noise,

$$n_2(t) = x_2(t) \cos 2\pi f_2 t - y_2(t) \sin 2\pi f_2 t \qquad (11\text{-}72)$$

After multiplying the f_1 and f_2 outputs by the carrier frequency, we shall have

$$v_1(t) = a(t) + x_1(t)$$

$$v_2(t) = x_2(t)$$

Since we must have $v_1 > v_2$ in order to avoid error, the probability of error will be

$$P_e = \text{prob}\,(v_1 < v_2) = \text{prob}\,(a + x_1 < x_2)$$

or

$$P_e = \text{prob}\,(a + x_1 - x_2 < 0)$$

But x_1 and x_2 have a gaussian distribution, each with zero mean and standard deviation σ^2. Therefore, the variate $z = a + x_1 - x_2$ will have a gaussian distribution with

$$\langle z \rangle = a \qquad \text{and} \qquad \sigma_z^2 = \langle (z - \langle z \rangle)^2 \rangle = \langle (x_1 - x_2)^2 \rangle = 2\sigma^2$$

Therefore

$$P(z) = \frac{1}{\sigma_z \sqrt{2\pi}} \exp\left[-\frac{(z - \langle z \rangle)^2}{2\sigma_z^2}\right]$$

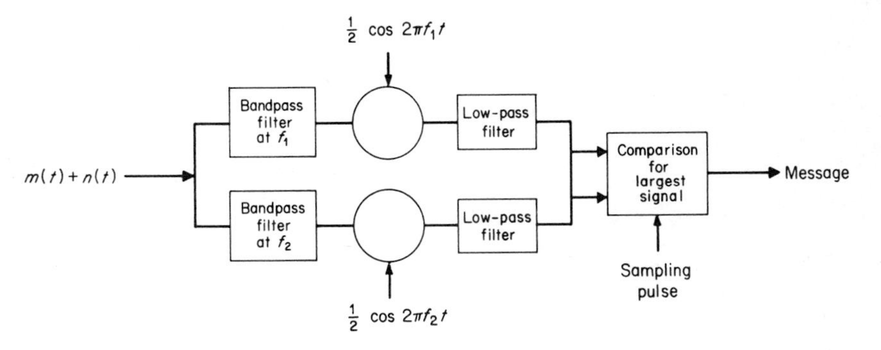

FIGURE 11-20 Coherent BFSK detector.

and

$$P_e = \int_{-\infty}^{0} P(z)\, dz = \tfrac{1}{2} \operatorname{erfc} \frac{a}{2\sigma_z}$$

or

$$P_e = \tfrac{1}{2} \operatorname{erfc} \sqrt{\frac{1}{2}\frac{S}{N}} = \tfrac{1}{2} \operatorname{erfc} \sqrt{\frac{1}{2}\frac{E_b}{n_0}} \qquad (11\text{-}73)$$

(Again for matched filtering, $S/N = E_b/n_0$ where E_b is the energy per bit and n_0 is the one-sided noise power density.)

Differential Coherent PSK

For the binary case, utilizing similar analyses as in the previous cases we derive

$$P_e = \tfrac{1}{2}e^{-S/N} = \tfrac{1}{2}e^{-Eb/no} \qquad (11\text{-}74)$$

NONCOHERENT DETECTION IN DIGITAL SIGNALING

The advantage of noncoherent detection is that, at the receiving end, one does not require a complete knowledge of the carrier signal, since synchronization is not necessary. After prefiltering and smoothing, an envelope detector is most often utilized for detection. Noncoherent detection is used for ASK and binary FSK, but not for PSK. A general analysis of noncoherent detection is presented in a previous section of this chapter, "Carrier System and Envelope Detection."

By applying this analysis to FSK and ASK transmission, after a number of simplifying assumptions, the following error probabilities are obtained. For noncoherent FSK,

$$P_e = \tfrac{1}{2}e^{-(1/2)S/N} = \tfrac{1}{2}e^{-(1/2)Eb/no} \qquad (11\text{-}75)$$

For noncoherent ASK with optimum threshold,

$$P_e \approx \tfrac{1}{2}e^{-(1/4)S/N} = \tfrac{1}{2}e^{-(1/4)Eb/no} \qquad (11\text{-}76)$$

PERFORMANCE OF DIGITAL SYSTEMS

The ratio E_b/n_0, that is, the ratio of energy per bit over the noise spectral density of white additive gaussian noise, is very important in determining the performance of a digital system.

Relationship of E_b/n_0 to C/N

Earlier in this chapter the following relationship was derived as Eq. (11-64):

$$\frac{E_b}{n_0} = \frac{A^2T}{2n_0} \tag{11-77}$$

where A is the peak amplitude of the carrier. Equation (11-77) can be rewritten as

$$\frac{E_b}{n_0} = \frac{1}{R}\frac{C}{n_0} = \frac{B}{R}\frac{C}{N} \tag{11-78}$$

where

$$R = \frac{1}{T} = \text{information or signaling rate, bits per second}$$

$$B = \text{RF bandwidth}$$

$$N = n_0B = \text{noise power}$$

$$\frac{C}{N} = \frac{A^2}{2N} = \text{rms carrier-to-noise ratio}$$

As an example, consider a satellite link in which $(C/N)_T$ is computed according to the discussion in Chap. 7. In the case of digital transmission, E_b/n_0 could be computed by using Eq. (11-78) if the transmission bandwidth B and the information rate R were known.

Symbol Energy vs. Bit Energy

In discussing the QPSK signaling concept, we indicated that each transmitted symbol contained 2 bits of information. Consequently, if E is the energy per symbol and E_b is the energy per bit, $E = 2E_b$.

This concept can be generalized by considering a symbol comprising a large alphabet. If we consider a symbol containing simultaneously M waveforms, with each pair of waveforms transmitting one bit of source information, then each symbol will transmit $\log_2 M$ bits. M-ary signaling, for example, could be accomplished by using M frequencies in an FSK system or M phases in a PSK system.

In the case of QPSK, we do not increase the bandwidth, but we double the transmitted energy because each orthogonal transmission requires as much energy as the energy of a BPSK signal. If T_M is the time required to transmit an M-ary symbol and R is the source information rate, the number of equally probable message sequences will be

$$M = 2^K \tag{11-79}$$

where

$$K = RT_M \tag{11-80}$$

The receiver, with each arriving waveform of duration T_M, will accept K source information bits and will act on this waveform as a whole. The result is that

$$E = KE_b = (\log_2 M)E_b \tag{11-81}$$

Equation (11-81) should not be confused with the relationships to be derived in the next chapter regarding algebraic codes. In algebraic code words, the detector examines the message bit by bit and does not consider the word as a symbol.

In obtaining the bit error rate at the output of the detector, having used an optimum filter, we express the error probability as a function of $\gamma = E/n_0$, where E is the energy per symbol. If we want to use the ratio $\gamma_b = E_b/n_0$, then we must make the substitution

$$\gamma = K\gamma_b = K\frac{E_b}{n_0} \tag{11-82}$$

The signaling rate $R_M = 1/T_M$ is related to the information rate R through Eq. (11-80), and therefore

$$R = KR_M \tag{11-83}$$

As an example, consider again the QPSK case in which the error probability P_e was found in Eq. (11-69b) to be

$$P_e = \text{erfc}\sqrt{\frac{1}{2}\frac{E}{n_0}}$$

But since $E_b = (1/K)E = \frac{1}{2}E$, Eq. (11-70) was derived to be

$$P_e = \text{erfc}\sqrt{\frac{E_b}{n_0}}$$

Comparison of Signaling Systems

The following list summarizes the results derived in a previous section regarding the error rate probability of various systems.

Coherent binary (antipodal) PSK (BPSK):

$$P_e = \frac{1}{2}\,\text{erfc}\sqrt{\frac{E_b}{n_0}}$$

Coherent four-phase (quadriphase) PSK (QPSK):

$$P_e = \left(\operatorname{erfc}\sqrt{\frac{E_b}{n_0}}\right)\left(1 - \tfrac{1}{4}\operatorname{erfc}\sqrt{\frac{E_b}{n_0}}\right)$$

$$\approx \operatorname{erfc}\sqrt{\frac{E_b}{n_0}} \qquad \frac{E_b}{n_0} \gg 1$$

Coherent differential PSK (DPSK):

$$P_e = \tfrac{1}{2}e^{-E_b/n_0}$$

Coherent FSK:

$$P_e = \tfrac{1}{2}\operatorname{erfc}\sqrt{\frac{1}{2}\frac{E_b}{n_0}}$$

Noncoherent FSK:

$$P_e = \tfrac{1}{2}e^{-(1/2)E_b/n_0}$$

Noncoherent ASK:

$$P_e \approx \tfrac{1}{2}e^{-(1/4)E_b/n_0}$$

Some of these functions are plotted in Fig. 11-21. In the case of coherent detection a perfect phase of the demodulating carrier was assumed. If a phase error $\Delta\phi$ is present, then the result should be multiplied by $\cos \Delta\phi$.

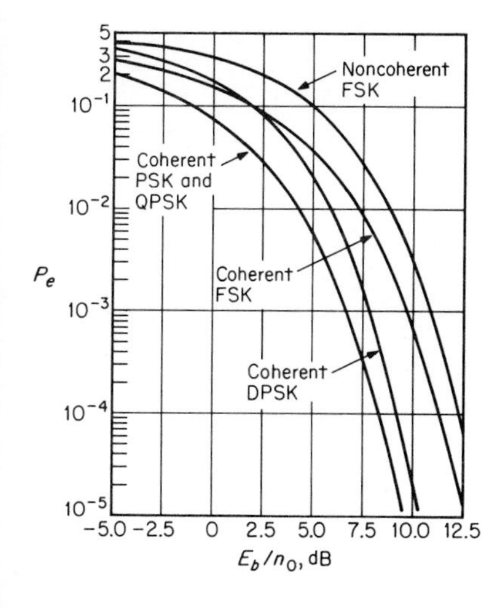

FIGURE 11-21 P_e versus E_b/n_o.

REFERENCES

Bennett, W. R., and J. R. Davey: *Data Transmission,* McGraw-Hill, New York, 1965.

Haykin, Simon: *Communication Systems,* John Wiley & Sons, New York, 1978.

Lucky, R. W., J. Salz, and E. J. Weldon, Jr.: *Principles of Data Communications,* McGraw-Hill, New York, 1968.

Schwartz, M., W. R. Bennett, and S. Stein: *Communication Systems and Techniques,* McGraw-Hill, New York, 1966.

Spilker, J. J., Jr.: *Digital Communications by Satellite,* Prentice-Hall, Englewood Cliffs, N.J., 1977.

Stein, S., and J. J. Jones: *Modern Communication Principles with Application to Digital Signaling,* McGraw-Hill, New York, 1967.

12
CODING AND FORWARD ERROR CORRECTION

THEORETICAL CHANNEL CAPACITY

The maximum binary rate that can be transmitted through a communications channel without an error, in the presence of white additive gaussian noise, over a bandwidth B is given by Shannon's formula

$$C = B \log_2 \left(1 + \frac{S}{N}\right) \qquad \text{bits/s} \tag{12-1}$$

where S/N is the signal-to-noise power ratio. This maximum binary rate can be transmitted without error when the message is appropriately coded. However, the coded message may also require an extremely long transmission time.

To approximate this maximum rate, the transmitted signals must approximate the statistical properties of white noise; that is, the power density spectrum $S(f)$ of the signal must be constant over the bandwidth under consideration.

In general, if $N(f)$ is the noise power spectrum and $S(f)$ is the signal power spectrum, we have

$$C = \int_0^\infty \log_2 \left[1 + \frac{S(f)}{N(f)}\right] df \tag{12-2}$$

When $S(f)$ is band-limited and the noise is white with one-sided spectral density $N(f) = n_0$, we obtain Eq. (12-1) from Eq. (12-2) if we also assume $S(f) = S$ = constant over the bandwidth B.

If we call C_∞ the value of C when the bandwidth increases to infinity, we get

$$C_\infty = \lim_{B \to \infty} C = \frac{S}{n_0} \frac{1}{\ln 2} = \frac{S}{n_0} \log_2 e \tag{12-3}$$

Now, if the signal S is transmitted over time T, the maximum transfer of information will be

$$C_\infty T = \frac{E}{n_0} \log_2 e \qquad \text{bits} \tag{12-4}$$

where $E = ST$, the signal energy. Equation (12-3) indicates that the maximum transmission rate does not depend on bandwidth.

Equation (11-76) was given as

$$\frac{E_b}{n_0} = \frac{B}{R}\frac{S}{N} = \frac{1}{R}\frac{S}{n_0} \tag{12-5}$$

But because the transmission rate R is less than C_∞, that is, $R \leq C_\infty$, from Eq. (12-3) we obtain

$$R \leq \frac{S}{n_0}\frac{1}{\ln 2} \tag{12-6}$$

Comparing Eqs. (12-6) and (11-76), we obtain

$$R \leq R\frac{E_b}{n_0}\frac{1}{\ln 2}$$

or

$$\frac{E_b}{n_0} \geq \ln 2 \doteq -1.6 \text{ dB} \tag{12-7a}$$

Equation (12-7a) provides a lower bound for E_b/n_0. Alternately, we have

$$R \leq C = B \log_2 \left(1 + \frac{S}{Bn_0}\right)$$

or

$$\frac{R}{B} \leq \log_2 \left(1 + \frac{R}{B}\frac{E_b}{n_0}\right)$$

and

$$\frac{E_b}{n_0} \geq \frac{e^{0.69R/B} - 1}{R/B} \tag{12-7b}$$

CODING CONCEPTS

In practice, we are always willing to accept a certain error rate, and as a result we deviate from the error-free channel capacity given by Shannon's formula.

However, in many cases, in order to achieve an acceptable error rate, codes allowing error detection and correction must be used. In previous discussion, we have seen that in digital transmission the error rate P_e is usually given as a function of the ratio E_b/n_0. As a matter of fact, in most practical schemes, the function $P_e = p(E_b/n_0)$ decreases monotonically with increasing E_b/n_0. In order to increase E_b/n_0 for a constant information transmission rate R, we must increase either the bandwidth B or the carrier-to-noise ratio C/N [see Eq. (11-76)]. In satellite applications, an increase in the carrier power means an increase in the total power capability of the spacecraft. This is an expensive proposition, since the cost per watt in orbit is substantial. Similarly, bandwidth allocations are scarce, and very often satellite transmission is bandwidth-limited. An effective technique in reducing the error rate of the transmission is to employ codes with error-correction capabilities.

In digital systems, error detection can be achieved by transmitting redundant digits which serve as parity checks at the receiving end. For example, at the end of a code word one adds, as necessary, a 1 or a 0 so that the total number of 1s in the code word is always even.

The redundant digits can be structured so that, besides identifying an error, they also can provide the information for correcting the same error. As a result, when a message is coded, not all the digits of a word are necessarily message digits. If the message is coded so that only k digits out of n in a code word are message digits and the remaining $n - k$ digits are control digits, the fraction $k/n \leq 1$ describes the number of message digits as a fraction of the total transmitted digits for the particular code. The ratio $R = k/n$ is called the code-rate efficiency. For example, if a code word of 30 digits contains 15 message digits, we have a rate 0.5 bit/bit.

If the $n - k$ parity bits check only the k information bits immediately preceding them in the transmitted sequence, the code is called a "block code" with an "n-bit code word." We usually refer to this code as an "(n, k) block code." If the $n - k$ parity bits check not only the k information bits immediately preceding them, but also information bits which appeared in the m preceding blocks, then the code is called a "convolutional code." Each convolutional code is associated with the "constraint length" L, which equals m. Most authors define L as a product:

$$L = km \tag{12-8}$$

In both block and convolutional codes, the code-rate efficiency R is given by

$$R = k/n \tag{12-9}$$

In addition, a code is called a "systematic code" when the k data bits transmitted are identical with the k bits generated by the message source and each of the $n - k$ control bits is a linear combination of the data bits.

In block codes, practical values for k range from 3 to several hundred, and for R from $\frac{1}{4}$ to $\frac{7}{8}$. In convolutional codes, practical values for n and k range from 1 to 8, and for R from $\frac{1}{4}$ to $\frac{7}{8}$. (Very often $k = 1$ and/or $n - k = 1$.)

In the case of linear block codes, the concept of Hamming distance is useful in determining the error-correcting ability of the codes. The "Hamming distance" between two words of equal length is defined as the number of positions in which the words differ. If a code is used as a straight error-detecting code, it is necessary and sufficient for the minimum Hamming distance between code words to be d so that $d - 1$ errors can be detected. Similarly, it is possible to correct all patterns of errors that do not exceed $(d - 1)/2$ positions. In general, if we want the code to correct all t error patterns or less, the Hamming distance must be at least $2t + 1$.

COMMUNICATIONS CHANNEL WITH ERROR CORRECTION

General Description and Coding Gain

A binary channel is defined as a channel utilizing two symbols for transmission, namely 0 and 1. Assume such a binary channel with k data symbols per n symbol words. If E_s is the energy per symbol, the receiver receives a total energy nE_s per word. Since there are only k bits per word, the energy expended per bit will be

$$E_b = \frac{n}{k} E_s$$

or

$$E_s = \frac{k}{n} E_b \tag{12-10a}$$

where $k/n < 1$.

Since all the expressions for the error-rate probability P_e have been derived in terms of the symbol energy, which constitutes the energy of each physical pulse at the input of the detector, we must also utilize Eq. (12-10) if we need to express P_e in terms of E_b.

For example, in the binary PSK (antipodal) case, Eq. (11-68) for the error-rate probability per symbol will become

$$P_e = \tfrac{1}{2} \operatorname{erfc} \sqrt{\frac{k}{n} \frac{E_b}{n_0}} \tag{12-10b}$$

E_b in Eq. (11-68) represents the uncoded message, and therefore the symbol energy is the same as the bit energy. In addition, Eq. (12-10) should not be confused with Eq. (11-81) of the M-ary system, in which the energy per symbol was larger than the energy per bit.

Coding gain is defined as the increase in E_b/n_0 by a transmission without coding that is needed to achieve the same bit error rate as the coded transmission (see Fig. 12-6).

The generalized block diagram of a system employing coding is depicted in Fig. 12-1. The channel within the dashed line is depicted as a satellite channel, but it could be any other discrete data channel with noise.

Encoding

In practice, two types of encoders have been used. The first type, the memoryless encoder, receives k symbols from the information source and derives the n-symbol word sequence. The $n - k$ control symbols are usually formed as linear combinations of the data symbols. This process will be discussed in more detail later. The result is a systematic block code which, in the binary channel, is often known as a "group code."

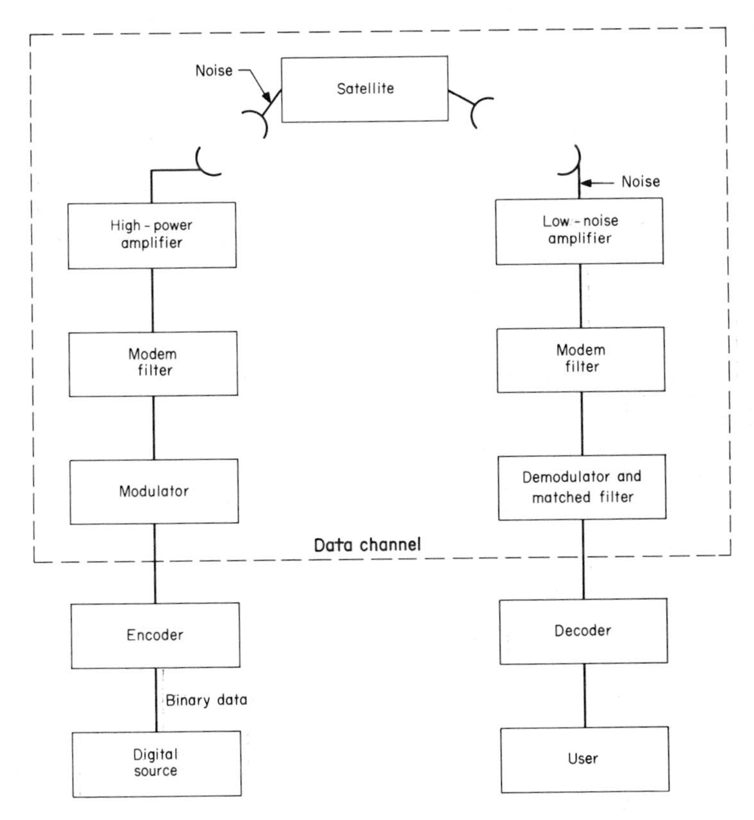

FIGURE 12-1 Digital communications process via satellite.

The second kind of encoder, which produces a convolutional code or "tree code," utilizes an encoder with memory. The k source symbols enter a binary shift register with km stages. A number n of modulo 2 adders are connected to certain of the shift-register stages. The output of the n adders is sampled sequentially, and the code word is produced. Figure 12-2 shows a block diagram of a convolutional coder.

The various codes could also be divided into two major categories: the random-error-correcting codes and the burst-error-correcting codes. Burst errors may occur in communication channels because of fading and other similar occurrences when a whole set of successive symbols is affected.

Decoding

Decoding Algorithms The decoding process is much more difficult than the encoding process. The symbols arrive in the decoder embedded in noise, and the decoder must determine what was the original symbol. Several algorithms have been developed for this process. If certain reliability or statistical information is used in the process, the decisions of the decoder are called "soft decisions" and the decoder is called a "soft-decision decoder." Decisions based on a methodology without reiteration or any statistical information are termed "hard decisions."

The various algorithms can be classified according to whether they apply to convolutional codes or to block codes.

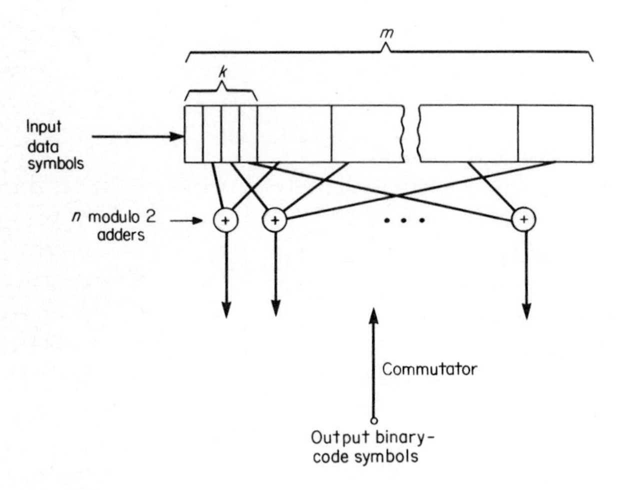

FIGURE 12-2 Convolutional coder.

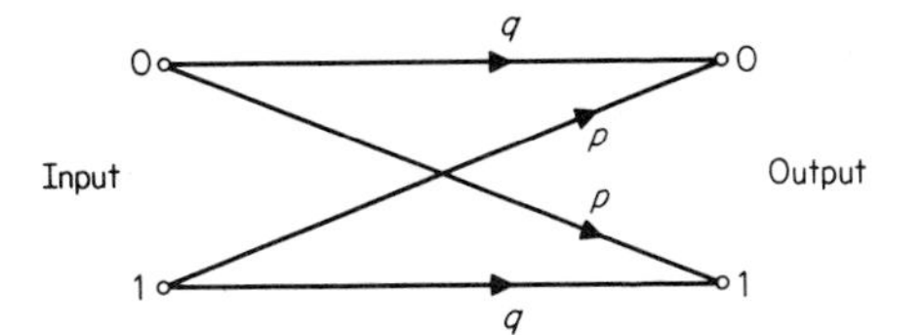

FIGURE 12-3 Symmetrical binary channel.

Some of the best-known algorithms for decoding convolutional codes are

1. Sequential algorithms
2. Viterbi algorithm
3. Threshold decoding algorithms

Block codes, for the most part, use algebraic procedures or certain other concepts based on the special structure of the codes, including threshold decoding.

Signal Quantization in a Symmetrical Channel The binary symmetrical channel is depicted in Fig. 12-3. The probability of correct reception is designated q, and the probability of incorrect reception is designated p. It is obvious that

$$p = 1 - q$$

As an example of a soft-decision data channel, consider a gaussian distribution of the incoming signal as described in Eq. (11-49). The probability of a symbol 1 is

$$p_1(v) = \frac{1}{\sqrt{2\pi\sigma^2}} e^{-(v-A)^2/2\sigma^2}$$

This probability density is shown in Fig. 12-4.

At the receiver, the output of the matched filter is further quantized to 8 levels before decision. Each level is coded in terms of 3 binary digits: for example, level 1, 000; level 2, 001; level 3, 010; level 4, 011; level 5, 100; level 6, 101; level 7, 110; level 8, 111. In this case the binary channel will have eight outputs. The transition probabilities from the two input states to the eight output states are computed as the areas under the corresponding segments of the probability density curve. The decoder must now decide what the output should be by applying certain criteria based on these transitional probabilities. When digital signaling is used, the symbol probability at the output of the matched filter is a function of the ratio E_b/n_0. By further quantizing the received symbol, a certain loss in the E_b/n_0 ratio takes place. A two-level quantization results in 2 dB loss.

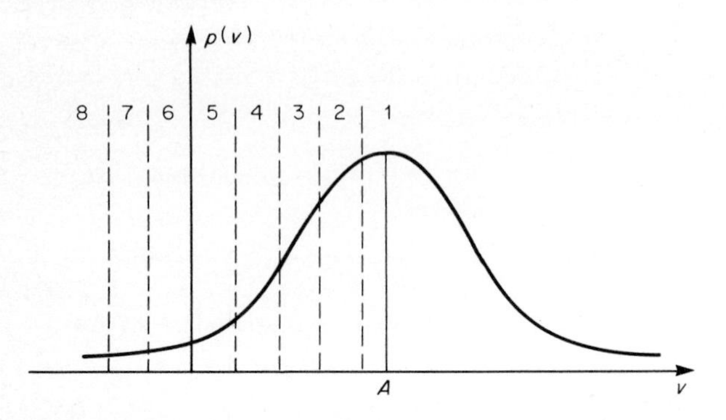

FIGURE 12-4 Symbol quantization.

This loss could be reduced by making the quantization finer. With an 8-level quantization, the loss is less than 0.25 dB. Soft-decision decoding could provide an overall improvement of 2 to 3 dB in coding gain.

Geometric Distances and Metrics The design objective for a decoder when decoding is to minimize the probability of error at the receiving end. Given a transmitted sequence S_1 of data symbols and the corresponding received sequence r, the decoder should be guided by the conditional probability $P(S_1/r)$. Under the assumption that all data sequences S_1 are equally likely, when the probability $P(S_1/r)$ is maximized, the probability $P(r/S_1)$ will also be maximized. The decoder is likely to lead to the sequence S_1, since it will sense that $P(S_1/r)$ is the maximum transitional probability.

Let us assume that the system operates in an environment of gaussian noise, utilizing a digital signaling technique. The difference $e_j = r_j - S_{1j}$ between the received symbol r_j and the transmitted symbol S_{1j} will be the interfering noise, which will have a probability density function $p = A \exp(-ae_j^2)$. The joint probability of the sequence symbols, if the symbols are independent, will be

$$P(r/S) = \prod_{j=1}^{n} Ae^{-a(r_j - S_{1j})^2} = A^n \exp\left[-a \sum_{j=1}^{n} (r_j - S_{1j})^2\right] \qquad (12\text{-}11a)$$

where S_{1j} is the jth symbol of the transmitted sequence.

The objective, then, is to minimize the sum

$$d^2 = \sum_{j=1}^{n} (r_j - S_{1j})^2 \qquad (12\text{-}11b)$$

The parameter d can be considered as the geometric distance (euclidean distance) between the presumed sequence S_1 and the received sequence r.

The soft-decision decoding process can be illustrated graphically by either a "code-tree" or a "trellis" diagram. Both of these diagrams depict the transition of the encoder from state to state in the coding process. The decoder, on the other hand, considers the probabilities of these transitions.

The code-tree diagram starts from the state 0. If the input is 0, an upper branch is drawn, whereas if the input is 1, a lower branch is drawn (Fig. 12-5). The output along each branch is the output of the coder when that branch is followed. For example, the input sequence 0110 will result in an output sequence 00, 11, 10, 10. Figure 12-5 represents such a code tree for a convolutional code.

From the quantization process (Fig. 12-4), a set of probabilities is derived, namely, $P(x_j|x)$, the probability that if x is transmitted, level x_j will be received. If now $S_1(S_{11} \cdots S_{1j} \cdots S_{1n})$ is the code vector describing the path and $r(r_1 \cdots r_j \cdots r_n)$ is the received vector, the path likelihood function to be computed by the decoder is the product of the individual likelihoods,

$$P(r/S_1) = \prod_{j=1}^{n} P(r_j|S_{1j})$$

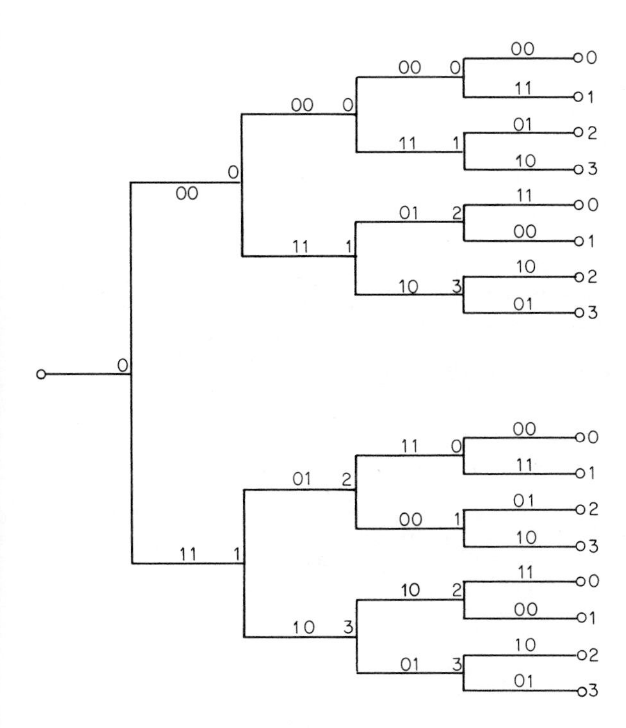

FIGURE 12-5 Code tree.

In order to convert the multiplication to addition, we take the logarithms of the probabilities and define these logarithms as "metrics":*

$$m_{1j} = \log P(r_j/S_{1j}) \tag{12-12a}$$

Consequently, the overall path metric M_1 will be

$$M_1 = \sum_{j=1}^{n} m_{1j} \tag{12-12b}$$

Equation (12-12*b*) is similar to Eq. (12-11*b*), and therefore the overall metric M_1 is equivalent to the geometric distance d.

The decoding process should maximize the likelihood of the path.

Symbol Error Probabilities

Consider an (n, k) error-correcting code in a channel with gaussian noise. Assume also that the code is capable of correcting errors up to t-error patterns. The number of j-symbol patterns within the n-tuple will be

$$1_j = \binom{n}{j} = \frac{n!}{j!(n-j)!} \tag{12-13}$$

Now let us assume that this code cannot correct j-error patterns when $j > t + 1$. (This is not strictly true for a code that corrects all t-error patterns, but it is a reasonable approximation.)

If p is the probability of error per symbol, then the average probability P_S of a sequence error after correction will be

$$P_S = \sum_{j=t+1}^{n} \binom{n}{j} p^j(1-p)^{n-j} \tag{12-14}$$

In the derivation of Eq. (12-14), it is assumed that the code corrects all t-error patterns or less.

For an error probability $p \ll 1$, the above relationship could be approximated as

$$P_S = \binom{n}{t+1} p^{t+1} \tag{12-15}$$

By entering the values of p corresponding to each signaling scheme, we can determine P_S by using Eq. (12-14) or (12-15). For example, in the case of BPSK, Eq. (12-10*b*) provides

$$p = P_e = \tfrac{1}{2} \operatorname{erfc} \sqrt{\frac{k}{n} \gamma_b}$$

*A "metric" can also be defined as a linear function of the logarithm; for example, $m = a + b \log p(r/s)$.

It is interesting to note that $(k/n)\gamma_b < \gamma_b$, and therefore there is a reduction in S/N because of the (n, k) coding. As a result, P_e has been increased by coding. Consequently, the improvement obtained by applying Eq. (12-15) must compensate for this decrease and produce additional improvement for the code to be effective.

Figure 12-6 shows, for certain specific block codes and for QPSK signaling, the sequence error probability as a function of γ_b. The derivation of the average bit error will be shown in a later section.

At high signal-to-noise ratios in codes with a minimum Hamming distance $d = 2t + 1$, this average bit error rate probability could be approximated as

$$P_b \approx \frac{d}{n} P_S \tag{12-16}$$

In the derivation of Eq. (12-16), it is also assumed that any received word containing $t + 1$ errors would be misinterpreted as containing t errors; thus an erroneous correction would be made.

BLOCK CODES

This section requires knowledge of fields, vector spaces, and matrix representation of vector spaces. Appendix D provides a summary of these topics and should be studied, if necessary, before this section is read.

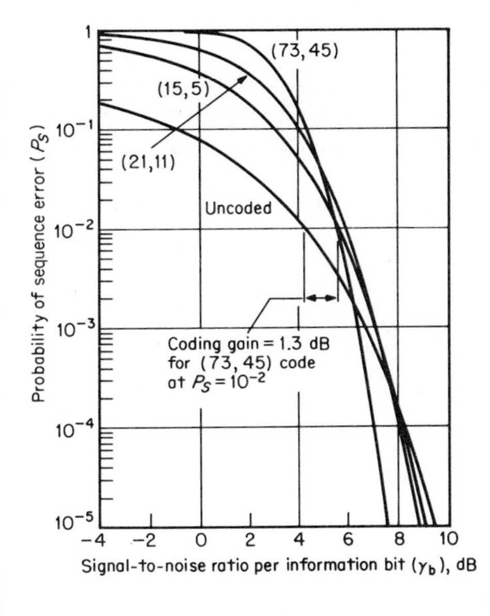

FIGURE 12-6 Coherent reception of QPSK.

Matrix Representation

The set of all possible n-tuples over a field forms a linear vector space, since an n-tuple is defined as a set of n ordered elements that satisfies all the axioms necessary to form such a space.

An (n, k) block code, where $k \leq n$, is defined as the collection of all the 2^k n-tuples, each with k data elements and the rest of the $n - k$ elements of each n-tuple serving as parity-check elements. The code word is the whole n-tuple.

Let us assume the k-row generator matrix G, representing the basis of the n-tuple space of the 2^k code words, written in a reduced echelon form:

$$G = [I_k P] = \begin{bmatrix} 1 & 0 & 0 & \cdots & 0 & a_{11} & a_{12} & \cdots & a_{1n-k} \\ 0 & 1 & 0 & \cdots & & a_{21} & \cdots & \cdots & a_{2n-k} \\ \cdots & \cdots & \cdots & \cdots & \cdots & \cdots & \cdots & \cdots & \cdots \\ 0 & 0 & 0 & \cdots & 1 & a_{k1} & \cdots & \cdots & a_{kn-k} \end{bmatrix} \tag{12-17}$$

A k-tuple represented by $v(x_1, x_2, \ldots, x_k)$, when multiplied by the matrix G, will give one of the n-tuple code words, say, code word w:

$$w = vG = (x_1, x_2, \ldots, x_k, y_1, y_2, \ldots, y_{n-k})$$

where

$$y_i = \sum_{j=1}^{k} a_{ji} x_j \tag{12-18}$$

In fact, y_i is a linear combination of the rows of G, where x_j is the coefficient of the jth row. This last relation indicates that the last $n - k$ elements of the n-tuple w are linear combinations of the first k elements of the code word. Such a code is called a "systematic code," with the first k elements being the information symbols chosen arbitrarily among the 2^k k-tuples and each of the last $n - k$ elements, called "check symbols," being a linear combination of the information symbols. Since every block code is represented by a vector space with a generator matrix that can be represented by a reduced echelon form, it can also be shown to be equivalent to a systematic code.

As shown in Appendix D, the matrix $H = (-P^T I_{n-k})$ has as a null space the vector space of the n-tuples; that is,

$$GH^T = 0 \tag{12-19}$$

This matrix H is called a parity-check matrix, and for any code vector w representing an n-tuple of the code, the parity check will be

$$wH^T = 0 \tag{12-20}$$

For any n-tuple (represented by the vector u) that is not a code vector, we will have

$$uH^T = S \neq 0 \tag{12-21}$$

The $(n - k)$-tuple S is called the syndrome of the n-tuple.

A vector v is a code only if its syndrome $S_v = vH^T$ equals zero.

The Standard Array

The standard array is a decoding table having as a first row all the code vectors starting with the identity vector **0** (0, 0, . . . , 0) at the left. There are 2^k such vectors.

The second row is a set of all vectors with the same syndrome, usually starting under the identity code vector with the vector that is most likely to be the identity vector if an error occurs in the transmission. The vectors in the third row also have the same syndrome with each other, and so on. Each row is called a "coset" and the first vector at the left is called a "coset leader."* In forming the array, we place under each code word a vector which is the sum of the coset leader and the code word. Thus, if w_{i1} is the coset leader in the ith row below the row of the code words, the vector in the jth column under the code vector V_j will be $w_{ij} = w_{i1} + V_j$. A received vector w_{ij} is decoded to the code word on the top of its column. If a vector V_j is transmitted and a vector w_{ij} is received, then this decoding array will decode correctly only if the difference $w_{ij} - V_j$ is a coset leader. This difference between received and transmitted vector is called the "error pattern."

A simple method of decoding can be derived that takes into account these considerations. Namely, when a vector is received, this vector is multiplied with the parity check matrix H^T, and thus the syndrome is computed. From the decoding table, the coset leader is found, i.e., the leader corresponding to this syndrome. This coset leader is presumably the error pattern. When this error pattern is subtracted from the received vector, the code word is obtained.

"Hamming weight" of a code vector is defined to be the number of the nonzero ordered elements in the vector. Since the Hamming distance between two vectors is defined as the number of elements that are different between the two vectors, the Hamming distance must be equal to the Hamming weight of the difference of the two vectors.

The minimum weight in a code (excluding the zero vector) equals the minimum Hamming distance of the code. This statement, although self-evident from the above definitions, is significant in the analysis of block codes.

Let us assume the n-tuple w consisting of two parts, the k-tuple w_d representing the data section and the $(n - k)$-tuple w_p representing the parity section, so that

$$w = (w_d w_p) \tag{12-22}$$

*Usually the minimum-weight n-tuple of the n-tuples in a coset is picked to be the coset leader.

The syndrome s can be computed as follows:

$$s = (w_d w_p) H^T \tag{12-23}$$

but, since, from the definition of H,

$$H^T = \begin{bmatrix} -P \\ I_{n-k} \end{bmatrix} \tag{12-24}$$

we have

$$s = [w_d w_p] \begin{bmatrix} -P \\ I_{n-k} \end{bmatrix} = w_d[-P] + [w_p] \tag{12-25}$$

The matrix P is a $k \times (n - k)$ matrix, and Eq. (12-25) shows that the syndrome s is an $(n - k)$-tuple. If we subtract this syndrome $(n - k)$-tuple from the w_p section of w, then the resultant n-tuple will be a code word. This is easily seen by considering a_j, the jth term of s, which, if it is not 0, indicates that the parity check with the jth column of H^T failed. By subtracting this element of the syndrome from the jth term of w_p, we will cause this check to become 0, since this subtracted element will be multiplied only by the corresponding 1 of the I_{n-k} matrix and added to the previous result, causing it to be 0.

We can further show that every $(n - k)$-tuple is a syndrome. Indeed, by considering the n-tuple w in which the first k elements are zero and the last $n - k$ elements are the $(n - k)$-tuple under consideration, one can easily see that the product

$$wH^T = w \begin{bmatrix} -P \\ I_{n-k} \end{bmatrix}$$

will be the $(n - k)$-tuple under consideration.

Consequently we can construct 2^{n-k} syndromes, and therefore the standard array will have 2^{n-k} rows. In addition, since we have k data bits, we will have 2^k columns,* resulting in $2^k \times 2^{n-k} = 2^n$ total entries, which is the total number of n-tuples. Since every n-tuple has exactly one syndrome and since there are 2^{n-k} syndromes, the standard array cosets are disjoint.

Binary Linear Codes and the Symmetrical Channel

A binary channel is defined as a channel utilizing two symbols for transmission, namely 0 and 1. The set of all n-tuples with elements 0 and 1 only constitutes a vector space over a field with two scalars, that is, 0 and 1. A set of vectors is called a "binary linear code" if and only if it is a subset of this vector space of the n-tuples. Binary linear codes are often referred to as group codes.

*Or, equivalently, we will have 2^k code words with n bits per code word.

The symmetrical binary channel is depicted in Fig. 12-3, where q is the probability of correct reception and p is the probability of incorrect reception. Clearly $p = 1 - q$.

If V is an (n, k) group code used over this channel and if the code vectors are equally likely to be transmitted, then the average probability P_{CD} of correct decoding will be

$$P_{CD} = \frac{1}{2^k} \sum_{i,j} p^{d_{ij}} q^{n-d_{ij}} \tag{12-26}$$

where d_{ij} is the Hamming distance between a received word V_{ij} (located at the ith row and the jth column of the standard array) and the code word V_{ij} into which it is decoded.

The summation is over all possible rows and columns, and this summation is divided by 2^k, the total number of k-tuples in the standard array representing the total number of code words which are assumed equally probable. This probability is maximized when each received word is decoded to the closest code word in a Hamming-distance sense. Decoding to the closest code word is ensured if each coset leader is chosen to have minimum weight in its coset, since the coset leader is assumed to be the error pattern and it is added to the received vector to find the transmitted code word. Of course, the choice of a different set of code words may result in an even higher average probability of correct decoding.

It is interesting to note that, when we consider only group codes, the minus signs can be omitted and a subtraction operation can be considered as an addition. (See the definition of $-c$ in Appendix D.) For example, the matrix $H = [-PI_{n-k}]$ can be written as $H = [PI_{n-k}]$ and Eq. (12-24) can be written as

$$H^T = \begin{bmatrix} P \\ I_{n-k} \end{bmatrix} \tag{12-27}$$

Error Rate Probabilities for Block Codes

Equations (12-14) and (12-16) provide an approximation for the average-error-rate probabilities for sequences and bits, respectively. A more accurate derivation of these probabilities for block codes will now follow.

For a block code with a minimum Hamming distance d, all t-error patterns for $t = (d - 1)/2$ can be corrected. Algorithms that correct all possible errors in a code word are called "complete." Algorithms that do not correct all received patterns but only flag the uncorrected ones are called "incomplete." Now assume the following:

- The code word v_i, corresponding to the ith column of the standard array, was transmitted.

- The word w_j, corresponding to the jth column of the standard array, was received.
- The standard array is divided by a horizontal line placed below the mth row ($m < n - k$, since there are $n - k$ syndromes and therefore $n - k$ rows). Received n-tuples falling below this line are not being corrected.

Then the received word w_j will be correctly decoded if and only if $i = j$, that is, if w_j also belongs to the ith column. Therefore the average probability for correctly decoding a sequence will be

$$P_{SC} = \sum_{i=1}^{2k} p(v_i)p(w_i/v_i) \tag{12-28}$$

(There are 2^k columns in the standard array.) The average error probability P_S for a sequence then will be

$$P_S = 1 - P_{SC} = 1 - \sum_{i=1}^{2k} p(v_i)p(w_i/v_i) \tag{12-29}$$

If we call a_{ij} the data symbols in error between the ith-column code word v_i and the jth-column received word w_j and call b_i the average number of errors for transmitting v_i and receiving w, an n-tuple that is not corrected by the code, i.e., one which falls below the line of the mth row, then the average bit error probability p_{be} will be

$$p_{be} = \sum_{i=1}^{2k} \sum_{\substack{j=1 \\ (j \neq i)}}^{2k} \frac{a_{ij}}{k} p(v_i)p(w_j/v_i) + \sum_{i=1}^{2k} \frac{b_i}{k} p(v_i)p(w/v_i) \tag{12-30}$$

Approximations of Eqs. (12-29) and (12-30) are given by Eqs. (12-14) and (12-16).

CYCLIC CODES

A linear block code is called a "cyclic code" if all the n-tuples $(a_0, a_1, \ldots, a_{n-1})$ of this code constitute a subspace of the n-tuple space, such that the subspace vectors are derived from each other by shifting the elements cyclically one position to the right. Consequently, the next n-tuple of the subspace will be $a_{n-1}, a_0, \ldots, a_{n-2}$, the next to the next will be $a_{n-2}, a_{n-1}, a_0, \ldots, a_{n-3}$, and so on.

We usually invert the sequence of the elements of the n-tuple when we describe the first n-tuple; consequently, we shall use for the general description the code word w, as in $w(a_{n-1}, a_{n-2}, \ldots, a_0)$, and the shifting to the left will yield $(a_{n-2}, a_{n-3}, \ldots, a_0, a_{n-1})$ and so on, the last term being $(a_0, a_{n-1}, \ldots, a_1)$.

It will become clear in the subsequent discussion that it is very useful to consider the elements a_i of the cyclic code n-tuple as coefficients of a polynomial of a variable x, namely,

$$a(x) = a_{n-1}x^{n-1} + a_{n-2}x^{n-2} + \cdots + a_1 x + a_0 \tag{12-31}$$

Of course in the case of linear binary codes, the a's are 1s or 0s. For simplicity we will consider in the following discussion only linear binary codes (group codes) and we will use the terms "code word" and "code polynomial" interchangeably.

If we multiply $a(x)$ by x we will get

$$\begin{aligned} xa(x) &= a_{n-1}x^n + a_{n-2}x^{n-1} + \cdots + a_1x^2 + a_0x \\ &= a_{n-1}(x^n - 1) + a_{n-2}x^{n-1} + \cdots + a_1x^2 + a_0x + a_{n-1} \end{aligned} \tag{12-32}$$

Therefore multiplication by powers of x, such as x^i, corresponds to a cyclic shift, in this case by i elements (modulo $x^n + 1$).

In the case of binary codes, where subtraction can be substituted for addition, the modulo $x^n - 1$ is often described as modulo $x^n + 1$.

The highest power of x in a polynomial is called the degree of the polynomial. If $g(x)$ is a code polynomial with the smallest possible degree, if it is also monic (the coefficient of the highest power of x is 1), which is certainly true for binary codes, and if it also exactly divides the polynomial $x^n - 1$, then $g(x)$ is called a "generator polynomial of the code." It is easy to derive this generator polynomial of a code by considering the generator matrix $G = [I_kP]$ of the code in its reduced echelon form. If we multiply the last column by x^0, the column next to the last by x^1, etc., each row gives the terms of a word polynomial. The polynomial with the smallest degree is formed in the kth row (last row), and its highest power is $n - k$. This is called the "generator polynomial $g(x)$."

The $k - 1$ row of the generator matrix can be formed by multiplying $g(x)$ by x, which shifts the kth row to the left. Now, if the x^{n-k-1} coefficient of $g(x)$ is 1, we add to $xg(x)$ the $g(x)$ itself; if it is 0, we keep the product $xg(x)$ as is. Therefore, the $k - 1$ row will be either $(x + 1)g(x)$ or $xg(x)$. The $k - 2$ row can be derived by the product $x^2g(x)$, adding none, one, or both of the previous rows, depending on the 0s or 1s of these rows that correspond to the first and second columns of the P matrix.

Since every code word is the sum of the rows of G, every code polynomial will be the sum of the row polynomials and therefore divisible by $g(x)$.

Now if we consider the H matrix

$$H = [P^T I_{n-k}]$$

and if we denote by $h(x)$ the polynomial of the top row of H with its coefficients reversed [i.e., the mth coefficient becomes the $(n - m)$th one], we will have

$$g(x)h(x) = 0 \bmod (x^n - 1) \tag{12-33}$$

since

$$GH^T = 0$$

The degree* of $h(x)$ is k, while $n - k$ is the degree of $g(x)$. Therefore the degree of their produce is n, but because of Eq. (12-33), we must have

$$g(x)h(x) = x^n - 1 \tag{12-34}$$

Therefore $g(x)$ exactly divides $x^n - 1$.

The above statements could be summarized in the following theorem:

> Every cyclic (n, k) code is generated by an exact divisor $g(x)$ of $x^n - 1$. Similarly, every divisor $g(x)$ of $x^n - 1$ generates a cyclic (n, k) code. The degree of $g(x)$ is $n - k$.

For every code subspace, there is only one $g(x)$ of degree $n - k$ because, if there were a second one, the addition of the two would give a code word of degree less than $n - k$, which is impossible. The $g(x)$ will also have a 1 in the last position; otherwise a shift will produce a code word with all zero information elements and at least one nonzero parity element, which is also impossible. Thus,

$$G = \begin{bmatrix} \overbrace{x^{n-1} & 0 & \cdots & 0}^{k} & \overbrace{\cdots & \cdots}^{n-k} \\ 0 & x^{n-2} & \cdots & 0 & \cdots & \cdots \\ 0 & 0 & \cdots & 0 & \cdots & \cdots \\ \cdots & \cdots & \cdots & \cdots & \cdots & \cdots \\ 0 & 0 & \cdots & 0 & \cdots & \cdots \\ 0 & 0 & \cdots & x^{n-k} & \cdots & 1 \end{bmatrix} \tag{12-35}$$

in which x's occur only in the main diagonal of the first k columns.

Encoding a data polynomial $d(x)$ of a degree less than k corresponds to the operation

$$x^{n-k}d(x) + r(x) \tag{12-36}$$

where $r(x)$ has a degree less than $g(x)$ and is the remainder of the division of $x^{n-k}d(x)$ by $g(x)$; that is,

$$\frac{x^{n-k}d(x)}{g(x)} = q(x) + \frac{r(x)}{g(x)} \tag{12-37}$$

The syndrome $s(x)$ of a polynomial $w(x)$ of the code space is given as the remainder of the division of $w(x)$ by $g(x)$,

$$s(x) = \text{remainder of } \frac{w(x)}{g(x)} \tag{12-38}$$

*The function $h(x)$ consists of the k symbols of the first row of P^T followed by the $n - k$ symbols of I_{n-k}, which are all 0s except for one.

As an example, consider $n = 7$. The polynomial $x^n - 1 = x^7 - 1$ has the following divisors:

$$(x^7 - 1) = (x - 1)(x^3 + x^2 + 1)(x^3 + x + 1)$$

Each of these divisors could be a generator polynomial. For the first divisor we have $n - k = 1$ or $k = 6$; for the second divisor we have $n - k = 3$ and $k = 4$, which is true for the third divisor also. In addition, the various products such as $(x - 1)(x^3 + x + 1)$ give a generator polynomial. For this particular polynomial, the code is a (7, 3) code, since $n - k = 4$. The minimum Hamming weight of the code words is the weight of $g(x)$ which also gives the minimum Hamming distance d of the code.

The polynomial $h(x)$ of Eq. (12-34), which is derived from the top row of the parity-check matrix H, is called the parity-check polynomial of the code generated by $g(x)$. The $h(x)$ could also be considered as a generator polynomial of an $(n, n - k)$ code. This $(n, n - k)$ code is called the "dual code" of the (n, k) code generated by $g(x)$.

MINIMUM HAMMING DISTANCE: HAMMING CODES

An n-tuple code word w of a group code with k data symbols and $n - k$ parity symbols was represented in Eq. (12-22) as

$$w = w_d w_p \tag{12-39}$$

where w_d represents the k data symbols and w_p the $n - k$ parity symbols. This word must satisfy the parity-check relationship:

$$wH^T = 0 \tag{12-40}$$

where H is the parity matrix $H = [P^T I_{n-k}]$.

An alternative way of writing Eq. (12-40) is

$$Hw^T = 0 \tag{12-41}$$

The vector w^T is a column vector with n elements and H is an $(n - k) \times n$ matrix. If the Hamming weight of w is d, then w will have d nonzero symbols. In forming the product $v = Hw^T$, the columns of the matrix H, whose elements multiply with these nonzero symbols, will be shifted out. In order for vector v to be zero, these shifted-out vector columns, when added, must also add to zero. For example, consider the H and w^T given by Eq. (12-42):

$$H = \begin{vmatrix} 1 & 0 & 1 & 1 & 0 \\ 1 & 1 & 0 & 0 & 1 \end{vmatrix} \qquad w^T = \begin{vmatrix} 1 \\ 0 \\ 1 \\ 0 \\ 1 \end{vmatrix} \tag{12-42}$$

For the product Hw^T to be zero, columns 1, 3, and 5 of H, corresponding to the nonzero elements 1, 3, and 5 of ω^T, must add to zero [see Eq. (12-40)].

$$\begin{vmatrix}1\\1\end{vmatrix} + \begin{vmatrix}1\\0\end{vmatrix} + \begin{vmatrix}0\\1\end{vmatrix} = 0 \qquad (12\text{-}43)$$

This leads to the conclusion that if d is the weight of w, H must have at least d column vectors that are linearly dependent; i.e., they must add up to zero. Furthermore, if d is the minimum Hamming distance of the code, $d - 1$ is the minimum number of linearly independent column vectors within the matrix H. Otherwise a code vector w' with a lesser distance than d will satisfy Eq. (12-41). (The minimum distance equals the minimum weight.)

This fundamental property of H provides a method for designing block codes with a desired minimum Hamming distance. In fact, by starting with two dissimilar columns in designing the matrix H, we can keep adding columns so that we never create a group of columns adding to zero that contains fewer than d columns. The null space G of this matrix H (that is, $H^TG = 0$) will be the generator matrix of the desired code.

Now consider a desired minimum distance $d = 3$. The parity matrix must not have any two columns that add to zero. This means that all columns must be distinct and nonzero. If $n - k = p$, there are $2^p - 1$ distinct p-tuples if the all-zero p-tuple is eliminated. Thus, the family of codes with $2^p - 1$ symbol code words ($n = 2^p - 1$) and $2^p - 1 - p$ data symbols ($k = 2^p - 1 - p$) has minimum distance $d = 3$. These codes are called "Hamming codes," and they are perfect codes in the sense that they have the minimum required redundancy for the number of errors that they are capable of correcting.

The two matrices G and H represent the null space of each other. If G is assumed to be the parity-check matrix and H the generator matrix, then we can derive a dual code. The dual code of a Hamming code is known as "maximum-length code."

The Bose-Chaudhuri-Hocguenhem (BCH) codes are a generalization of the Hamming codes. They are the most powerful codes known for multiple-random-error correction. These codes are cyclic codes and are best defined in terms of the roots of the generator polynomials. The reader can find a detailed discussion of these codes in the references at the end of this chapter.

CONVOLUTIONAL CODES

If m is the number of n'-tuples checked by the parity symbols of a convolutional code n'-tuple, then the k' data bits in each n'-tuple block must be taken over a number m of blocks (n'-tuples) to find the data checked by the parity symbols. If d_0 is the k' data bits of the current n'-tuple, d_1 is the k' data bits of the previous n'-tuple, and so on, then the data sequence

$$d = d_0d_1 \cdots d_{m-1}$$

is the data stream being checked. All the k''s have the same number of data bits, and therefore

$$k = mk'$$

The overall n-tuple is similarly

$$n = mn' \tag{12-44}$$

If we denote by $0_{k'}$ and $I_{k'}$ the zero and identity matrices of order $k' \times k'$, the generator matrix of the convolutional code will be

$$G = \begin{bmatrix} I_{k'}P_0 & 0_{k'}P_1 & \cdots & 0_{k'}P_{m-1} \\ & I_{k'}P_0 & \cdots & 0_{k'}P_{m-2} \\ & & & \\ & & & I_{k'}P_0 \end{bmatrix} \tag{12-45}$$

and the parity check matrix will be

$$H = \begin{bmatrix} P_0^T I & & & \\ P_1^T 0 & P_0^T I & & \\ & & & \\ P_{m-1}^T 0 & P_{m-2}^T 0 & \cdots & P_0^T I \end{bmatrix} \tag{12-46}$$

where I and 0 are the identity and 0 matrices of order $n' - k'$.

The matrices P_r^T or P_r are arbitrary $(n' - k') \times k'$ or $k' \times (n' - k')$ matrices. The element of the ith row and jth column of the matrix P_r^T checks the jth information bit which precedes the current block by r blocks.

We again have

$$GH^T = 0 \tag{12-47}$$

and a code vector w is a code word if and only if

$$wH^T = 0$$

The syndrome of a code vector v is defined as

$$s = vH^T \tag{12-48}$$

With these definitions, one can derive properties for convolution codes in the same way that the properties of the block codes were derived.

ELEMENTARY LINEAR SWITCHING CIRCUITS

The three basic elements of a linear switching circuit that are used to generate, multiply, and divide the various code polynomials are the following:

1. The designation of the EXCLUSIVE OR (addition), where the output is the sum modulo 2 of the inputs, given in Fig. 12-7a.

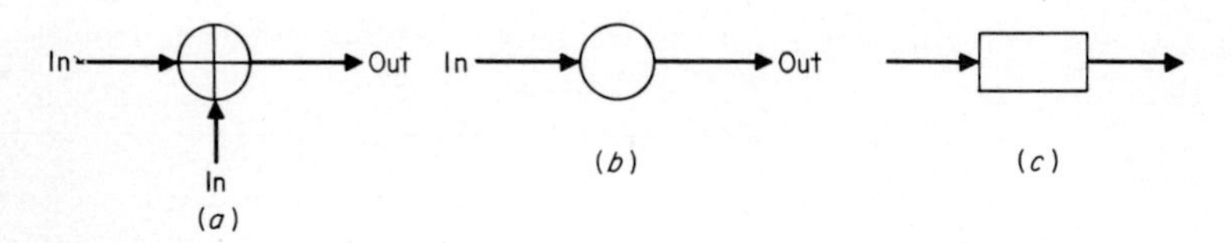

FIGURE 12-7 The basic elements of a linear switching circuit. (*a*) EXCLUSIVE OR. (*b*) Multiplier. (*c*) Storage element.

2. The designation of the multiplier, multiplying the input with a constant, the constant being 0 or 1 for binary codes, given in Fig. 12-7*b*.
3. The storage element which stores the input for one unit of time, depicted in Fig. 12-7*c*. The storage element is a delay device representing a single stage of a shift register.

The multiplication of a random polynomial (such as a data stream),

$$d(x) = d_0 + d_1x + \cdots + d_kx^k$$

by a standard multiplier polynomial (such as the parity check),

$$h(x) = h_0 + h_1x + \cdots + h_rx^r$$

can be achieved by inserting the coefficients of $d(x)$ one at a time (the highest one, d_k, first) in the circuit depicted in Fig. 12-8, and shifting $r + k$ times.

A circuit for dividing $d(x)$ by

$$g(x) = g_0 + g_1x + \cdots + g_rx^r$$

is depicted in Fig. 12-9, in which we again insert $d(x)$ highest-power first. For the first r shifts, the output is 0.

After k shifts, the output is the quotient and the shift registers contain the remainder.

Various combinations and alternative forms of the above circuits are employed for encoding and decoding processes.

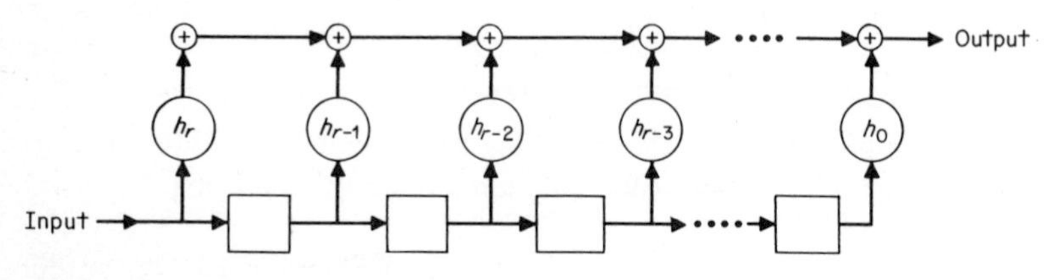

FIGURE 12-8 Multiplication circuit.

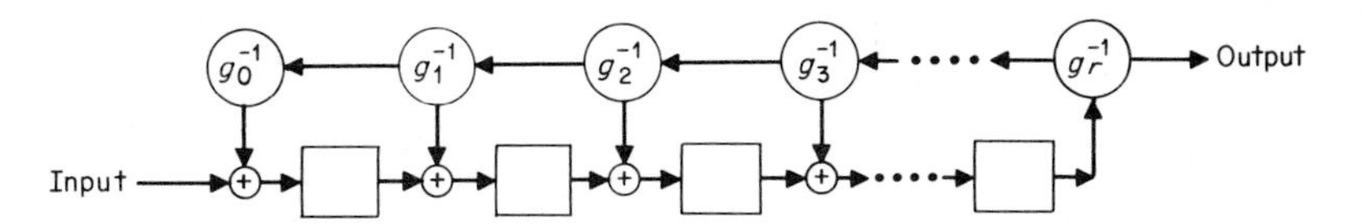

FIGURE 12-9 Division circuit.

DECODING TECHNIQUES AND PERFORMANCE OF ERROR-CORRECTING CODES

In satellite communications, as in most modern digital communications systems, coding is used to improve the performance in the presence of noise. For each signaling scheme, the bit error probability is expressed in terms of E_b/n_0. The performance of error-correcting codes is expressed in terms of the coding gain in decibels. Coding gain is defined as the required increase in E_b/n_0 by an uncoded message in order to maintain the same bit error probability as the coded one. The two most widely used types of codes are the block codes and the convolutional codes. Block codes include a very important class of codes, namely, the cyclic codes. The code rate $R = k/n$, where k is the number of data symbols and n is the number of code word symbols, measures the redundancy employed by a code. The constraint length applies to convolutional codes and measures the interrelationship between blocks.

The decoding technique is also very important in the overall performance. Two classes of decoders are widely used: hard-decision and the soft-decision decoders. Soft-decision decoders can provide an additional coding gain in the range of 2 dB. Some decoding algorithms are applicable to both hard- and soft-decision decoding, whereas other decoding algorithms are applicable to only one type of decoding. Soft decisions in block codes result in an additional coding gain close to 2 dB. Algorithms such as the Viterbi algorithm or sequential decoding could provide as much as a 6-dB additional coding gain for convolutional codes.

Decoding of Group Codes

Group codes can utilize either hard-decision or soft-decision decoding techniques. In this section we will discuss only hard-decision decoding. Soft-decision decoding will be discussed in connection with convolutional codes.

Hard-decision techniques for group codes can be divided into two categories: algebraic and nonalgebraic decoding. Algebraic techniques determine errors by the simultaneous solution of a set of algebraic equations. Nonalgebraic techniques depend on the structure of the code and determine the error with a more direct method, utilizing the specific structure of the code.

Algebraic Techniques These techniques are capable of correcting multiple errors and basically involve the simultaneous solution of a set of algebraic equations. The algebraic structure of the applicable codes allows the development of efficient decoding algorithms. It is easy to illustrate this technique by multiplying the received sequence by the parity-check matrix H^T and obtaining the syndrome S. Equation (12-25) of the section on standard arrays gives the result of this multiplication. By solving the resulting algebraic equations, the symbols which are in error in the received sequence can be identified. Of course, if there are more errors than equations, the solution is not deterministic.

Meggit Decoding Meggit decoding is a hard-decision nonalgebraic decoding technique. It belongs to the general category of the table look-up decoding techniques because it utilizes a stored table containing all the correctable-error patterns.

Meggit decoding is applied to cyclic codes of relatively short length. When the block sequence is received, it is stored in a register and the syndrome is computed. There is a one-to-one correspondence between each syndrome of the code and a correctable-error pattern taken from the set of all correctable error patterns stored in the table. Once the corresponding error pattern is identified, the proper correction is applied to the received code-word sequence. Coding gains of about 2.5 dB at $P_e = 10^{-5}$ are achievable with Meggit decoders.

Information-Set Decoding A nonalgebraic hard-decision technique, information-set decoding, applies to a large number of group codes. The decoder selects a certain number of information sets, assumes that they are error-free, and reencodes these information sets, producing a set of hypothesized code words. It then selects from the hypothesized code words the one which is closest to the received sequence. The selection of the information sets is made by utilizing a specific rule such as a random search or a predetermined set. The decoder could also consider the information set of the received sequence to be error-free. Utilizing this information set, it reencodes and produces a new sequence. This new sequence is compared with the received one, and the error pattern of the parity symbols is derived. Subsequently, the decoder must determine the parity-check set that contains this error pattern. Several strategies have been developed for performing this task.

Threshold Decoding Threshold decoding can be applied with both hard and soft decisions. It is also used extensively with convolutional codes, and it will be discussed in the section on convolutional-code decoding.

Decoding of Convolutional Codes

Convolutional codes as originally introduced by Elias do not lend themselves to algebraic techniques for multiple-error correction. Most of the convolutional codes are the result of computerized searches for codes with good distance properties. Both hard decisions and soft decisions are used with convolutional codes. Soft-decision techniques could be divided into two categories. The first category uses decoders that seek to minimize the average error of each sequence. The second category seeks to minimize the distance between the code word and the received sequence. As a result, the probability of sequence error is minimized and leads to the maximum-likelihood sequence decoding. Soft-decision decoders provide approximately 2-dB greater gain than do the corresponding hard-decision decoders.

Threshold Decoding The major advantage of threshold decoding is that it is easily instrumented for both block and convolutional codes. Coding gains in the range of 2 to 5 dB and 1 to 4 dB can be achieved for block codes and convolutional codes, respectively. Threshold decoding is limited to binary codes that possess the necessary structure. *Threshold Decoding* by James L. Massey (MIT Press, Cambridge, Mass., 1963) treats the subject of threshold decoding in depth.

Threshold decoding is applicable to both hard and soft decisions. For convolutional codes, the performance of threshold decoding is satisfactory for codes up to 100 symbols in length. In addition, given a fixed rate of data transmission, the error probability cannot be made arbitrarily small. Threshold-decoding algorithms could provide efficient decoding for several interesting classes of block codes. For both convolutional and block codes, the error-correction capability is not limited to the minimum distance of the code; many error patterns of weight greater than $t = (d - 1)/2$, where d is the minimum Hamming distance, can also be corrected. If s_i is the transmitted symbol of a sequence S_1 and r_i is the ith received symbol of the corresponding received sequence r, we form the difference $e_i = r_i - s_i$, where e_i is the error or noise symbol introduced by the communications channel.

The parity-check equations for the code under consideration are transformed (mapped) to a set of parity-check equations that are orthogonal to the noise symbol e_i. A set of equations that contains only once each error term being checked—except e_i, which appears in all of these equations—is said to be orthogonal on e_i. From this set of J orthogonal parity-check equations, one can obtain J estimates A_i orthogonal to the noise symbol e_i. A decision function is formed on the basis of these estimates. This decision function is compared with a threshold, and if it is greater than the threshold, e_i is assigned the value 1; otherwise it is assigned the value 0.

Two distinct forms of threshold decoding have been developed from this concept, majority decoding and a posteriori probability (APP) decoding.

"Majority decoding" is a hard-decision method utilized with block codes. In effect, the majority decoder adds up all the estimates A_i of e_i and compares the sum to a predetermined threshold. If

$$\sum_{1}^{J} A_i > \tfrac{1}{2}T \qquad \text{for } T = J$$

where $\frac{1}{2}T$ is the threshold, e_i is assigned the value 1; otherwise e_i is assigned the value 0.

"APP decoding" forms the soft-decision function by utilizing probability metrics.* APP assigns to e_i the value that is more probable, given the set of values of the orthogonal parity checks. The decision relationship becomes

$$\sum_{1}^{J} m_i A_i > \frac{T}{2}$$

Here, the m_i's are probabilistic parameters and T is the sum of these parameters. It can be seen that when the m_i's become all 1s, we obtain the majority-decoding formula.

By not considering the entire set of ordinary parity checks and limiting the decoding process only to orthogonal parity checks, we can reduce the decoding problem to a simple form. Of course, the structure of the code must be such that the mapping from ordinary parity checks to orthogonal parity checks can be carried out in an efficient manner.

Sequential Decoding J. M. Wozencraft presented the technique of sequential decoding at the 1957 National Convention of the Institute of Radio Engineers and, subsequently, in a book coauthored with B. Reiffer, *Sequential Decoding* (Technology Press/John Wiley & Sons, New York, 1961). The sequential decoder compares the received sequence with the possible transmitter sequences and selects the one that seems to be the best. To illustrate the process in a simple manner, we will divide the possible transmitted sequences into two subsets, namely, the subset with sequences starting with 1 and the subset of sequences starting with 0. When the decoder finds that the received sequence differs substantially from every sequence in one of these subsets, it assumes that the transmitted sequence is in the other subset. Consequently, the first symbol of the sequence is determined. Subsequently, the first symbol is dropped from the sequence and the process is repeated. Thus, sequentially, the

*See the footnote on p. 242 for the definition of "metric."

symbols of the process are determined. However, as soon as the decoder determines that a certain sequence is beginning to diverge substantially from the received sequence, it drops the sequence and does not complete the comparison. The sequential decoding process with soft decision is a trial-and-error process. This can be illustrated by the code tree in Fig. 12-5. The decoder examines the various paths (branches) of the tree and extends (or updates) only the paths that appear to be the most probable. The decoder will backtrack if it realizes that it has followed the wrong path.

The sequentially decoding algorithm used most widely to date was introduced by Fano. The principal rule of this algorithm is that the decoder will not proceed along a path in either direction if the path falls below the current metric threshold. From the current node, the decoder proceeds along the branch with the maximum likelihood of occurrence as long as the metric of the path to be followed is higher than the current threshold. If the decoder cannot move forward, it searches laterally or backward without lowering the threshold. If no paths can be found that meet this requirement, the threshold is lowered and the decoder moves forward. If, in a forward move, the decoder crosses a higher threshold, then the threshold is raised to this new threshold. Each time the decoder moves forward, a tentative decision is made. If the correct path is selected, the number of required computations is substantially reduced. To illustrate the utilization of metrics with the sequential decoding technique, we will utilize the code tree depicted in Fig. 12-5, which corresponds to a rate $R = \frac{1}{2}$ convolutional code. Assume that the received sequence is r = 01, 10, 00, 11, which corresponds to an information sequence 1100. There are errors in the first and third branches. The branch metric is defined by Eq. (12-12*a*) as a logarithmic function. However, every time that the signal-to-noise ratio changes, we will have to adjust the metric values. Instead, we can assign a fixed set of metric values that provides a good approximation over a wide range of signal-to-noise ratios. A scheme often used is to define $m_j = j$. For the case depicted in Fig. 12-4, with 8 levels, $j = 7$ and all the other metrics take the values 0, 1, . . . , 7. For a large range of codes, this selection of metrics causes only a small degradation of performance. In our example, let us assign $+1$ for each symbol that agrees and -4 for each symbol that disagrees with the received sequence r. The decoder will compute the metrics per branch as depicted in Fig. 12-10. Starting from state A, the decoder is presented with two branches with the same metric -4. The convention is to try the upper branch first. As the decoder tries this branch, it fails the threshold test according to the Fano algorithm. As a result the decoder comes one step back and lowers the original zero-value threshold. Each time a new sequence is received, the branch metric computer can easily determine the metric values of each of the different branches. The metric of a certain path will be the sum of the metrics of the branches of the path.

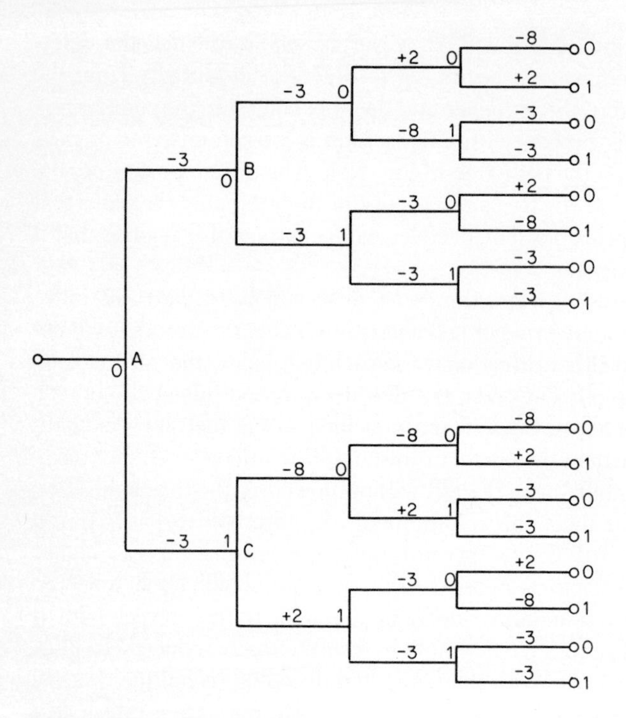

FIGURE 12-10 Branch metrics.

Viterbi Algorithm The Viterbi algorithm is a maximum-likelihood algorithm for convolutional codes. Like the sequential decoder, it can achieve a substantial improvement with coding gains in the range of 4 to 7 dB. The decoder searches all possible paths in the decoding sequence and selects the path with the best metric. The number of paths grows exponentially as the sequence increases. The decoder, however, does not have to examine all of them; it can discard a number of paths at every node by eliminating duplicate paths and making a judgment based on a maximum-likelihood criterion.

EXAMPLE 12-1 $\frac{1}{2}$ RATE CONVOLUTIONAL CODER: VITERBI DECODING

The $\frac{1}{2}$ rate convolutional coder depicted in Fig. 12-11*a* is often used as an example in the literature. It consists of two adders that produce the two symbols of the sequence and a shift register with three stages. The constraint length is 3, since there are three shift-register stages. For each data symbol entering the shift register ($k = 1$), the two adders produce a sequence of two symbols ($n = 2$), resulting in a rate $R = k/n = \frac{1}{2}$.

The symbols stored in the two leftmost stages of the register will be shifted to the right as a new data symbol enters from the left. Thus, the last symbol in the third stage will be lost. Therefore the two symbols in the two leftmost stages determine the state of the register. For a binary code we have four possible states, namely, 00 (number 0 state), 01 (number 1 state), 10 (number 2 state) and 11 (number 3 state).

The code tree in Fig. 12-5 represents the possible paths this encoder may take. The branches of the tree continuing from any state, say 3, are identical with the branches emanating from any other state 3. This indicates that these identical branches could be merged. This merging of the branches is used by the Viterbi algorithm. As a result, the potential exponential increase of the number of the branches is contained and the implementation of a decoder with soft decision becomes manageable.

The "trellis" diagram representation of this coding scheme is depicted in Fig. 12-11*b*. Again, there are four states (00, 01, 10, 11), and from each state the emanating upper branch corresponds to a 0, which results in a transition, while the lower branch corresponds to a 1. The output is designated along the branch. It can be seen that there is a maximum of two paths entering each node. As a result, the decoder at each node has to decide only among the past path combinations that end with these two paths. This selection is made on the basis of maximum likelihood and the decision is final, since beyond the node both paths entering the node have the same opportunities. From the paths examined, the decoder discards the nonqualifying paths and retains the "survivor" paths. For the $k = 3$ case, a maximum of four survivor paths will exist. For example, an input sequence of 1011 will provide the output 11, 01, 00, 10.

The number of the nodes in the trellis diagram does not increase as the number of the input symbol increases. When the constraint length $k = 3$, we have a depth of 4 ($= 2^2$). In general, this depth will be 2^{k-1}. After that depth, the trellis repeats itself.

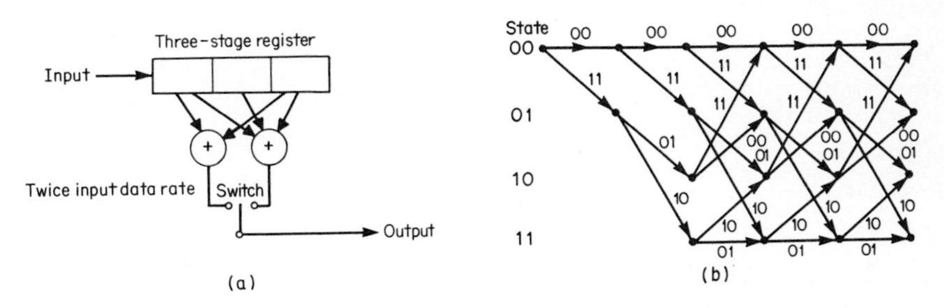

FIGURE 12-11 **(*a*) Convolutional coder ($R = \frac{1}{2}$). (*b*) Viterbi decoding (trellis diagram).**

TABLE 12-1 APPROXIMATE CODING GAINS

Type of code	Coding gain, dB		Data rate capability
	$P_e = 10^{-5}$	$P_e = 10^{-8}$	
Two-level (concatenated) Reed-Solomon and Viterbi	6–7	8–9	Moderate
Sequential decoding with soft decision	6–7	8–9	Moderate
Block codes with soft decision	5–6	6–7	Moderate
Viterbi decoding	4–6	5–7	High
Sequential decoding with hard decision	4–5	6–7	High
Block codes with hard or threshold decision	2–4	3–5	High
Convolutional codes (table or threshold)	1–3	2–4	High

Coding Gain of Various Codes

The coding-gain capability of each code depends on the level of the error probability P_e at which the code is asked to perform. Forney introduced a technique utilizing multiple levels of coding for codes with very long block lengths. These codes, known as "concatenated codes," most often use a Reed-Solomon code as a first level (outer code).

Table 12-1 gives an approximate value of the coding gain to be expected from various codes for two values of error-rate probability. A coherent PSK channel is assumed.

REFERENCES

Clark, G. C., Jr.: *Error-Correction Coding for Digital Communications*, Plenum Press, New York, 1981.

Forney, G. D., Jr.: "The Viterbi Algorithm," *Proceedings of the IEEE*, vol. 61, March 1978, pp. 268–278.

Heller, J. A., and I. M. Jacobs: "Viterbi Decoding for Satellite and Space Communications," *IEEE Transactions on Communications*, vol. COM-19, October 1971, pp. 835–848.

Lucky, R. W., J. Salz, and E. J. Weldon, Jr.: *Principles of Data Communications*, McGraw-Hill, New York, 1968.

Peterson, W. W.: *Error-Correcting Codes*, MIT Press/John Wiley & Sons, New York, 1961.

Shannon, C. E., and W. Weaver: *The Mathematical Theory of Communications*, The University of Illinois Press, Urbana, Ill., 1959.

13 TRANSMISSION IMPAIRMENTS

NONLINEARITIES

A number of nonlinearities are present in every communications system. Of course, there are certain networks and components that must be nonlinear in order to perform their functions, such as limiters and certain types of modulators. However, the nonlinearities discussed here are unwanted ones, introducing spurious components of signals that disturb the fidelity of the system as noise does.

Several of these nonlinearities are the result of operating a given component close to its saturation level in order to extract maximum power output. An example of such nonlinearity, discussed in Chap. 5, is the one introduced by the saturation characteristics of a traveling-wave-tube (TWT) amplifier.

The Nonlinear Power Amplifier

The idealized voltage-transfer characteristic of the nonlinear amplifier to be examined in this section is shown in Fig. 13-1. This type of amplifier is known as a "limiter." The TWT amplifier of a satellite repeater is a limiter.

For an input $x = x(t)$ with a maximum amplitude $x_{max} \leq b$, the operation is strictly linear with a voltage gain $G = a$. When the amplitude of the input signal becomes larger than b, we have nonlinear operation.

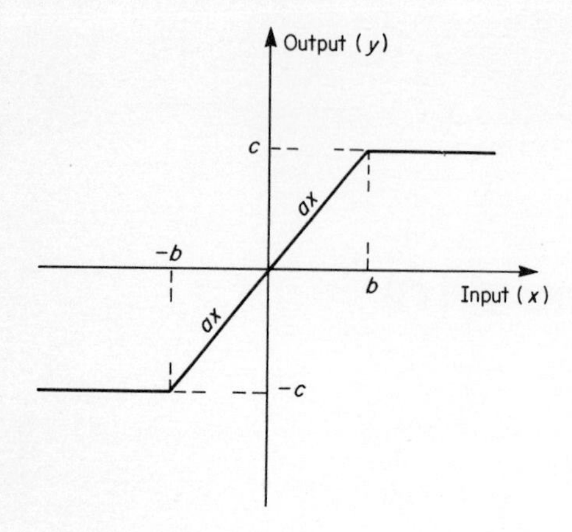

FIGURE 13-1 Voltage characteristic of a nonlinear amplifier.

Assume a sinusoidal input voltage

$$x(t) = A \cos(\omega_c t + \phi) = A \cos \theta \tag{13-1}$$

in which $x(t_1) = b$ for $t = t_1$. At $t = t_1$, the angle θ becomes $\theta_1 = \omega_c t_1 + \phi$, so that $b = A \cos \theta_1$, or

$$\cos \theta_1 = \frac{b}{A} \tag{13-2}$$

The output $y(t)$, shown in Fig. 13-2*a*, is saturated for values of y above c. This output is also a periodic function, with the amplitude B_1 of the fundamental given by the Fourier series coefficients:

$$B_1 = \frac{1}{\pi} \int_0^{2\pi} y(t) \cos \theta \, d\theta = \frac{4}{\pi} \int_0^{\pi/2} y(t) \cos \theta \, d\theta \tag{13-3}$$

We can evaluate the integral by splitting it into two parts:

$$\begin{aligned} B_1 &= \frac{4}{\pi} \int_0^{\theta_1} c \cos \theta \, d\theta + \frac{4}{\pi} \int_{\theta_1}^{\pi/2} aA \cos^2 \theta \, d\theta \\ &= \frac{4}{\pi} \left| c \sin \theta \right|_0^{\theta_1} + \frac{aA}{2} \left| (\theta + \tfrac{1}{2} \sin 2\theta) \right|_{\theta_1}^{\pi/2} \end{aligned} \tag{13-4}$$

But

$$\cos \theta_1 = \frac{b}{A}$$

$$\sin\left(\frac{\pi}{2} - \theta_1\right) = \cos \theta_1 = \frac{b}{A}$$

$$a = \frac{c}{b} = \text{voltage gain } G$$

Consequently,

$$B_1 = \frac{2c}{\pi}\left[\sqrt{1-\left(\frac{b}{A}\right)^2} + \frac{A}{b}\sin^{-1}\frac{b}{A}\right] \tag{13-5}$$

or

$$B_1 = GA\,\frac{2}{\pi}\left[\frac{b}{A}\sqrt{1-\left(\frac{b}{A}\right)^2} + \sin^{-1}\frac{b}{A}\right] \tag{13-6}$$

By defining g as the "gain-suppression" parameter, we obtain

$$B_1 = gGA$$

where

$$g = \frac{2}{\pi}\left[\frac{b}{A}\sqrt{1-\left(\frac{b}{A}\right)^2} + \sin^{-1}\frac{b}{A}\right] \tag{13-7}$$

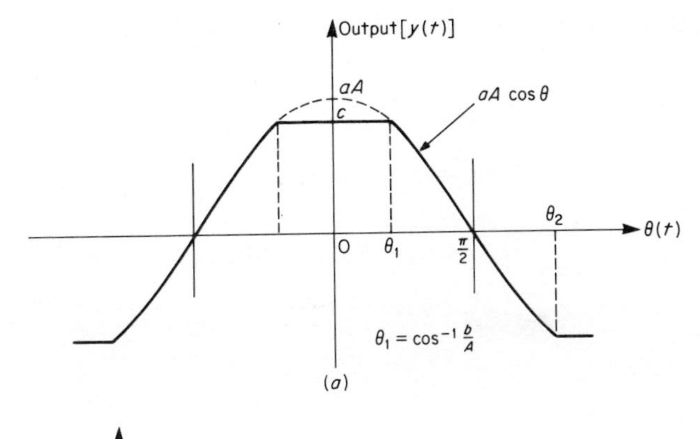

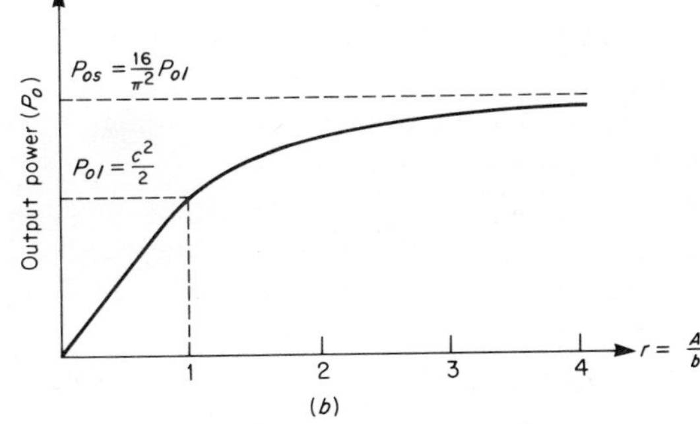

FIGURE 13-2 Idealized nonlinear-amplifer operation. (*a*) Saturated output waveform. (*b*) Output-input power characteristic.

The rms input power (for a 1-Ω load) is $P_1 = A^2/2$, and the rms output power of the fundamental is

$$P_o = \frac{B_1^2}{2} \tag{13-8}$$

The saturated output power P_{os} (for $A/b \to \infty$) is expressed as

$$P_{os} = \frac{8}{\pi^2}(Gb)^2 = \frac{8}{\pi^2}c^2 \tag{13-9}$$

and if the maximum linear output power P_{ol} is

$$P_{ol} = \tfrac{1}{2}(Gb)^2 = \frac{c^2}{2} \tag{13-10}$$

we obtain

$$\frac{P_{os}}{P_{ol}} = \frac{16}{\pi^2} = 1.62 \tag{13-11}$$

The plot of the output power P_o versus the ratio $r = A/b$ is given in Fig. 13-2*b*. The *n*th harmonic amplitude can be obtained by evaluating the integral

$$B_n = \frac{1}{\pi}\int_0^{2\pi} y(t)\cos n\theta\, d\theta = \frac{4}{\pi}\int_0^{\pi/2} y(t)\cos n\theta\, d\theta \tag{13-12}$$

We can evaluate this integral by splitting it into two parts:

$$B_n = \frac{4}{\pi}\int_0^{\theta_1} c\cos n\theta\, d\theta + \frac{4}{\pi}\int_{\theta_1}^{\pi/2} aA\cos\theta\cos n\theta\, d\theta \tag{13-13}$$

where

$$\cos\theta_1 = \frac{b}{A} \tag{13-14}$$

Equation (13-13) is easy to evaluate for various values of n, and all the harmonics can thus be obtained.

The process for deriving the output fundamentals and harmonics for an input consisting of the sum of two sinusoids, $x(t) = A_1\cos\theta_1 + A_2\cos\theta_2$, where $\theta_1 = \omega_1 t + \phi_1$ and $\theta_2 = \omega_2 t + \phi_2$, will be the same as the process used for one sinusoid. The integral must be evaluated piecewise with the linear region considered separately from the saturated region. However, the mathematics becomes quite complicated and is usually performed with the aid of computers. The case of a large number of inputs could be approximated by a continuous function of frequency, such as noise, and analyzed by assuming an approximate distribution function. A number of analytical models have been developed by assuming that the distribution is a gaussian distribution. For carriers with equal amplitude, one can also assume a white-noise-like spectrum. In many cases both amplitude and phase nonlinearities must be combined to provide an acceptable model.

Taylor Series Expansion

A common method of treating nonlinear transfer characteristics of the type represented in Figs. 13-1 and 13-2*b* is to express the output in terms of a Taylor series expansion including enough terms to provide the desired accuracy. If v_o is the output voltage for a v_1 input, we have

$$v_o = k_1 v_1 + k_2 v_1^2 + k_3 v_1^3 + \cdots + k_m v_1^m \tag{13-15}$$

Such a representation can be used over a limited range, especially when performance is highly sensitive to the frequency and phase of the signals.

Consider the case in which v_1 is a simple sinusoid:

$$v_1(t) = A \cos at \tag{13-16}$$

By inserting Eq. (13-16) into Eq. (13-15), we obtain

$$\begin{aligned} v_o &= k_1 A \cos at + k_2 A^2 \cos^2 at + k_3 A^3 \cos^3 at + \cdots \\ &= k_1 A \cos at + k_2 A^2 (\tfrac{1}{2} + \tfrac{1}{2} \cos 2at) \\ &\quad + k_3 A^3 (\tfrac{3}{4} \cos at + \tfrac{1}{4} \cos 3at) + \cdots \end{aligned}$$

By considering only the first three terms of Eq. (13-15), we obtain

$$\begin{aligned} v_o &= \tfrac{1}{2} k_2 A^2 + (k_1 A + \tfrac{3}{4} k_3 A^3) \cos at \\ &\quad + \tfrac{1}{2} k_2 A^2 \cos 2at + \tfrac{1}{4} k_3 A^3 \cos 3at \end{aligned} \tag{13-17}$$

One can see that, by use of proper filtering, the dc as well as the higher-frequency terms can be eliminated, whereas the desired frequency term is multiplied by a factor $k_1[1 + \frac{3}{4}(k_3/k_1)A^2]$. The inverse of the factor $[1 + \frac{3}{4}(k_3/k_1)A^2]$ is called a "suppression factor." The other terms of Eq. (13-17) are called "intermodulation products."

One can perform the same operation for an input consisting of more than one sinusoid, for example, three:

$$v = A \cos at + B \cos bt + C \cos ct \tag{13-18}$$

A detailed treatment of this case with three sinusoids is given in "Intermodulation Products" in Chap. 5. There, the output v_o was found to contain a dc factor and all possible combinations* of frequencies such as $\pm\, p_a a \pm p_b b \pm$

*As long as $p_a + p_b + p_c \leq 3$.

$p_c c$, where p_a, p_b, and p_c are integers with values 0, ±1, ±2, ±3. A term containing the sum of the fundamentals will be present at the output. This term was found to be

$$k_1 A(1 + L) \cos at + k_1 B(1 + M) \cos bt + k_1 C(1 + N) \cos ct$$

where L, M, N are functions of A, B, C, and the ratio k_3/k_1.

The rest of the terms represent intermodulation products of the second and third orders, such as $\cos 2at$, $\cos (b + c)t$, and $\cos (a - c)t$ for second-order products and $\cos 3bt$, $\cos (2a + b)t$, and $\cos (2c + b)t$ for third-order products.

Now, assume the general case in which there are n sinusoidal inputs

$$v_1(t) = \sum_{i=1}^{n} C_i \cos (\omega_i t + \phi_i) \tag{13-19}$$

into a nonlinear system with an input-output characteristic given by Eq. (13-15). The output will be given as a sum of sinusoids, each one of the form

$$v_{oj}(t) = K_j \cos \left(\sum_{i=1}^{n} p_{ij} \omega_i t + \theta_j \right) \tag{13-20}$$

where the coefficients p_{ij} will take the values 0, ±1, ±2, . . . , $\pm m$. The sum of the coefficients p_{ij} in Eq. (13-20) will take all possible values up to $\Sigma p_{ij} = m$, where m is the maximum power of the series in Eq. (13-15). The sum of the fundamentals will appear in the output as

$$v_o(t) = \sum_{i=1}^{n} k_i \cos (\omega_i t + \theta_i) \tag{13-21}$$

where θ_i is a function of the nonlinearities and input parameters. The term $v_{oj}(t)$ represents an intermodulation product of the order

$$m_j = p_{ij} + p_{2j} + \cdots + p_{xj} + \cdots + p_{nj} \tag{13-22}$$

The maximum value that m_j can take is m.

As was previously indicated, a complete analysis of the general case requires computer aid. Several approaches utilizing either time- or frequency-domain methods are discussed in the references at the end of this chapter.

AM-PM Conversion

The analysis in the previous section utilizing Eq. (13-15) and the single sinusoid of Eq. (13-16) does not take into account the phase characteristics of the network transfer function; it assumes nonlinearities in the amplitude gain only.

In general, the transfer characteristic of a linear two-port network can be expressed as

$$H(f) = |H(f)|e^{j\theta} \tag{13-23}$$

where θ is a function of the frequency.

However, in the case of a nonlinear network, θ could also be a function of the amplitude of the input signal. If such is the case, the phase of the output signal will be affected and a certain phase distortion will be introduced. In the case of an angle-modulated signal, this phase distortion will result in a degradation of the fidelity of the transmitted message. This effect is commonly known as the "AM-PM conversion effect." AM-PM conversion is usually measured in degrees of phase shift per decibel of amplitude change.

In many cases, the phase nonlinearity is approximated by

$$\theta(a) \approx ka^2(t) \tag{13-24}$$

where $a(t)$ is the amplitude of the input and k is the AM-PM conversion factor in degrees per decibel. This is a good approximation for TWT-amplifier satellite repeaters.

Now if the input signal is

$$V_i = a(t) \cos(2\pi f_c t + \phi_1) \tag{13-25}$$

the output due only to this AM-PM nonlinearity will be

$$v_o = a(t) \cos[2\pi f_c t + \phi_0 + \theta(a)] \tag{13-26}$$

If we assume that $a(t)$ is an amplitude-modulated signal with a small modulation index m, we can represent $a(t)$ as

$$a(t) = A(1 + m \cos 2\pi f_0 t) \tag{13-27}$$

and because of Eq. (13-24),

$$\begin{aligned}\theta(a) &= kA^2(1 + m \cos 2\pi f_0 t)^2 \\ &= kA^2(1 + 2m \cos 2\pi f_0 t + m^2 \cos^2 2\pi f_0 t)\end{aligned}$$

For small m,

$$\theta(a) \approx kA^2(1 + 2m \cos 2\pi f_0 t) \tag{13-28}$$

Therefore the peak error deviation will be

$$\theta_m = kA^2(1 + 2m) \tag{13-29}$$

For many sinusoidal signals, we can carry out a similar analysis, assuming an input of

$$v_i = \sum_{i=1}^{n} a_i(t) \cos (2\pi f_i t + \phi_i) \tag{13-30}$$

which can be expressed with an equivalent

$$v_i = a(t) \cos (2\pi f_i t + \phi) \tag{13-31}$$

where $a(t)$ and ϕ are functions of the i components. By calling $a_i(t) = C_i$ and $2\pi f_i = \omega_i$, we can derive Eq. (13-19) and carry on the generalized analysis, including both the AM-PM conversion as well as the amplitude nonlinearities.

INTERSYMBOL INTERFERENCE

In digital transmission systems, the information is carried by discrete pulses. For the purpose of our discussion, we will call each pulse a symbol. Both the transmitting and the receiving ends employ filters that limit the bandwidth of the communications channel. Each filter distorts the shape of a pulse, and, as a result, a trailing waveform appears at the output instead of a sharp pulse cutoff. This pulse trail interferes with the adjacent pulses. The tails produced from a series of signaling pulses, added together, can deform the next incoming pulse to such a degree that the receiver detector may erroneously interpret the presence or absence of the signaling pulse. This type of interference is known as "intersymbol interference."

Ideal Low-Pass Filter and Impulse Signaling

Consider a transmitter having as an output filter the ideal low-pass-filter transfer characteristic shown in Fig. 13-3*a*. The transfer characteristic is $|H(f)| = 1$

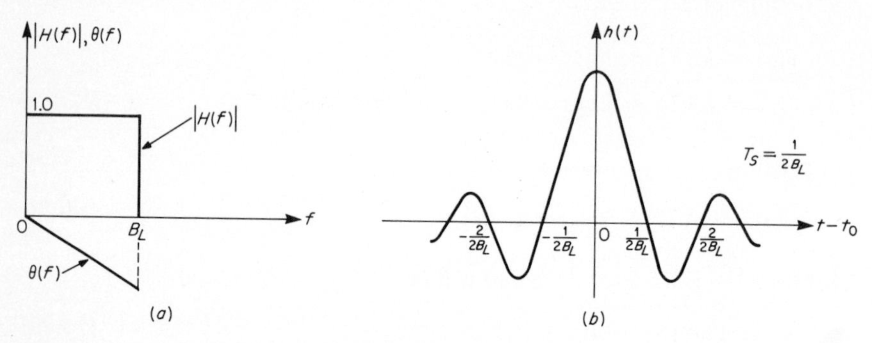

FIGURE 13-3 (*a*) Ideal low-pass filter. (*b*) Impulse response of ideal low-pass filter.

within the passband 0 to B_L. The phase characteristic $\theta(f)$ is linear with frequency. The response of this filter to a unit-impulse excitation $\delta(t)$ is a $(\sin x)/x$ function, shown in Fig. 13-3*b*. This response is given as

$$h(t) = B_L \frac{\sin 2\pi B_L(t - t_0)}{2\pi B_L(t - t_0)} \tag{13-32}$$

where t_0 is the time delay introduced by the low-pass filter [see Eq. (8-10)].

The response crosses the time axis and it is zero for $t = t_1$ when $(\sin x)/x$ becomes zero. Consequently,

$$h(t_1) = 0 \qquad \text{for } t_1 - t_0 = \frac{n}{T} = \frac{n}{2B_L} \tag{13-33}$$

where $n = 1, 2, 3, \ldots$. If we now consider the transmitter signaling with a continuous sequence of impulses (symbols), each one centered at a time determined by the values of $n = 0, 1, 2, \ldots$, we will have a signaling rate $1/T_S$, where

$$T_S = \frac{1}{2B_L} \tag{13-34}$$

There will be no intersymbol interference if the detector at the receiver end is also sampling at exactly the same rate and at the zero-crossing times $t_1 = n/2B_L$. The interval $T_S = 1/2B_L$, known as the "Nyquist interval," corresponds to a signaling rate twice the cutoff frequency B_L of the ideal low-pass filter.

In practice, an ideal low-pass filter is not physically realizable. In addition, if the receiver detector samplings are not performed at the exact zero-crossing times t_1, the result will be intersymbol interference. This interference will be caused by the sum of all the $(\sin x)/x$ waveform tails at the sampling time $t_1 + \Delta t$. It can be proved that this sum does not converge, and as a result, the interference could be substantial.

Low-Pass Filter with Gradual Roll-Off

Figure 13-4*a* shows the transfer characteristic, marked *b*, of an ideal low-pass filter at the output of a transmitter. The impulse response of this filter is marked in Fig. 13-4*b* by the same letter *b*. If we design the low-pass filter with a more gradual cutoff, the oscillatory nature of the tails of this response is reduced. In addition, the realization of such a filter becomes more practical.

The roll-off depicted by curve *a* in Fig. 13-4*a* has been synthesized by utilizing an upper part and a lower part that have a certain symmetry. Specifically, the upper part is identical with the dashed curve marked *a′*. This dashed curve has an odd symmetry, with the lower part of curve *a* about the cutoff frequency f_1. Nyquist has shown that when such a symmetry exists, the zero-crossing points

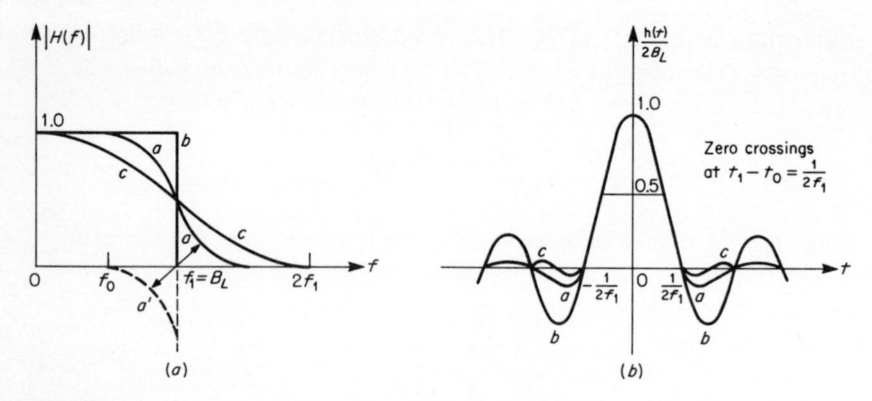

FIGURE 13-4 **(*a*) Symmetrical gradual roll-off in a low-pass filter. (*b*) Impulse responses to gradual roll-off.**

at $t_1 = n/2B_L$ of the impulse response are retained. However, the amplitude of the oscillations is substantially reduced. If we consider the response of this filter (called the "Nyquist filter") to a train of impulses, the sum of the overlapping tails forms a converging series. As a result, an error by the receiver detector in the sampling time is much more tolerable, in contrast with error when an ideal filter is used.

A sinusoidal function is often used to form the roll-off designated by curve *a*.

The transfer characteristic of a filter with sinusoidal roll-off is given by the expression

$$H(f) = \begin{cases} 1 & \text{for } 0 < f < f_0 \\ \frac{1}{2}\left[1 + \cos\dfrac{\pi(f - f_0)}{2(f_1 - f_0)}\right] & \text{for } f_0 < f < 2f_1 - f_0 \\ 0 & \text{for } f > 2f_1 - f_0 \end{cases} \tag{13-35}$$

where f_0 is the frequency at which roll-off starts and $f_1 = B_L$. The factor $p = 1 - f_0/f_1$ is called the "roll-off factor." For $p = 0$ we have $f_1 = f_0$; that is, we have ideal low-pass-filter behavior. For $p = 1$ we have $f_0 = 0$. This is depicted in Fig. 13-4*a* by curve *c*. The transfer characteristic for $f_0 = 0$ $(p = 1)$ is

$$H_1(f) = \begin{cases} \frac{1}{2}\left(1 + \cos\dfrac{\pi f}{2f_1}\right) & \text{for } 0 < f < 2f_1 \\ 0 & \text{for } f > 2f_1 \end{cases} \tag{13-36}$$

The impulse response $h(t)$ of the filter represented by the transfer function in Eq. (13-35) is shown in Fig. 13-4*b* as plot *a* and is given by the relationship

$$h(t) = 2B_L\,(\text{sinc}\ 2f_1t)\,\frac{\cos 2\pi p f_1 t}{1 - 16p^2f_1^2t^2} \qquad (13\text{-}37)$$

where the sinc function is defined as sinc $x = (\sin \pi x)/\pi x$ and, for simplicity, $t - t_0$ has been replaced by t. The filter represented by Eq. (13-36) is called the raised-cosine roll-off filter. The impulse response of this filter with $p = 1$ will be

$$h_1(t) = h(t)_{p=1} = 2B_L\,\frac{\text{sinc}\ 4f_1t}{1 - 16f_1^2t^2} \qquad (13\text{-}38)$$

The response $h_1(t)$, shown in Fig. 13-4*b* as curve *c*, has additional zeros midway between the zeros of the ideal low-pass-filter response. In addition, $h_1(t)$ has half the maximum amplitude for $t = 1/4f_1$, that is, halfway between 0 and the first zero crossing. This property could be used by a slicer in producing a pulse resembling a square pulse. Consequently, the raised-cosine filter is a good answer to the problem. The penalty is that it extends its zero cutoff point to $2f_1$.

In practice, for digital signaling, we use rectangular pulses instead of impulses. A rectangular pulse has a frequency spectrum of the $(\sin x)/x$ type. In order to compensate for this, when a Nyquist filter with gradual roll-off is used, the transfer characteristic of the filter is equalized; i.e., it is multiplied by $x/(\sin x)$. The aperture equalization by $x/(\sin x)$ will provide a filter which will respond to a square pulse in the same way the Nyquist filter will respond to an impulse. This concept could be generalized for a channel with several filters in series, such as transmitting filters and receiving filters. If the pulse $p(t)$ is transmitted through a series of filters $H_1(f)$, $H_2(f)$, . . . , $H_n(f)$, we form the product

$$P_o(f) = P(f)H_1(f)H_2(f) \cdots H_n(f)$$

where $P(f)$ is the spectrum of $p(t)$. We then select the equalized functions $H_1(f)$, $H_2(f)$, . . . , $H_n(f)$ in such a way that $P_o(f)$ has the desired spectrum. For example, the response $p_o(t)$ has zeros equally spaced, with period $T_S = 1/2B_L$, where B_L is the baseband width of the transmission channel.

Nyquist described three criteria for the transmission of digital pulses:

1. Signaling without intersymbol interference is possible at a rate $1/2f_1$.
2. Intersymbol interference can be removed halfway between adjacent impulses by a transmittance function of the form

$$H(f) = \begin{cases} \cos \dfrac{\pi}{2}\dfrac{f}{f_1} & \text{for } 0 < f < f_1 \\ 0 & \text{for } f > f_1 \end{cases}$$

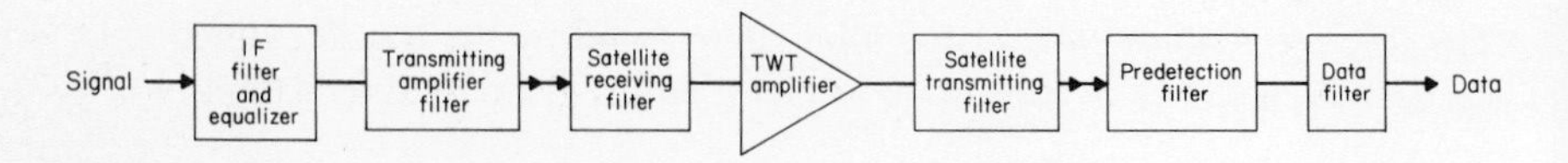

FIGURE 13-5 Filtering sequence in a satellite channel (block diagram).

3. The area of the response to an impulse will be zero for every signaling interval except its own when the transmittance function is of the form

$$H(f) = \begin{cases} \dfrac{(\pi/2f_1)f}{\sin(\pi/2f_1)f} & \text{for } 0 < f < f_1 \\ 0 & \text{for } f > f_1 \end{cases}$$

Criteria 1 and 2 can be satisfied simultaneously by the raised-cosine function.

Filtering Sequence of the Digital Channel

The block diagram in Fig. 13-5 depicts the sequence of filtering in a satellite channel. The IF or RF bandwidth B is usually twice the baseband width B_L because of the imaging effect when a signal is translated from baseband to IF.

If T is the width of a symbol pulse, the signaling rate R_T will be $1/T$. If we signal with a Nyquist rate $1/2B_L = 1/B$, we have

$$TB = 1 \qquad R_T = B \tag{13-39}$$

The bandwidth of a Nyquist filter placed in the IF section must be at least B, where B is twice the baseband width B_L. It has been shown that for a nonlinear channel, such as the one in Fig. 13-5, improved performance is obtained with

$$1 < BT < 1.2 \tag{13-40}$$

Consequently, the signaling rate should be

$$B > R_T > \frac{B}{1.2} \tag{13-41}$$

The filtering arrangement is described in Chap. 14 under "The Digital Data Channel."

REFERENCES

Bennett, W. R., and J. R. Davey: *Data Transmission*, McGraw-Hill, New York, 1965.

Devieux, C., and M. E. Jones: "A Practical Optimization Approach for QPSK/TDMA Satellite Channel Filtering," *IEEE Transactions on Communications*, vol. COM-29, no. 5, 1981.

Members of the Technical Staff of Bell Telephone Laboratories: *Transmission Systems for Communications*, rev. 4th ed., Bell Telephone Laboratories, Winston Salem, N.C., 1971.

Shimbo, Osamu: "Effects of Intermodulation, AM-PM Conversion and Additive Noise in Multicarrier TWT Systems," *Proceedings of the IEEE*, vol. 59, February 1971.

Spilker, J. J., Jr.: *Digital Communications by Satellite*, Prentice-Hall, Englewood Cliffs, N.J., 1977.

Sunde, E. D.: "Intermodulation Distortion in Multi-Carrier FM Systems," *1965 IEEE International Convention Record*, pt. 2, 1965.

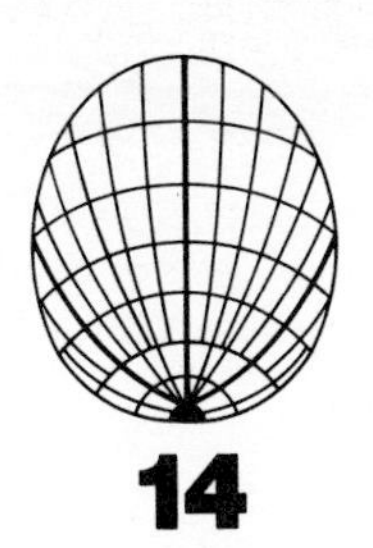

14

NETWORKS AND SYSTEMS

A satellite communications network usually includes a number of earth stations communicating with each other via the satellite channels, referred to as "transponders." Transponders channelize the satellite capacity both in frequency and in power. A transponder may be accessed by one or by several carriers, but since it also exhibits strong nonlinear characteristics, multicarrier operation—unless properly balanced—may result in unacceptable interference. In addition, since each uplink requires expensive equipment, it is usually advantageous to combine several messages in one carrier. In this chapter, we will examine techniques used in satisfying the network requirements described above, as well as the requirements imposed by the specific nature of various services such as voice, television, and data. Some of the system requirements stem from the geometry of the network as defined by the various nodes and the capacity allocation between nodes.

In summary, this chapter covers the following

- General techniques useful in solving network systems problems
- The handling of the various types of services

MULTIPLEXING

Multiplexing techniques are used for the transmission of a large number of messages over the same carrier. For example, a single 36-MHz satellite transponder can be used for approximately 1200 one-way simultaneous telephone messages on a single carrier. The two most commonly used multiplexing techniques are frequency-division multiplexing (FDM) and time-division multiplexing (TDM). The multiplexed messages are assumed to be band-limited. For example, a voice message occupies frequencies from 200 to 3300 Hz. Such a message in an FDM system is allocated a 4000-MHz channel, which includes a "guard band" of frequencies so that proper frequency separation between messages is ensured.

Frequency-Division Multiplexing

In an FDM system, each message, with the aid of filters and the appropriate product modulator, is translated in frequency to the band within the channel that is specifically allocated for the message. Usually a single-sideband (SSB) modulation process is used in which only the lower part of the message spectrum is employed. Quite often, the ring modulator depicted in Fig. 10-4 is used for this purpose.

Figure 14-1*a* depicts the original spectrum of a message limited to f_m, and the spectrum of the same message at the output of a product modulator. Figure 14-1*b* is the lower sideband of this spectrum, isolated for transmission in a multiplexed channel. At the receiving end, the demodulator must utilize subcarriers with accurate frequency and phase for the demodulation process. Figure 14-2 shows the block diagram of an FDM channel.

In telephony, a single voice message is multiplexed to a 4-kHz SSB channel.

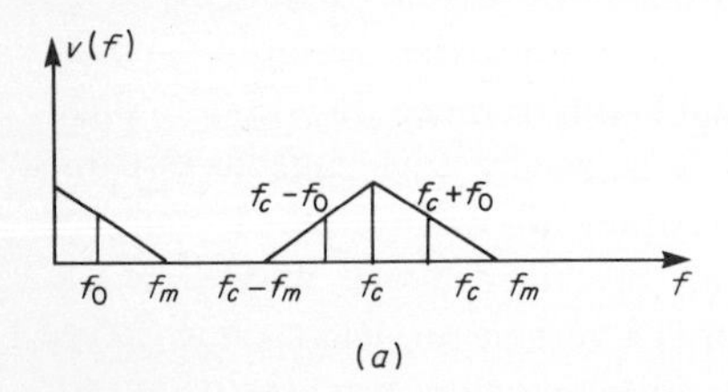

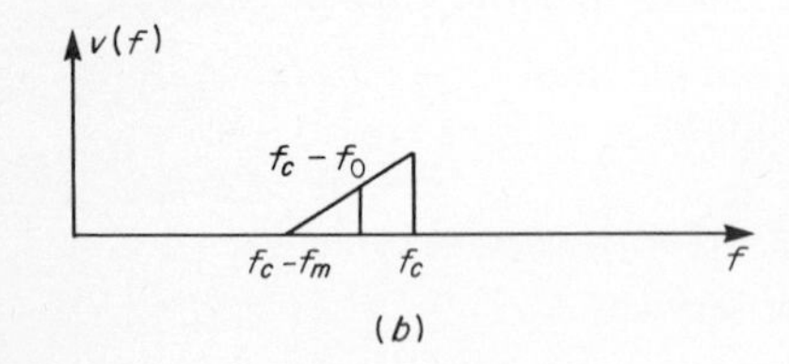

FIGURE 14-1 Frequency spectrum of SSB transmission of a multiplexed channel. (*a*) Spectrum of message and DSBSC modulated signal. (*b*) Spectrum of SSB signal.

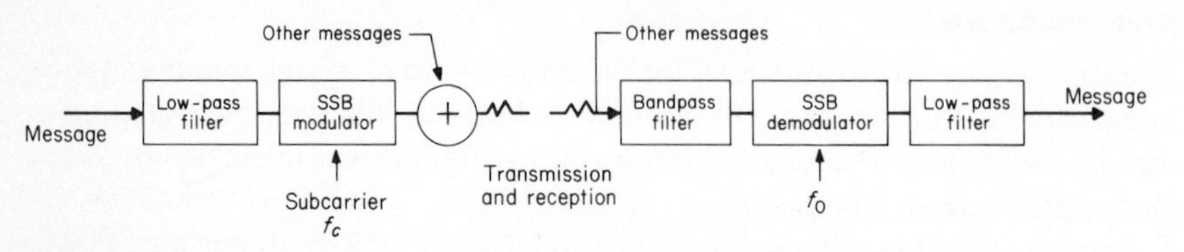

FIGURE 14-2 Frequency-division multiplexing and demultiplexing.

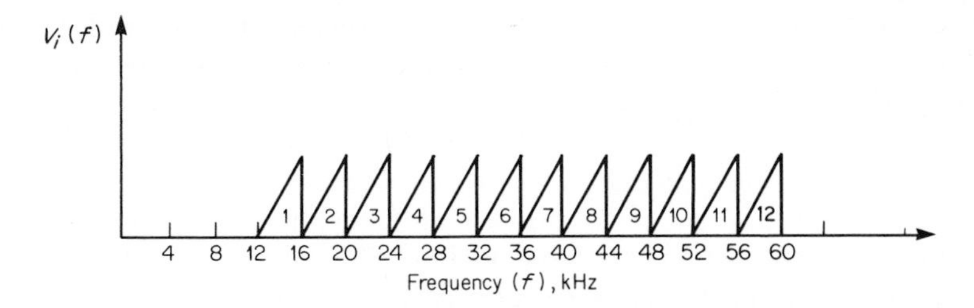

FIGURE 14-3 Group of 12 FDM voice channels.

Twelve of these channels make a group. In a second level of multiplexing, five of these groups are multiplexed to a supergroup of 60 channels. Ten supergroups multiplexed together result in a mastergroup of 600 voice channels. Figure 14-3 represents a typical frequency assignment of 12 multiplexed voice channels.

A group occupies 48 kHz of bandwidth, including the frequency guards. A supergroup occupies a $5 \times 48 = 240$-kHz bandwidth, and a mastergroup occupies a band of 2520 kHz.

Time-Division Multiplexing

Whereas FDM is used mainly for analog transmission, TDM is used for digital transmission. In TDM, a message is sampled and the sampled value is transmitted in a time slot; the next time slot is occupied by a sample of the next message, and so on. When all messages have been sampled and time-slotted sequentially, the system cycles to the first message and the process is repeated. Of course, the sampling rate for each message must be high enough to satisfy the criteria of the sampling theorem, that is, $T \leq 1/2f_m$, where T is the time between samples of a single message. Each sample must have a duration much shorter than T, so that T is long enough for all the messages to be sampled once. The bandwidth requirements of the RF channel increase in proportion to the number of multiplexed channels; if a single message requires a bandwidth B, the multiplexed sample will require bandwidth nB, where n is the number of messages. The increased bandwidth is offset by the noise performance advantage of digital transmission. Again, guard-time slots must be allowed. At the receiving end, by proper gating, each channel receives the sample pulses corresponding only to that channel. This implies adequate synchronization capability. Usually, the stream of transmitted pulses includes timing pulses so that the receiving end can update its clock continually. Again, proper low-pass filtering is necessary in both ends.

Every sample of each message corresponds to a number of code-modulated pulses and is transmitted in the allocated time slot. If we need to multiplex channels having nonidentical pulse rates, extra pulses can be stuffed in the slowest channel to equalize the pulse rates. These stuffed pulses must be subtracted at the receiving end.

The hierarchy of multiplexing for the U.S. public network has been standardized at certain transmission rates. Twenty-four digitized voice channels utilizing 7-bit pulse-code modulation (PCM) are multiplexed for transmission over a 1.544-Mb/s line. Each channel is rated at 64 kb/s. The carrier system for this 1.544-Mb/s transmission is known as the T1 carrier.

Presently this system is updated with channel banks which encode 96 voice channels into 8-bit binary PCM, resulting in a rate that is approximately 4 times the T1 rate. Four T1 pulse streams are combined in a T2 line operating at 6.3 Mb/s. Pulse stuffing is necessary, since the T2 rate is slightly higher than 4 times the old T1 rate. Seven T2 lines are multiplexed to a 46.3-Mb/s line, the T3 pulse-stream rate.

Commercial color television is sampled and encoded into 9-bit words, resulting in 92.6-Mb/s pulse streams.

With the appropriate compression techniques, a voice channel could be compressed to require 32 kb/s or an even smaller data rate, instead of 64 kb/s. The two most frequently used compression techniques are near-instantaneous companding (NIC) and variable-slope delta modulation (VSDM).

Digital voice via satellite is usually transmitted at a rate of 32 kb/s per voice channel. Adaptive PCM techniques of 32 kb/s are under consideration.

Baseband Noise

The baseband signal is the signal used to modulate the RF carrier and occupies the baseband channel. The basic thermal noise of a baseband channel, such as a multiplexed channel, is called "idle noise." To measure noise, a noise generator is connected to the input of the channel, providing the signal over the channel bandwidth. Then the overall power P_T in the channel will be

$$P_T = P_S + P_{IN} + P_{IM}$$

where

P_S = signal power
P_{IN} = idle noise
P_{IM} = intermodulation noise

Now, if a selective filter with a narrow passband is introduced between the signal (noise) generator and the receiver, the only remaining power in the channel within this narrow band will be the sum $P_R = P_{IN} + P_{IM}$ of the idle noise and the intermodulation noise. The ratio P_R/P_T is called the "noise power ratio" (NPR).

When the signal (noise) generator is removed and the channel is terminated in its characteristic impedance, all noise is removed except for the idle noise P_{IN}. The ratio P_{IN}/P_T is called the "baseband intrinsic noise ratio" (BINR).

SATELLITE MULTIPLE ACCESS

The Requirements for Multiple Access

In general, the various earth stations accessing a satellite are not similar to each other. Their size, capacity, and frequency of operation depend on the specific requirements of the network node which they serve. A single earth station serving a network node may be accessing one or more transponders, or even a fraction of a transponder. It may utilize one carrier per transponder or multiple carriers per transponder. Correspondingly, it may receive only one carrier or several carriers. As a result, each satellite transponder may be accessed by only one carrier or by several carriers. But because each transponder is a nonlinear repeater with limited power and bandwidth, a serious network problem exists if the system is to operate optimally. In addition, the problem can be further complicated if the communications requirements of each node change dynamically, that is, from second to second. Certain networks have been designed with a demand-assignment multiple-access (DAMA) capability. These networks are able to instantaneously reconfigure on demand the access of each node to satellite transponders as well as the connectivity arrangements with other nodes. To satisfy these access requirements, a number of modulation techniques have been developed. All these techniques have been significantly influenced by the nonlinearity of the transponder as well as by the nature of the baseband signals and their characteristics. The most frequently used techniques are the following:

Frequency-division multiple access (FDMA): FDMA is characterized by the allocation of a certain frequency band for access. This frequency band may be only a fraction of the frequency band of a transponder, or it may occupy a whole transponder. This access is continuous with time, with the allocated frequency band assigned permanently.

Time-division multiple access (TDMA): TDMA is characterized by the allocation of a certain time slot for access. Each carrier or node occupies a different time slot. The carriers use the transponder sequentially; when each has had a turn to use it, the use cycle begins over again. As a result, the entire satellite-channel frequency spectrum and the power are utilized by only one carrier at any given time. Usually this satellite channel is a whole transponder, but it need not be.

Code-division multiple access (CDMA): CDMA has two major subdivisions, spread-spectrum multiple access (SSMA) and pulse-address multiple access (PAMA). The CDMA method allocates separate codes to each user;

SSMA utilizes angle-modulation coding, and PAMA utilizes amplitude-modulation coding. CDMA can be characterized as a random-access technique; FDMA and TDMA are controlled-access techniques.

FREQUENCY-DIVISION MULTIPLE ACCESS

System Considerations for FDMA Networks

The major advantage of the FDMA technique is its simplicity of implementation. Each access is preassigned and therefore permanent, each is continuous in time, each requires no synchronization or central timing, and each is almost independent of all other accesses. However, if more than one FDMA carrier accesses the same transponder, the transponder nonlinearities dictate a back-off (discussed in Chap. 5) and therefore reduce the total transponder capacity. The back-off of the operating point toward the linear portion is necessary to avoid excessive intermodulation products. Each carrier requires individual power-level control if one is to achieve appropriate power allocation per carrier.

In summary, the transponder capacity is a function of the number of carriers per transponder. In addition, each carrier may be suppressed and interfered with by intermodulation products. To improve multicarrier operation, hard-limiting transponders could be used. A hard limiter has a voltage output-input characteristic like the one shown in Fig. 13-1 except that the slope of the linear portion is 90°, that is, the output is always saturated. Usually a frequency-division-multiplexed signal is frequency-modulated and then becomes a part of an FDMA network. Such a transmission is called FDM-FM-FDMA transmission. FDMA most frequently is used with analog multiplexing and modulation techniques. However, a special case of FDMA, single-channel-per-carrier (SCPC) transmission, to be discussed later, has been extensively used with digital signals.

Figure 14-4 shows a 36-MHz transponder frequency assignment with four FDMA carriers. The receiving earth stations separate the carriers by utilizing the appropriate filters.

In deciding on the frequency assignments for each carrier, one must consider three main constraints:

1. Total useful bandwidth per transponder
2. Available power and uplink power control capability
3. Interference

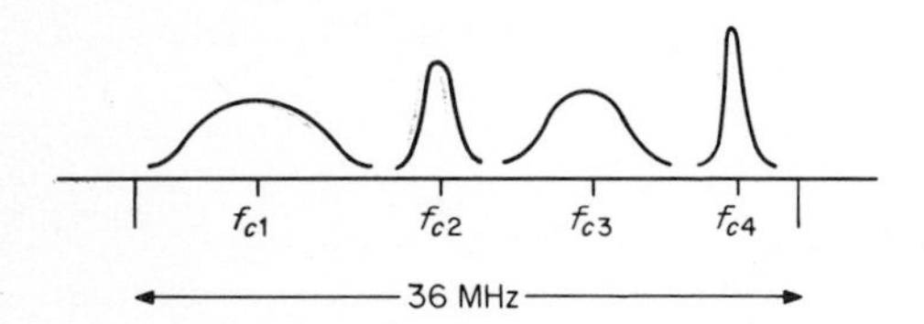

FIGURE 14-4 Frequency spectrum of a 36-MHz transponder with four FDMA carriers.

We have two types of interference, self-induced interference due to the system nonlinearities and interference induced by adjacent channels.

Self-Induced Interference Due to System Nonlinearities The main source of nonlinearities is the power amplifier and the in-band intermodulation products it causes. Both phase and amplitude nonlinearities cause this type of interference. Interference causing intelligible cross talk is important in voice communications. It is distinguished from the general type of interference because it becomes extremely disturbing; hearing extraneous talk while trying to conduct a two-way conversation is very disconcerting. This type of interference in angle-modulated carriers is introduced by AM-PM conversion.

Adjacent-Channel Interference Certain types of adjacent interference are also caused by adjacent transponders operating in a nonlinear mode. However, there are other causes, and in general, one can list the following types of interference:

1. Frequency-spectrum overlap between adjacent carriers in the same transponder
2. Intermodulation products due to an adjacent transponder spilling over these products
3. Interference due to cross-polarization imperfections
4. Other emissions due to satellites and earth stations
5. Dual-path distortion caused by recombining signals that have followed alternate paths to arrive at the receiving antenna

TWT Operating Point and Back-off

In Chap. 7, we derived the communications link equations considering the carrier-to-noise ratio $(C/N)_U$ for the uplink, the carrier-to-noise ratio $(C/N)_D$ for the downlink, and the carrier-to-intermodulation ratio $(C/N)_I$. This last ratio $(C/N)_I$ is either computed with the aid of computer programs or measured, and it is then introduced in the link analysis. In a few simple cases, a longhand computation of $(C/N)_I$ is feasible.

The total link carrier-to-noise ratio $(C/N)_T$ was derived as

$$\left(\frac{C}{N}\right)_T^{-1} = \left(\frac{C}{N}\right)_U^{-1} + \left(\frac{C}{N}\right)_D^{-1} + \left(\frac{C}{N}\right)_I^{-1} \tag{14-1}$$

The uplink carrier power must be properly controlled when a multicarrier operation is desired. Let us call ϕ_S the flux density that just saturates the satellite

TWT amplifier, measured at the satellite receiving antenna. This flux density, called "saturation flux density," is measured at the center of the beam of a single carrier in decibels, referred to 1 W, per square meter (dBW/m²). In Chap. 7, Eq. (7-4) expressed, in general, the flux density at the satellite receiving antenna as

$$\phi = \frac{G_{TE} P_{TE}}{4\pi R_U^2}$$

Consequently, if P_{TES} is the transmitting-earth-station power resulting in a saturation flux density ϕ_S, we will have

$$P_{TES} = L_U \frac{4\pi R_U^2}{G_{TE}} \phi_S \tag{14-2}$$

The L_U represents a factor necessary to compensate for the transmission-medium losses. By introducing Eq. (14-2) into Eq. (7-8) for $(C/n_0)_U$, we will get the saturation $(C/n_0)_{US}$ as

$$\left(\frac{C}{n_0}\right)_{US} = \phi_S \frac{G_{RS}}{T_{RS}} \frac{\lambda^2}{4\pi} \frac{1}{k} \tag{14-3}$$

Now, if from Fig. 5-5 we select an operating point with a back-off BO_i relative to the single-carrier saturation, the unsaturated uplink $(C/N)_U$ can be derived from Eq. (14-3) as

$$\left(\frac{C}{n_0}\right)_U = \frac{\phi_S}{BO_i} \left(\frac{G}{T}\right)_S \frac{\lambda^2}{4\pi} \frac{1}{k} \tag{14-4}$$

The same relationship expressed in decibels will be

$$\left(\frac{C}{n_0}\right)_U = \phi_S + \left(\frac{G}{T}\right)_S - 10 \log \frac{4\pi}{\lambda^2} - 10 \log k - BO_i \tag{14-5}$$

where

ϕ_S = saturation flux density, dBW/m²

$\left(\frac{G}{T}\right)_S$ = satellite receiving gain–to–noise temperature ratio, dB/K

λ = wavelength, m

k = Boltzmann's constant ($10 \log k = -228.6$ dBW/K·Hz)

BO_i = TWT input back-off, dB relative to single-carrier saturation power

The transmitting earth-station power should be

$$P_{TE} = P_{TES} - BO_i \qquad \text{dB} \tag{14-6}$$

If BO_o is the output back-off relative to satellite EIRP, then by using Eq. (7-9) for the downlink, we obtain

$$\left(\frac{C}{n_0}\right)_D = \text{EIRP}_{\text{sat}} - BO_o + \left(\frac{G}{T}\right)_E + 20 \log \frac{\lambda}{4\pi R_D} - 10 \log L_D - 10 \log k \tag{14-7}$$

where

EIRP_{sat} = satellite EIRP at beam center for single-carrier saturation
BO_o = TWT output back-off relative to single-carrier saturation
$\left(\frac{G}{T}\right)_E$ = gain–to–noise temperature ratio for the receiving earth station
λ = wavelength, m
R_D = distance from satellite to earth station, m
L_D = medium loss (power ratio)
k = Boltzmann's constant

Figure 14-5 shows the plots of the three C/n_0 ratios of Eq. (14-1) as a function of the back-off BO_i. The total $(C/n_0)_T$ is also plotted, and the optimum operating point which maximizes the $(C/n_0)_T$ is determined. When we have a multicarrier operation, this optimization process can be applied by utilizing for the

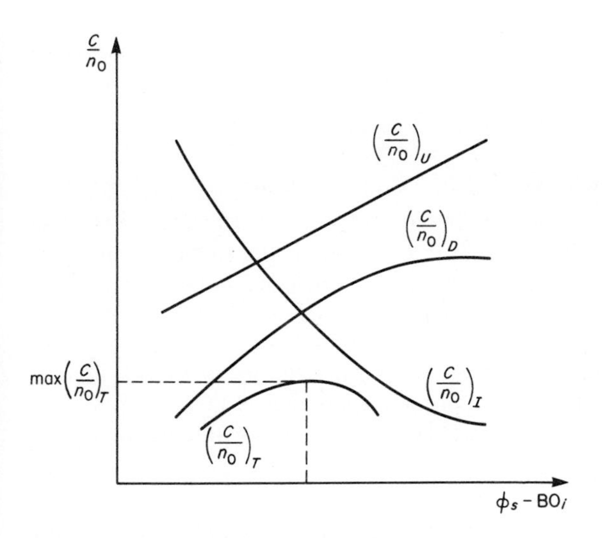

FIGURE 14-5 FDMA transponder: operating-point optimization.

uplink and the downlink the C/n_0 resulting from the addition of the various carriers. We add C/N's by using the general relationship

$$\left(\frac{C}{n_0}\right)_T^{-1} = \sum_i \left(\frac{C}{n_0}\right)_i^{-1} \tag{14-8}$$

When the intermodulation $(C/n_0)_I$ is computed, both the amplitude nonlinearity and the AM-PM conversion must be combined. This means that the phase nonlinearity must also be included. In addition, while a large number of carriers could be treated in the analysis as gaussian noise, a small number of carriers—say three or four—must be analyzed as individual interfering carriers.

Carriers located near the edge of the transponder bandwidth may also experience an additional degradation due to the gain slope at each transponder end. A group-delay distortion will be introduced, and as a result, the end channels must be designed to accommodate this additional interference.

When adjacent carriers are not equal, they may interfere with each other. When the interfering carrier is of a lesser amplitude, we have what is termed "convolution noise." When the desired carrier is greater than the interfering carrier, we have "Impulse Noise." An appropriate guard band should be allowed in order to bring impulse and convolution noise within acceptable limits. A typical 36-MHz satellite transponder, with 5-W TWT amplifier, can provide 1200 voice-grade telephone channels when it operates with 10-m uncooled earth stations and a single carrier. If the number of FDMA carriers increases to two, because of the back-off BO_i, the number of voice-grade telephone channels reduces to 600. If the number of carriers increases to 4, the number of telephone channels decreases to 450. This reduction in capacity in a multicarrier operation is the major disadvantage of an FDMA system.

SINGLE CHANNEL PER CARRIER

FDMA modulation can be utilized with a carrier that carries only one channel with digital information. This technique is known as the single-channel-per-carrier (SCPC) technique. For example, a voice channel can be digitized and encoded utilizing 8-bit PCM. With a bit rate of 8 kb/s, we obtain a 64-kb/s rate per voice channel.

Each earth station is given a frequency assignment, or an FDMA carrier, upon request. A common channel is utilized to keep track of the occupied channels and the requests. Since it is very unlikely that all users will request channels simultaneously, and since at least the return channel of a two-way conversation is empty most of the time, one can obtain very good utilization of the transponder capacity by also using the idle slots. This technique was first utilized by the Intelsat system. A 45-kHz RF band was assigned per voice channel, including a 7-kHz guard band. Approximately 800 such channels were derived from a 36-MHz transponder. A common TDMA, 128-kb/s channel was utilized by Intelsat IV as order wire. In addition, each channel was voice-activated. The number of one-way voice channels—that is, 800—compared very favorably with a normal FDM-FM-FDMA multicarrier use of the same transponder.

SCPC transmission can also be used very effectively to transmit data at rates as high as 56 kb/s or even several megabits per second. Usually a QPSK signaling scheme with forward error correction is used for this kind of application.

The system design for SCPC transmission with analog modulation is identical with the process described in the previous section for FDMA systems. In the case of digital transmission, the equation used for system design is Eq. (11-78). Restating Eq. (11-78), we obtain

$$\frac{C}{N} = \frac{E}{n_0}\frac{R}{B} \qquad \text{or} \qquad \frac{C}{n_0} = R\frac{E}{n_0} \tag{14-9}$$

where

$\dfrac{E}{n_0}$ = ratio of symbol pulse energy to noise density; when forward error correction is used, $E/n_0 = (k/n)(E_b/n_0)$

R = signaling rate

B = RF bandwidth

For a desired performance, that is, for a given bit error rate probability P_e, the required E/n_0 is obtained from the equation for the digital-modulation scheme under consideration. For example, in the case of BPSK, $P_e = \frac{1}{2}\,\text{erfc}\,\sqrt{E/n_0}$. In the case of power-limited transmission, a certain threshold carrier-to-noise ratio exists,* and by assuming a design margin M in decibels we obtain

$$\frac{E}{n_0} + 10 \log R = \left(\frac{C}{n_0}\right)_{\text{threshold}} + M \tag{14-10}$$

If an error-correcting code is used, the signaling rate is a function of the desired bit error rate R_b and the code structure. For an (n, k) block code,

$$R = \frac{n}{k} R_b$$

For a desired transmission rate R, the threshold carrier-to-noise ratio can be determined, and therefore trade-offs between required satellite EIRP, transponder back-off BO_i, and number of channels per transponder can be performed.

*The threshold carrier-to-noise ratio is computed by utilizing the available satellite downlink EIRP [see Eq.(14-18)].

TIME-DIVISION MULTIPLE ACCESS

In general terms, a TDMA system is a true orthogonal system, since only one carrier operates at a given time. As a result, the nonlinear nature of the transponder is minimized, no multicarrier interference is generated, and the full power of the transponder is available, as well as the full bandwidth. In a sense, this technique is an energy-efficient technique. TDMA, however, requires network synchronization, coded messages, and buffer storage from burst to burst. The message must be coded in order to provide address, timing, and other pertinent information. Sufficient guard time must be allowed from burst to burst so that any possible interference will be avoided. In order to accommodate these overhead functions, a certain loss of efficiency occurs. A TDMA system can attain better than 90 percent efficiency of satellite power utilization. In comparison, an FDMA system may lose 3 to 6 dB of the available power.

The frame rate, i.e., the time required to include in sequence all the bursting stations, constrains the data rate per burst. Each burst duration is a fraction of the frame duration. The maximum burst rate is also limited by the satellite EIRP and the earth-station G/T. All information, including voice, must be digitized, since TDMA works only with digital signals.

Figure 14-6 depicts the TDMA concept. The bursts of three earth stations combine to form a frame. The frame, consisting of bursts 1, 2, and 3, repeats itself after burst 3. To ensure that overlap between successive bursts does not occur, a guard time is allowed between bursts.

As an example, consider a 60-Mb/s system. A guard time of 200 ns will be adequate to assure that overlap does not take place. This 60-Mb/s system can

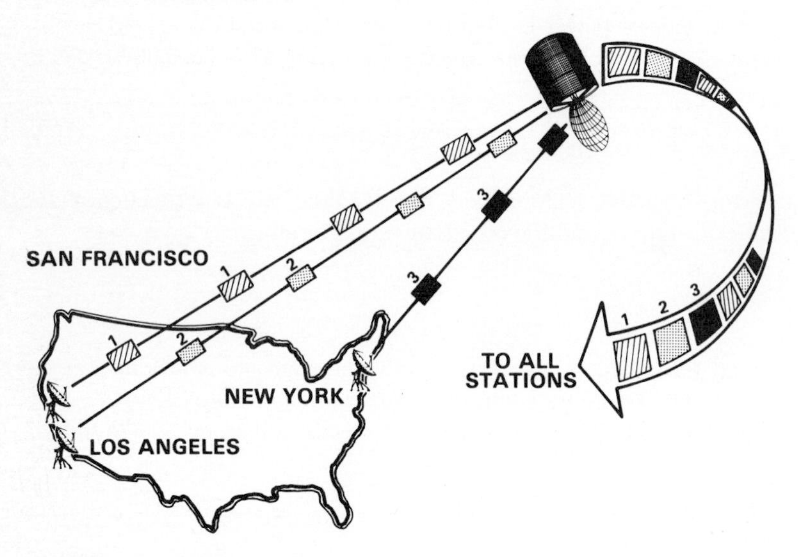

FIGURE 14-6 TDMA network.

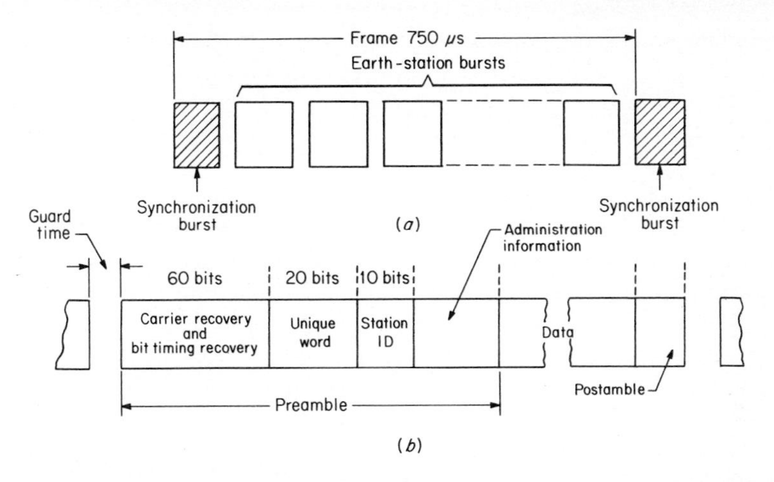

FIGURE 14-7 TDMA bursts. (*a*) TDMA format. (*b*) Earth-station burst organization.

typically accommodate 10 to 15 earth stations. As many as 50 earth stations could be accommodated, if necessary. A station, designated as the reference station, provides a control burst per frame, usually known as a synchronization burst. The other stations utilize this reference to place their bursts in their allocated time slots. A typical synchronization burst includes three pieces of information:

1. Carrier recovery (CR) and bit timing recovery (BTR), used by the demodulator for coherent detection
2. Unique word (UW), used to establish an accurate time reference for the received bursts and the location of each bit within the burst
3. Transmitting-station identifying code

Each earth-station transmission includes a preamble at the beginning. This preamble carries the same three messages as the synchronization pulse. After the data transmission, a "postamble" is transmitted to designate the end of the burst. The preamble may include other information, such as a specific address and an order-wire request. A typical earth-station burst is depicted in Fig. 14-7. The times and lengths of preamble words in Fig. 14-7 are taken from a specific 60-Mb/s system. Each system has different frame times and different preamble lengths. With four carriers per transponder, the 60-Mb/s system, with the type of frame shown in Fig. 14-7, derives 800 voice channels per 36-MHz transponder. As was previously discussed, under the same conditions the FDMA system would be capable of deriving only 450 voice channels.

Frame times may vary from a few hundred microseconds (as low as 100 μs) to several milliseconds. Low-speed data networks may even utilize frame times in the range of seconds. The number of the data bits in a frame, divided by the number of the total bits, including the overhead bits, determines the efficiency of the TDMA system. Efficiencies better than 95 percent have been achieved in certain special designs.

The most difficult problem in any TDMA system is synchronization. Each earth station must identify its time slot in the frame sequence, acquire it, and maintain accurate synchronization so that the station burst arrives at the satellite at the predetermined time. Slant range and satellite motion must be taken into account. For acquisition and synchronization, a variety of concepts have been implemented utilizing open-loop, closed-loop, or some kind of feedback control.

The data rate that can be supported by a single transponder is limited by either bandwidth or power. In bandwidth-limited transmission, the number of pulses (i.e., the number of symbols) that can be transmitted per second is limited by intersymbol interference. A rule of thumb is that a 1-MHz bandwidth can support $1/1.2 = 0.83$ "megasymbols" per second. Therefore a 36-MHz transponder could transmit approximately $36 \times 0.83 = 30$ megasymbols. For a QPSK system with 2 bits per symbol, this rate corresponds to 60 Mb/s.

In a power-limited case, the relationship

$$\frac{C}{n_0} = R\frac{E_b}{n_0}$$

must be used. For a desired error probability, E_b/n_0 can be defined. By using the link equations for the downlink, one can compute the attainable C/n_0 plus required margin from the available satellite EIRP. The range of the required margin is in the neighborhood of 2.5 dB. Consequently, the data rate R can be determined. If a transmission rate derived on the basis of bandwidth limitation is the same as the rate derived on the basis of power limitation, then this rate makes optimum use of the transponder.

One of the main advantages of a TDMA system is the ability to change the length of the bursting time allocated for each earth station. By changing the length of the bursting time, the system changes the capacity allocated per station and thus can easily respond to changing traffic demands. On the other hand, each earth station must have the ability to occupy the full transponder and also accept a full transponder transmission. As a result, even if the station represents a small user, it must have full EIRP and G/T capability. In addition, sophisticated synchronization and timing capability must be included in every TDMA earth station.

CODE-DIVISION MULTIPLE ACCESS

Systems employing CDMA transform the transmission of each station and spread it over the frequency and/or time axis by using a coded transformation.

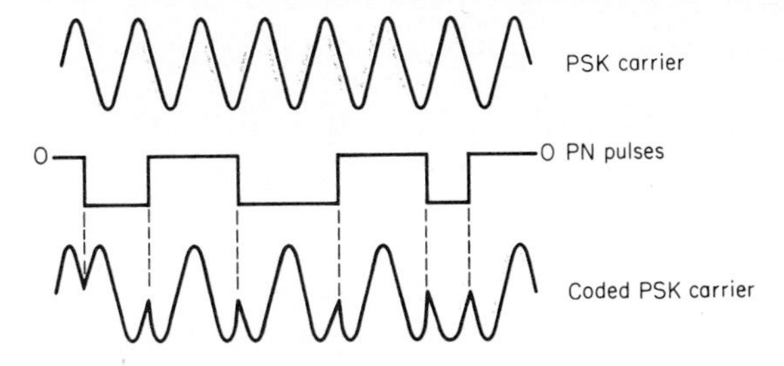

FIGURE 14-8 PN-coded multiple-access carrier.

A typical example of a CDMA system is one utilizing a pseudo-random noise (PN) code to transform (encode) a PSK signal. A PN code consists of a series of pulses generated in random by a special generator. The transmitted signal is multiplied by the PN pulses before transmission. Figure 14-8 shows the waveforms of an unmodulated PSK carrier before and after multiplication by a PN pulse train. The receiver, knowing the exact PN-code pulse train, can reverse the process and decode the message. Of course, the PSK carrier will also be modulated by the message.

A technique commonly used in CDMA is frequency hopping. The transmitter changes carrier frequency at random. The receiver, knowing the hopping code, follows this change.

A number of stations can operate at the same frequency band, utilizing orthogonal codes. As long as the codes are orthogonal, interference is avoided.

In order to avoid intentional jamming, the military has developed spread-spectrum techniques. Each user uses the whole transponder spectrum under a specially coded message. Again, as long as the coded messages over the same frequency band utilize orthogonal codes, separation of messages at the receiving end is possible. Several references in the bibliography at the end of the book cover the subject of CDMA in great depth.

VARIABLE-ACCESS DEMAND ASSIGNMENT

The multiple-access schemes described in the previous sections assume a fixed network in which the access requirements remain relatively constant. This type of access is called "scheduled variable multiple access." Scheduled variable multiple access is used when the traffic requirements are known in advance and the network configuration remains fixed for some period of time. However, if a rapid reassignment of resources is necessary, then a demand-assignment multiple-access (DAMA) technique may provide a more efficient use of the available resources.

To request channel space, networks with DAMA usually provide an order-wire channel for common use by the various users. The network takes advan-

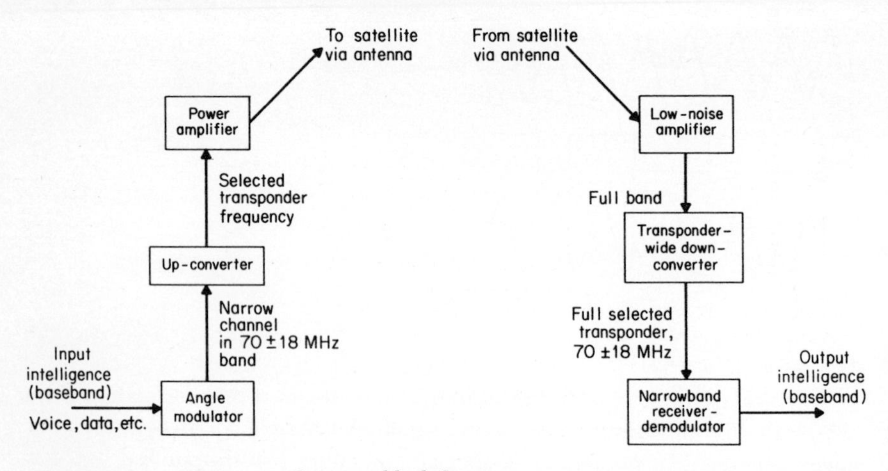

FIGURE 14-9 FDMA ground station (block diagram).

tage of the noncontinuous user requirement for capacity. It assigns capacity to a specific user only on demand. As a result, the network does not have to be sized to accommodate all users simultaneously. Such a simultaneous demand from every user is rare and probably never occurs. As a result, the system is designed for a reasonable "blocking" probability based on the statistics of the traffic. If a user receives a busy signal, the user will be queued for the first available opening. The result is that, for a desired grade of service, a substantial reduction of maximum capacity is possible. A master station usually provides

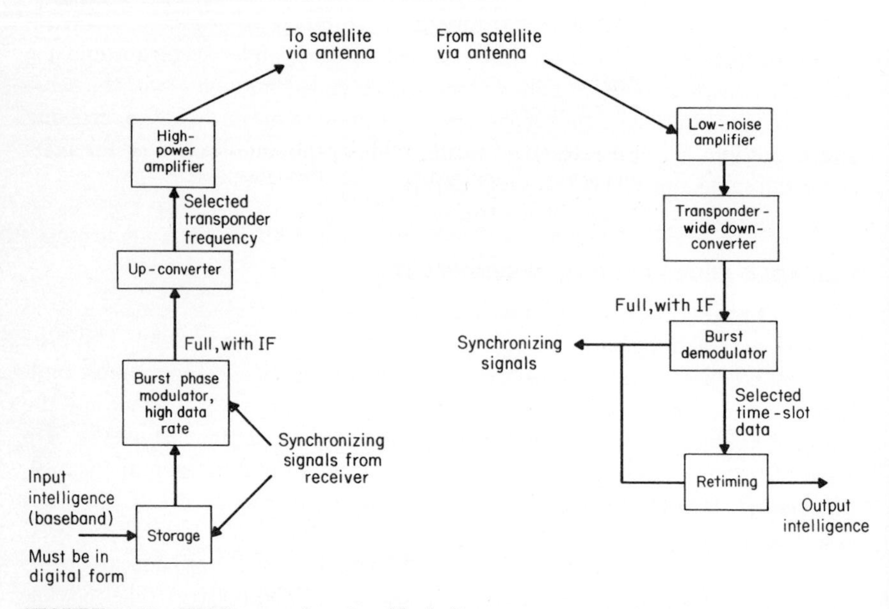

FIGURE 14-10 TDMA ground station (block diagram).

instant control over the network via the network facilities. DAMA systems work best with TDMA-operated networks. A typical example of a DAMA system is the Intelsat Spade system which has been in operation for several years. The Spade system uses a single channel per carrier concept. Each subscriber utilizes its channel approximately 40 percent of the time. Consequently, a substantially higher number of carriers—over 800—can be accommodated per transponder.

COMPARISON OF MULTIPLE-ACCESS TECHNIQUES

Figures 14-9 and 14-10 are block diagrams of typical FDMA and TDMA earth stations, respectively.

The major difference in the TDMA station is the incorporation of sophisticated equipment for burst generation and synchronization. This equipment is expensive and delicate. But of course, on the positive side, a TDMA system introduces substantial flexibility into the network. Table 14-1 tabulates the main advantages and disadvantages of the multiple-access techniques.

TABLE 14-1 ADVANTAGES AND DISADVANTAGES OF MULTIPLE-ACCESS TECHNIQUES

Access technique	Advantages	Disadvantages
FDMA	Channel availability is fixed. No central control is required. Users with various capacity needs are easily accommodated. Relatively unsophisticated earth stations.	Intermodulation requires back-off, reducing transponder throughput. Rigid system; reassigning resources to reflect traffic change is difficult.
TDMA	Transponder bandwidth and power fully utilized. No complicated frequency assignments. Users with different capacity requirements are easily accommodated. Flexible system: resources are easily reassigned.	Guard times and headers reduce throughput. Central timing and synchronization control are required. High-cost earth terminals. Even low-traffic users must have full EIRP and G/T terminals.
CDMA	No central control; fixed channels. All earth terminals interchangeable. Earth-station sophistication is at baseband. Relatively immune to external interference.	Transponder must be backed off; reduced throughput. Only a limited number of orthogonal codes exist. Works efficiently with preselected data rates.

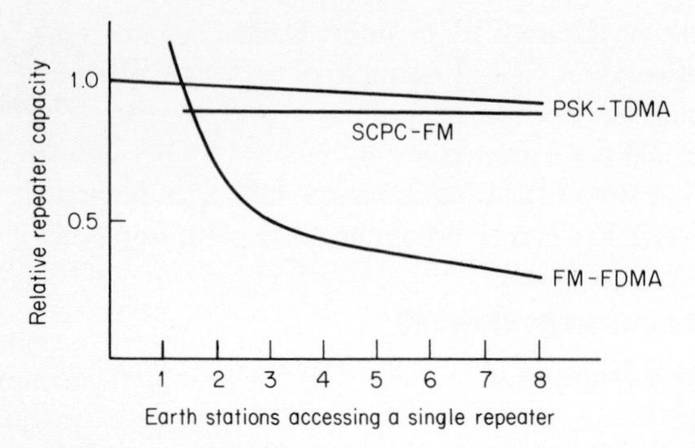

FIGURE 14-11 Comparison of multiple-access systems.

Figure 14-11 shows the relative capacity for the various accessing modes of a satellite transponder as a function of the number of the nodes accessing the transponder.

MULTIPLEXED VOICE

Standards of Performance

The quality of a single telephone channel is specified by CCITT standards. Two quantities are specified:

1. Signal level standardized at 1 mW at 1000 Hz as applied and received at a point of the user.
2. The total noise, in picowatts (10^{-12} W), referred to this signal level. The weighted mean noise power in any one hour should not exceed 10,000 pW, psophometrically weighted (pWp). In addition, the mean noise power over 1 min should not exceed 10,000 pWp by more than 20 percent in any one month.

Although higher levels of noise are allowed over small percentages of the time, 10,000 pW noise is used in the design of satellite channels. The noise budget is usually divided as follows:

Satellite link (uplink plus downlink and transponder intermodulation noise)	8000 pW
Earth station (intermodulation and equipment)	1000 pW
Interference (intermodulation and thermal)	1000 pW

Consequently, if the postdetection signal is 1 mW and the allowed satellite link noise level (before further degradation by the earth-station equipment) is 8000 pW, the postdetection signal-to-noise ratio should be at least

$$\left(\frac{S}{N}\right)_0 = \frac{1 \times 10^{-3}\ \mathrm{W}}{8000 \times 10^{-12}\ \mathrm{W}} = \frac{10^6}{8} \doteq 51\ \mathrm{dB}$$

If an FM detector with threshold extension is being used, the minimum allowed signal-to-noise ratio is 10 dB. In addition, a certain margin should be included in the link over and above this threshold. For example, in the case of a 24-voice channel, the practice is to allow at least a 2.5-dB margin. Consequently, the system should be designed for a threshold $C/N = 12.5$ dB.

The postdetection signal-to-noise ratio (51 dB) and the threshold carrier-to-noise ratio (12.5 dB) of the link are also connected through a relationship that depends on the modulation scheme. The parameters involved in this modulation-dependent relationship can now be evaluated if the above values are used for the threshold carrier-to-noise ratio and the postdetection signal-to-noise ratio.

Multiplexed Voice: FDM-FM-FDMA

The signal modulating the FM carrier, being the sum of all the speech channels, is random and very similar to a gaussian signal. As a result, the frequency deviation required to transmit this multichannel signal cannot be easily defined.

In practice, a single-frequency test tone (in the neighborhood of 1000 Hz) is used for carrying out the computations and evaluation measurements. The required peak frequency deviation ΔF_{mcp} for this multichannel (mc) voice transmission and the corresponding Carson bandwidth B are computed from empirical data.

Let us define the various parameters of the FM multichannel transmission as follows:

b = single-channel bandwidth for voice = 3.1 kHz ($b' = 4$ kHz, including guards)

n_v = the number of multiplexed voice channels

f_m = maximum baseband frequency $= n_v b + f_0$

f_0 = lowest frequency in baseband to be chosen arbitrarily (usually $f_0 = 12$ kHz for voice channels)

ΔF_r = rms deviation of the test-tone frequency

g = peak-to-rms ratio for the test-tone frequency deviation ($g = 3.16 \doteq 10$ dB for a large number of voice channels; for a small number of channels, g may be as high as $8.5 \doteq 18.6$ dB)

l = multichannel load capacity factor multiplying ΔF_r in order to adjust for the number of channels n_v

$$L = 20 \log l = \begin{cases} -15 + 10 \log n_v & \text{for } n_v \geq 240 \\ -1 + 4 \log n_v & \text{for } 12 \leq n_v < 240 \\ & \text{(empirical formula)} \end{cases}$$

ΔF_{mcp} = multichannel peak frequency deviation $= gl\,\Delta F_r$

B = RF bandwidth $= 2(\Delta F_{\text{mpc}} + f_m) = 2(gl\,\Delta F_r + f_m)$

B_S = satellite bandwidth
p = psophometric noise weighting factor = 1.78 ≐ 2.5 dB (see "Signal and Noise Transmission Levels in Transmission Systems," Chap. 6)
w = preemphasis improvement factor = 2.5 ≐ 4.0 dB (see "Preemphasis and Deemphasis," Chap. 10)

The postdetection signal-to-noise ratio $(S/N)_0$ per channel (the top channel is the worst) is given by the relationship

$$\left(\frac{S}{N}\right)_0 = \frac{C}{N}\frac{B}{b}\left(\frac{\Delta F_r}{f_m}\right)^2 pw \qquad (14\text{-}11)$$

As an example, assume the following system parameters

$$n_v = 24 \text{ voice channels}$$
$$f_0 = 12 \text{ kHz}$$
$$f_m = f_0 + n_v \times 4 \text{ kHz} = 12 \text{ kHz} + 24 \times 4 \text{ kHz} = 108 \text{ kHz}$$
$$l = \text{antilog}\,\frac{-1 + 4\log n_v}{20} = \text{antilog}\,\frac{-1 + 4\log 24}{20} = 1.68$$
$$g = \text{peak-to-rms ratio} = 10 \text{ dB} \doteq 20 \log 3.16$$

Therefore, Eq. (14-11) will become (units in decibels)

$$\left(\frac{S}{N}\right)_0 = \frac{C}{N} + 20\log\frac{\Delta F_r}{108} + 10\log\frac{2[\Delta F_r(3.16)(1.68) + 108 \times 10^3}{3.1 \times 10^3} + 2.5 + 4 \qquad (14\text{-}12)$$

The performance of the system requires:

$$\left(\frac{S}{N}\right)_0 = 51 \text{ dB}$$
$$\frac{C}{N} = 12.5 \text{ dB}$$

(see "Standards of Performance," this chapter). By substituting these values in Eq. (14-12) and solving for ΔF_r, we obtain

$$\Delta F_r = 164 \text{ kHz}$$

We also obtain

$$\Delta F_{\text{mc}} = l\,\Delta F_r = (1.68)(164 \text{ kHz}) = 275 \text{ kHz}$$
$$\Delta F_{\text{mcp}} = g\,\Delta F_{\text{mc}} = (3.16)(275 \text{ kHz}) = 870 \text{ kHz}$$

The radio bandwidth (Carson's bandwidth) will be

$$B = 2(\Delta F_{\text{mcp}} + f_m) = 2(870 \text{ kHz} + 108 \text{ kHz}) = 1.96 \text{ MHz}$$

The occupied satellite bandwidth is 1.96 MHz, but the required satellite bandwidth corresponds to $g = 13$ dB or a ratio of 4.47.

Therefore

$$B_S = 2[(4.47)(275 \text{ kHz}) + 108 \text{ kHz}] = 2.67 \text{ MHz}$$

THE FM-FDMA TELEVISION CHANNEL

The recommendation of the International Radio Consultative Committee (CCIR) is that the signal-to-noise ratio for television signals be expressed as

$$\frac{S_{pp}}{N} = \frac{\text{peak-to-peak luminance signal}}{\text{weighted rms noise}} \tag{14-13}$$

If we were to relate this ratio to the normal S/N, where S is the mean square of a sinusoid, we would obtain

$$\frac{S_{pp}}{N} = 4\left(\frac{S}{N}\right)$$

Indeed, for a sinusoid, the ratio of the mean square to the peak-to-peak value is 1:4.

Therefore, for an FM signal, we will have

$$\frac{S_{pp}}{N} = 4\left(\frac{S}{N}\right) = 4\left(\frac{3}{2}\right)\left(\frac{C}{N}\right)\left(\frac{\Delta F}{f_m}\right)^2 \frac{B}{f_m} \text{W}$$

$$= 6\left(\frac{C}{N}\right)\left(\frac{\Delta F}{f_m}\right)^2 \frac{B}{f_m} \text{W} \tag{14-14}$$

where

ΔF = peak frequency deviation
f_m = maximum baseband frequency in the video signal
B = radio bandwidth $= 2(\Delta F + f_m)$
$\frac{C}{N}$ = carrier-to-noise ratio
W = noise weighting improvement

If preemphasis-deemphasis is used, the deemphasis will modify the noise spectrum, and the combined effect of preemphasis and noise weighting will be

$$q = q(w, p)$$

where p is the preemphasis-deemphasis factor. This combined effect is provided in the CCIR standard as a function of the video frequency.

THE DIGITAL DATA CHANNEL

The design objective of a data transmission system is to transmit the maximum bit rate R_b through a transponder with the minimum possible bit error rate probability P_e. With the introduction of forward error correction, depending on the coding gain of the particular code, the required E_b/n_0 for a desired P_e could be substantially reduced. Table 12-1 shows that anywhere between 2 and 9 dB improvement is feasible with well-proven techniques. Of course, there is an additional cost associated with the introduction of encoding and decoding equipment. Consequently, the first step should be to examine if there is a digital-modulation scheme that could possibly result in the desired P_e for a reasonable value of E_b/n_0. The performance curves for various digital schemes such as PSK and FSK provide P_e as a function of E_b/n_0. Examination of these plots makes it immediately clear that some kind of coding will be required for $P_e > 10^{-4}$.

Once the appropriate coding scheme is selected (and this selection could be a reiterative process), the required E_b/n_0 can be easily determined. The per-symbol E/n_0 can be derived from the relationship

$$\frac{E}{n_0} = \left(\frac{k}{n}\right)\left(\frac{E_b}{n_0}\right) \tag{14-15}$$

where k and n are the typical parameters of an (n, k) code. Depending on the signaling scheme, a symbol may contain one or more transmitted bits. For example, a BPSK system transmits 1 bit per symbol whereas a QPSK system transmits 2 bits per symbol. It must be noted that these bits are not data-source bits, but instead are one of the n bits of the code word. Consequently, the signaling rate R_T is related to the data-source bit rate R_b by the equation

$$R_b = m\frac{k}{n}R_T \tag{14-16}$$

where m is the number of transmitted bits per symbol and k/n is the code efficiency rate. The link is operating at a transmission rate R_T. This rate may be constrained either by bandwidth or by power.

Power Constraint

Equation (14-15) provides the required E/n_0 for the desired bit error rate probability. In addition, as was shown in a previous chapter,

$$\frac{C}{N} = \frac{R_T}{B}\left(\frac{E}{n_0}\right) \qquad \text{or} \qquad \frac{C}{n_0} = R_T\left(\frac{E}{n_0}\right) \tag{14-17}$$

where B is the transmission bandwidth. C/N is related to the satellite transmitted power through the link equations derived in Chap. 7. Specifically, for the downlink,

$$\left(\frac{C}{n_0}\right)_D = \text{EIRP}_S - M - 10 \log \left(\frac{G}{T}\right)_R - P_L - 10 \log k \qquad (14\text{-}18)$$

where

EIRP_S = satellite EIRP for the channel under consideration, dB
M = required margin, dB
$\left(\frac{G}{T}\right)_R$ = figure of merit of receiving station
P_L = total path losses, dB
k = Boltzmann's constant

If there are also uplink or intermodulation noise limitations, the overall C/N can be computed by the general relationship

$$\left(\frac{C}{N}\right)^{-1} = \sum_i \left(\frac{C}{N}\right)_i^{-1} \qquad (14\text{-}19)$$

By constraining the EIRP_S of the channel, we constrain C/n_0, and therefore the transmission rate R_T will be constrained as indicated by Eq. (14-17).

Bandwidth Constraint

If B is the RF bandwidth allocated to the channel, the transmission rate R_T over this channel will also be constrained because of limitations imposed by the bandwidth. Design considerations such as intersymbol interference have resulted in the relationship

$$B \geq 1.2R_T \qquad \text{or} \qquad R_T \leq \frac{B}{1.2} \qquad (14\text{-}20)$$

In the discussion of TDMA, this relationship was used to derive a maximum transmission rate R_T of 30 Mb/s for each 36-MHz transponder.

If R_T in a given channel approaches both limits, that is, the bandwidth-imposed limit as well as the power-imposed limit, then optimum utilization of the channel resources is being made.

In the case of a linear channel, a matched filter such as the integrate-and-dump filter is a desirable solution for optimizing the predetection signal-to-noise ratio. However, the satellite channel depicted in Fig. 13-5 for a TWT amplifier is not a linear channel. Consequently, the received signal will be distorted. The integrate-and-dump filter is not the best choice for distorted signals. Experiments combined with analysis show that improved performance can be obtained with the following:

1. A gradual-roll-off Nyquist or a wider non-Nyquist (such as Chebyshev or Butterworth) transmitting filter ahead of the channel nonlinearities.
2. A Nyquist filter after demodulation but before sampling when a wider non-Nyquist filter is used in the transmitting end. Otherwise, a flat-passband noise-limiting filter will be adequate.
3. A signaling rate $R_T = 1/T$ such that

   $$1.0 < BT < 1.2$$

   where T is the symbol pulse width and B is the noise bandwidth (i.e., RF bandwidth). This conclusion differs from the linear-channel optimum solution $BT = 1$. The $BT = 1$ solution in a linear channel is optimum from the standpoint of avoiding intersymbol interference.

REFERENCES

Dicks, J. D., P. H. Schultze, and C. H. Schmidtt: "Systems Planning," *COMSAT Technical Review*, vol. 2, Fall 1972, pp. 452–469.

Dicks, J. L., and M. P. Brown, Jr.:"Frequency Division Multiple Access (FDMA) forSatellite Communications Systems," *IEEE-EASCON'74*, IEEE Press, New York; October 7–9,1974, pp. 167–178.

Gabbard, O. G., and P. Kaul: "Time-Division Multiple Access," *IEEE-EASCON '74*, IEEE Press, New York; October 7–9, 1974, pp. 179–184.

Houston, S. W.: "Modulation Techniques for Communications," *IEEE 1975 National Aerospace and Electronic Conference*, IEEE Press, New York; June 1975, pp. 51–58.

Members of the Technical Staff of Bell Telephone Laboratories: *Transmission Systems for Communications*, rev. 4th ed., Bell Telephone Laboratories, Winston Salem, N.C., 1971.

Reference Data for Satellite Communications and Earth Stations, ITT Space Communications, Ramsey, N.J., 1972.

Rinstenbatt, M. P., and J. L. Daws, Jr.: "Performance Criteria for Spread Spectrum Communications," *IEEE Transcripts on Communications*, vol. COM-25, August 1977.

Spilker, J. J., Jr.: *Digital Communications by Satellite*, Prentice-Hall, Englewood Cliffs, N.J., 1977.

Stein, S., and J. J. Jones: *Modern Communication Principles with Application to Digital Signaling*, McGraw-Hill, New York, 1967.

APPENDIX A

THE APPARENT MOTION OF THE SUN WITH RESPECT TO AN EARTH-SATELLITE VEHICLE

DEFINITION OF COORDINATE-SYSTEM ANGLES

The coordinate system is chosen so that the angles used to describe the position of the sun are analogous to the angles used to describe the position of a celestial body in certain terrestrial astronomic situations. As a result, common names and symbols for the terrestrial angles will be applied to this space-satellite case.

The pole axis of the polar-coordinate system is the satellite local vertical. The angle measured around the local vertical on the local horizontal plane between the orbit axis and the line of sight to the sun projection is called the "azimuth" Z. The angle that the line of sight to the sun makes with the satellite local horizontal plane is called the "altitude" h. Figure A-1 shows a sketch of the celestial sphere with an example of the measurement of the position of the sun with these angles. The orbit axis is perpendicular to the orbital plane.

The azimuth is measured in the local horizontal plane. It increases with the sun's apparent clockwise motion, as viewed downward from the satellite (toward the center of the earth).

The altitude is the angle that the line of sight to the sun makes with the satellite local horizontal plane. The altitude, therefore, is 0° when the sun lies on the local horizontal plane. The range of altitude is -90 to $+90°$ ($+90°$ is the direction directly away from the center of the earth).

MERIDIAN ANGLE AND DECLINATION

In order to determine the altitude and azimuth of the sun for any particular inertial space-fixed orbit, it is necessary to describe the position of the satellite in its orbit and the position of the orbit relative to the sun using another polar-coordinate system. The coordinate system is sketched in Fig. A-2. The polar axis

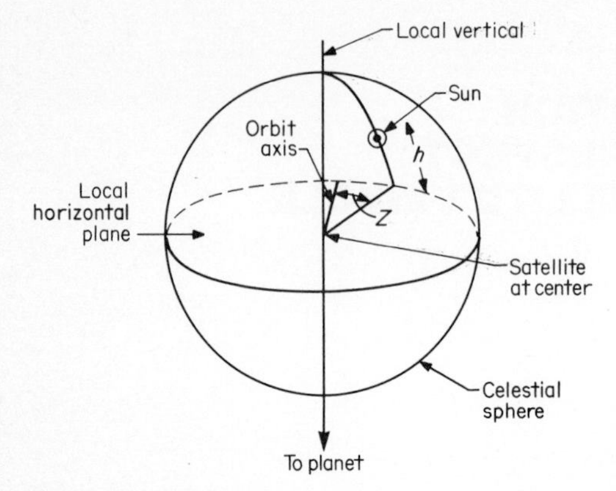

FIGURE A-1 Altitude and azimuth measurement.

of the coordinate system is the orbit axis. The angle that the line of sight to the sun makes with the orbit plane is called the "declination" d. The declination is equal to 0° when the sun is on the orbit plane. Its range is −90 to +90°. In the earth coordinate system, $d = 90°$ is the direction of the north pole. Looking down on the earth from the north pole, one would see the earth turn counterclockwise. In accordance with earth practice, the positive-declination hemisphere will be defined as the hemisphere from which the satellite motion (relative to inertial space) is counterclockwise. The satellite motion about the orbit axis is measured with the "meridian angle" t. The meridian angle is measured in the orbit plane and is 0° (or 180°) when the local vertical of the planet lies in the plane that contains the orbit axis and the line of sight to the sun. (We resolve the ambiguity by requiring that the altitude reach a positive value when the meridian angle is equal to 0°.) The meridian angle is undefined when the declination is equal to either +90° or −90°, but since the meridian angle has no significance for these conditions, the problem can be ignored.

The existing 180° ambiguity in azimuth can also be resolved. It can be stated that the azimuth is equal to 0° when the direction to the sun (measured in the local horizontal plane) coincides with the +90° direction of the declination.

SUN-MOTION EQUATIONS

With the quantities defined above, the following equations describing the motions of the sun may be derived

$$\sin d = \cos h \cos Z$$

$$\sin t = \tan Z \tan d$$

$$\sin h = \cos d \cos t$$

The sun-position information given by these equations is shown in Fig. A-3. The ordinate values of Fig. A-3 are the sun altitudes; the abscissa values are the sun azimuths. Against these values are plotted constant-declination lines and constant-meridian-angle lines. The full range of declination is covered so that the sun's motion for every possible orbit is covered. Slight discrepancies will be introduced in the curves by parallax due to the nonzero diameter of the satellite orbit. This effect is usually quite small. As will be discussed later, the declination will usually change during an orbit. This effect, too, is usually quite small.

The constant-declination lines of Fig. A-3 are continuous curves labeled along the 0° altitude axis. The constant-meridian-angle curves are shown for positive values of declination only. They are labeled along the 0° declination curve.

SUN-RATE CURVES

The sun-motion equations given above may be differentiated with respect to the meridian angle to obtain the rate of change of altitude and azimuth with respect to the change in meridian angle. Plots of these functions are shown in Figs. A-4 and A-5. For a circular orbit, the meridian angle will change linearly with time. These curves, therefore, show approximately the time rate of change of altitude and azimuth. Inspection of the dh/dt function shows that the maximum value that dh/dt can reach is 1, that is 1° change in altitude for 1° change in meridian angle. On the other hand, dZ/dt becomes infinite when the declination is equal to 0° and the meridian angle is equal to 0°. This high rate of azimuth change presents a design problem that will be discussed later.

THE INITIAL VALUE AND TIME VARIATION OF THE DECLINATION ANGLE

The original value of the declination angle depends on the way that the satellite is placed in orbit. For earth orbits, the angle between the orbit plane and the ecliptic determines the range of the possible initial values of the declination angle. For example, since the sun must lie in the ecliptic plane, if the orbit plane

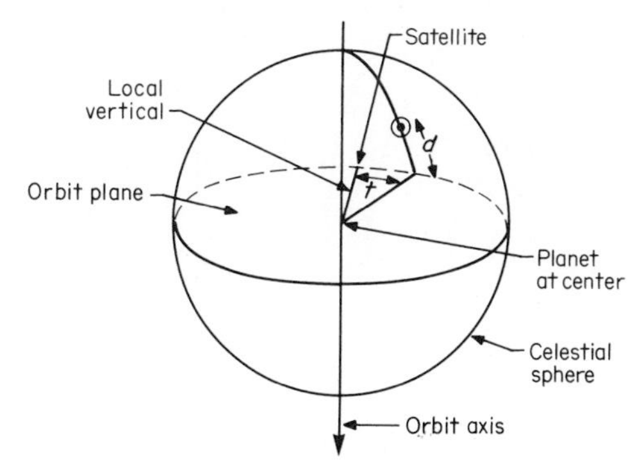

FIGURE A-2 Declination and meridian angle measurement.

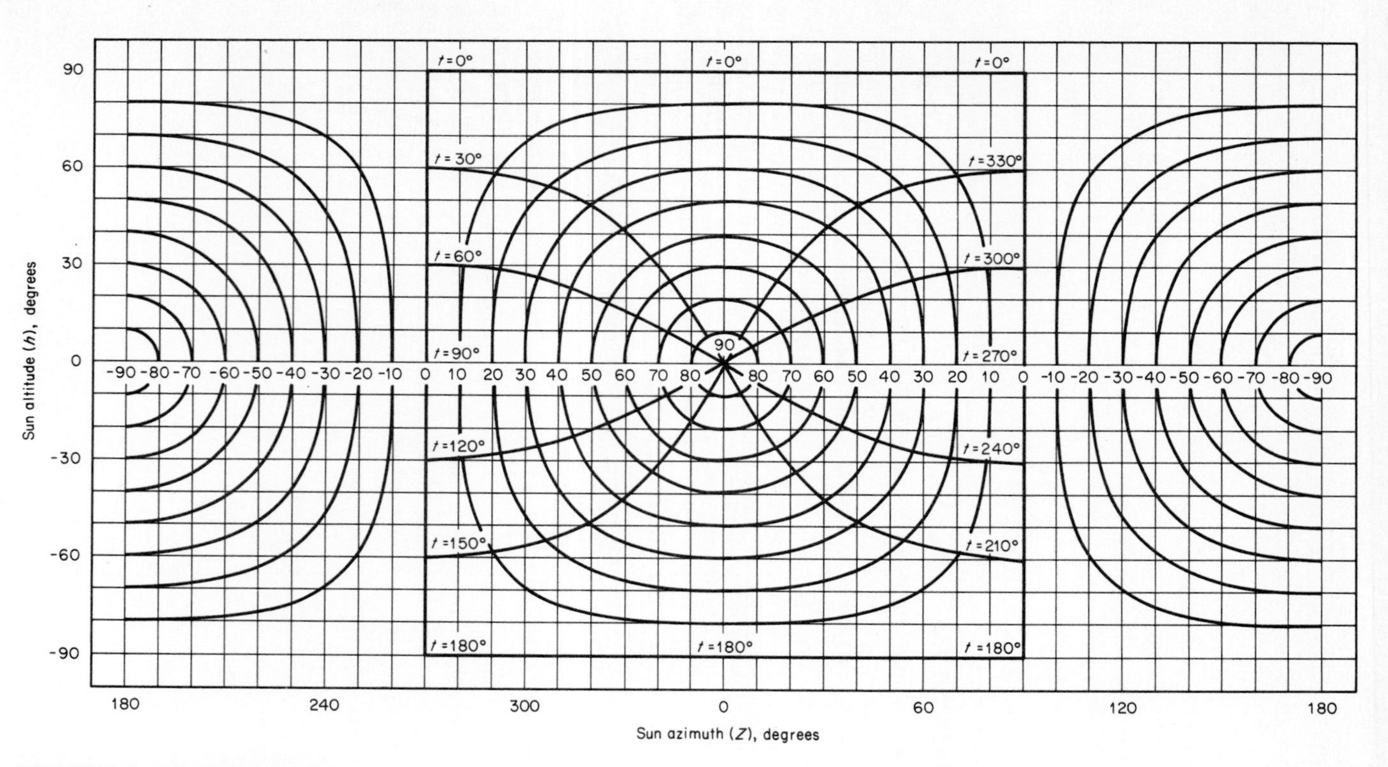

FIGURE A-3 Sun-position angles.

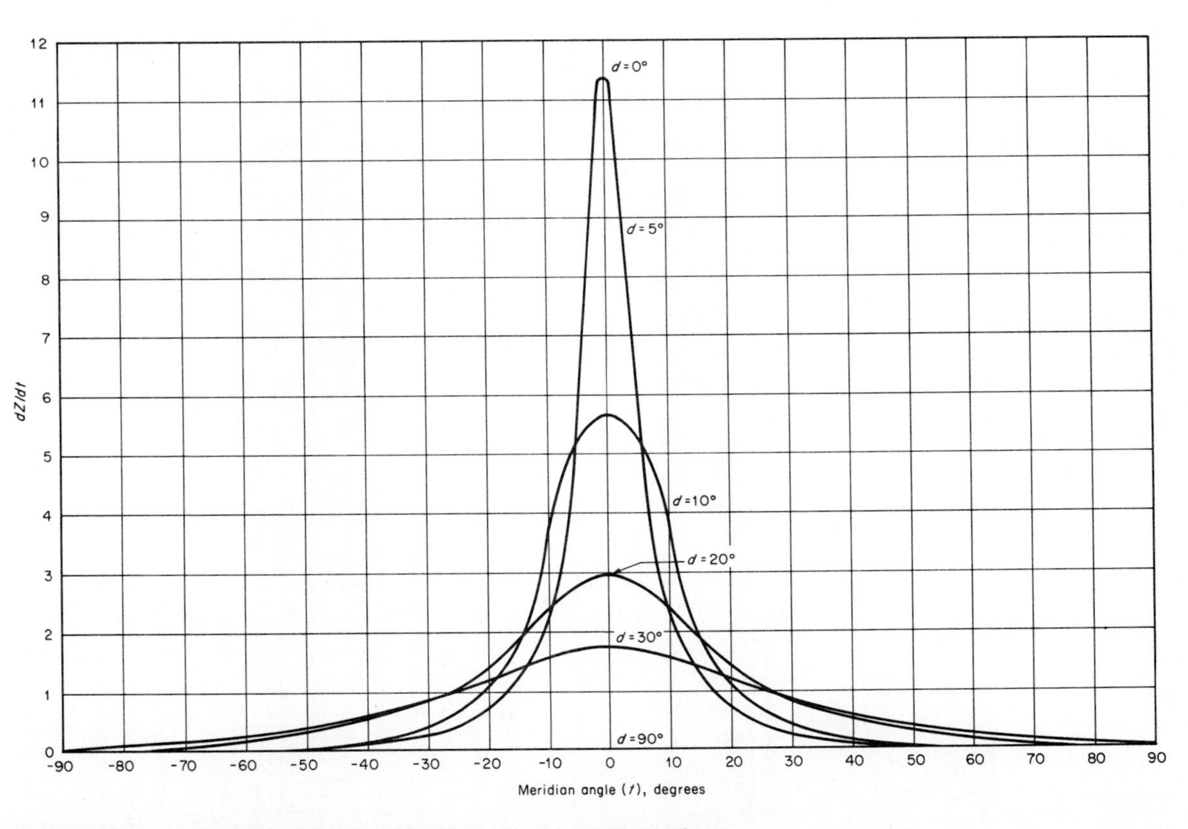

FIGURE A-4 Rate of change of azimuth with respect to meridian angle.

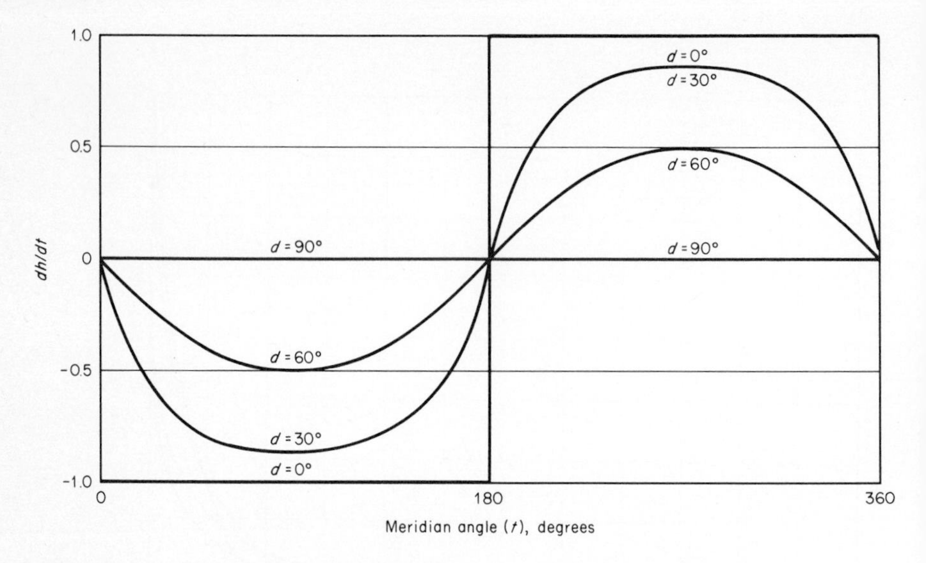

FIGURE A-5 Rate of change of altitude with respect to meridian angle.

coincides with the ecliptic plane, the declination angle can be nothing other than 0°.

Figure A-6 shows the possible values of declination for orbits with various inclinations to the ecliptic plane. Note that the constant-inclination lines are plotted against the value of the angle between the line from the center of the earth to the center of the sun (which is contained in the ecliptic plane) and the projection on the ecliptic plane of the satellite orbit axis. If there is no regression of the orbit plane, this angle will vary nearly linearly through 360° each year. Over that period of time, the declination will vary according to the inclination of the orbit plane to the ecliptic plane. If the direction of the orbit plane in inertial space is affected by orbit regression, the abscissa of Fig. A-6 will probably not change linearly with time. In addition, the tilt of the orbit plane to the ecliptic plane will probably change. To the extent that the regression effects can be predicted, the value of the declination angle can be determined and the operation of the power and orientation system can be evaluated for the life of the satellite.

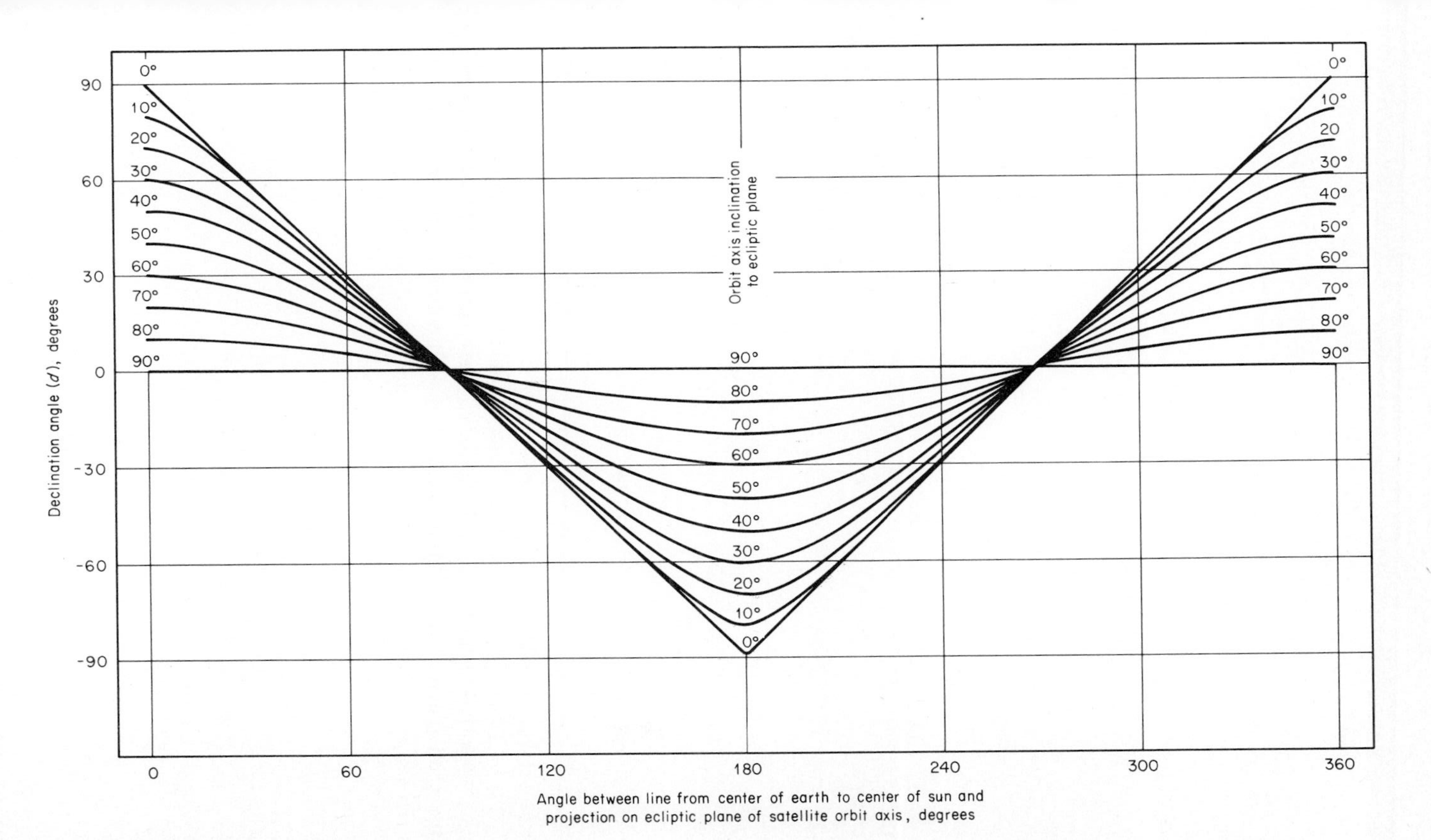

FIGURE A-6 Declination angles.

APPENDIX B

SUMMARY OF FOURIER ANALYSIS

FOURIER SERIES REPRESENTATION OF A PERIODIC FUNCTION

A function $g(t)$ is said to be periodic if it satisfies the relationship

$$g(t) = g(t + T) \qquad \text{for all } t \tag{B-1}$$

where T is defined as the period. The angular frequency ω is defined by the relationship

$$\omega = 2\pi f = \frac{2\pi}{T} \qquad \text{rad/s} \tag{B-2}$$

where f is the frequency in hertz. If in addition, $g(t)$, within the period T, satisfies the Dirichlet conditions—

1. $g(t)$ is a single-valued function of t
2. $g(t)$ has a finite number of discontinuities
3. $g(t)$ has a finite number of maxima and minima
4. $g(t)$ is bounded; that is, there always exists a noninfinite number A such that $|g(t)| < A$

then $g(t)$ can be represented by a series, called a Fourier series,* of sinusoids,

$$g(t) = \frac{A_0}{2} + \sum_{n=1}^{\infty} (A_n \cos n\omega t + B_n \sin n\omega t) \tag{B-3}$$

*For a detailed treatment of Fourier series, see R. V. Churchill, *Fourier Series and Boundary Value Problems*, McGraw-Hill, New York, 1963.

where

$$A_n = \frac{2}{T} \int_0^T g(t) \cos n\omega t \, dt$$

$$B_n = \frac{2}{T} \int_0^T g(t) \sin n\omega t \, dt \qquad \text{(B-4)}$$

$$A_0 = \frac{2}{T} \int_0^T g(t) \, dt$$

An alternative way of representing $g(t)$ is

$$g(t) = \sum_{n=-\infty}^{\infty} a_n e^{jn\omega t} \qquad \text{(B-5)}$$

where

$$a_n = \frac{1}{T} \int_0^T g(t) e^{-jn\omega t} \, dt = \frac{\omega}{2\pi} \int_0^T g(t) e^{-jn\omega t} \, dt \qquad \text{(B-6)}$$

It is obvious from Eq. (B-6) that if $g(t)$ is a real function of t,

$$a_n = a_{-n}^{*} \qquad \text{(B-7)}$$

and it can easily be proven that

$$A_n = a_n + a_{-n} = 2 \operatorname{Re} a_n \qquad B_n = j(a_n - a_{-n}) = -2 \operatorname{Im} a_n \qquad \text{(B-8)}$$

From Eq. (B-3), one can also derive

$$g(t) = \frac{C_0}{2} + \sum_{n=1}^{\infty} C_n \cos (n\omega t + \phi_n) \qquad \text{(B-9)}$$

where

$$C_n = \sqrt{A_n^2 + B_n^2} = 2|a_n| \qquad \text{(B-10)}$$

$$\phi_n = \tan^{-1} \frac{B_n}{A_n} \qquad \text{(B-11)}$$

By plotting the magnitude of amplitude of the Fourier coefficient a_n and C_n versus frequency, we obtain the two-sided and one-sided frequency spectra, respectively. The one-sided frequency spectrum has double the amplitude of the two-sided spectrum.

APERIODIC FUNCTIONS AND FOURIER TRANSFORMS

If $g(t)$ is an aperiodic function satisfying the previously mentioned Dirichlet conditions and if, in addition, $g(t)$ is absolutely integrable [that is, if

$\int_{-\infty}^{\infty} |g(t)|\, dt < \infty$], then the following relationships are true:

$$g(t) = \int_{-\infty}^{\infty} G(\omega)e^{j\omega t}\, d\omega$$

$$G(\omega) = \frac{1}{2\pi}\int_{-\infty}^{\infty} g(t)e^{-j\omega t}\, dt \tag{B-12}$$

where $g(t)$ and $G(\omega)$ are Fourier transform pairs. By using f instead of ω, we can rewrite the above relationships as follows:

$$g(t) = \int_{-\infty}^{\infty} G(f)e^{j2\pi ft}\, df$$

$$G(f) = \int_{-\infty}^{\infty} g(t)e^{-j2\pi ft}\, dt \tag{B-13}$$

where $G(f)$ is redefined to equal $2\pi G(\omega)$.

Equations (B-12) could also be written in a symmetrical form:

$$g(t) = \frac{1}{\sqrt{2\pi}}\int_{-\infty}^{\infty} G(\omega)e^{j\omega t}\, d\omega \tag{B-14a}$$

$$G(\omega) = \frac{1}{\sqrt{2\pi}}\int_{-\infty}^{\infty} g(t)e^{-j\omega t}\, dt \tag{B-14b}$$

where the Fourier transform is $\sqrt{2\pi}$ times the Fourier transform defined by Eq. (B-12).

It is easy to see from Eq. (B-14) that, for $g(t)$ real,

$$G(\omega) = G^*(-\omega) \tag{B-15}$$

If

$$G(\omega) = a(\omega) - jb(\omega) \tag{B-16}$$

from Eq. (B-14) we get

$$a(\omega) = \frac{1}{\sqrt{2\pi}}\int_{-\infty}^{\infty} g(t)\cos\omega t\, dt$$

$$b(\omega) = \frac{1}{\sqrt{2\pi}}\int_{-\infty}^{\infty} g(t)\sin\omega t\, dt \tag{B-17}$$

Therefore

$a(\omega)$ is an even function of ω.
$b(\omega)$ is an odd function of ω.

Substituting the value of $G(\omega)$, as given by Eq. (B-16), into Eq. (B-14*a*) and observing that the integral of the odd function vanishes, we obtain

$$g(t) = \frac{1}{\sqrt{2\pi}} \int_{-\infty}^{\infty} [a(\omega) \cos \omega t + b(\omega) \sin \omega t] \, d\omega$$

$$= \frac{2}{\sqrt{2\pi}} \int_{0}^{\infty} [a(\omega) \cos \omega + b(\omega) \sin \omega t] \, d\omega \tag{B-18}$$

But Eq. (B-18) can be rewritten as

$$g(t) = \frac{2}{\sqrt{2\pi}} \int_{0}^{\infty} C(\omega) \cos [\omega t + \phi(\omega)] \, d\omega \tag{B-19}$$

where

$$\begin{aligned} 2C(\omega) &= 2\sqrt{|a(\omega)|^2 + |b(\omega)|^2} = 2|G(\omega)| \\ \phi(\omega) &= \tan^{-1} \frac{b(\omega)}{a(\omega)} \end{aligned} \tag{B-20}$$

Again, the amplitude of the one-sided spectrum represented by $2C(\omega)$ is double the amplitude of each corresponding frequency of the two-sided spectrum represented by $G(\omega)$. The frequency spectrum of the function $g(t)$ is defined as the plot of $G(f)$ versus frequency, where $G(f) = |G(f)|e^{j\theta}$. In general, there are two such plots, one for $|G(f)|$ and one for the phase $\theta(f)$. The spectrum of the function $G(f)$ is extended over both the negative and positive sides of the f axis, while the spectrum of the function $C(f)$ extends over only the positive side of the f axis. The spectrum of a periodic function described by a Fourier series is not a continuous one but instead is represented by discrete values. Since, for a real function $g(t)$, we have $G(f) = G^*(-f)$, the amplitude spectrum will be an even function of the frequency f while the phase spectrum will be an odd function of f.

EXAMPLE B-1 FOURIER TRANSFORM OF A RECTANGULAR PULSE

Consider a rectangular pulse of duration T and amplitude A. This pulse is depicted in Fig. B-1*a*. the mathematical expression for this pulse is

$$g(t) = \begin{cases} A & \text{for } -\frac{T}{2} < t < \frac{T}{2} \\ 0 & \text{for } |t| > \frac{T}{2} \end{cases} \tag{B-21}$$

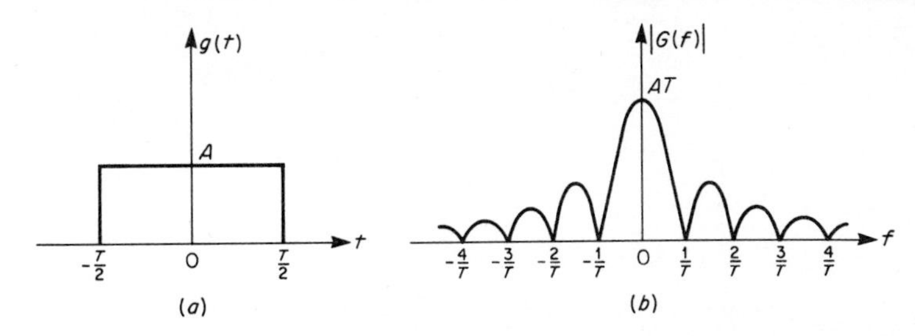

FIGURE B-1 **(a) Rectangular pulse. (b) Amplitude spectrum $|G(f)|$ of the rectangular pulse.**

The Fourier transform of this pulse is given by the relationship

$$G(f) = \int_{-T/2}^{T/2} Ae^{-j2\pi ft}\,dt = AT \operatorname{sinc} fT \tag{B-22}$$

where $\operatorname{sinc} fT = (\sin \pi fT)/\pi fT$. The amplitude spectrum of $g(t)$ is depicted in Fig. B-1*b*. Whereas sinc fT takes alternately positive and negative values, the frequency spectrum of $g(t)$ will be always positive, since it represents the absolute value of $G(f)$.

EXAMPLE **B-2** PERIODIC PULSE TRAIN

A periodic train of rectangular pulses $g(t)$ of duration τ and repetition rate T is shown in Fig. B-2*a*. This train of pulses could be expressed in terms of a Fourier series with coefficients

$$a_n = \frac{1}{T}\int_{-\tau/2}^{+\tau/2} Ae^{-j2\pi nt/T}\,dt = \frac{A}{n\pi}\sin\frac{n\pi\tau}{T} \tag{B-23}$$

The discrete amplitude and phase spectra are depicted in Fig. B-2*b* and *c*. The phase spectrum takes the values 0 and $\pm 180°$, depending on the polarity of sinc $(n\tau/T)$, since

$$a_n = \frac{\tau A}{T}\operatorname{sinc}\frac{n\tau}{T} \tag{B-24}$$

SOME IMPORTANT FOURIER TRANSFORM PROPERTIES

The properties listed below can be easily proved by utilizing the fundamental relationships, Eqs. (B-13), defining the Fourier transform pairs.

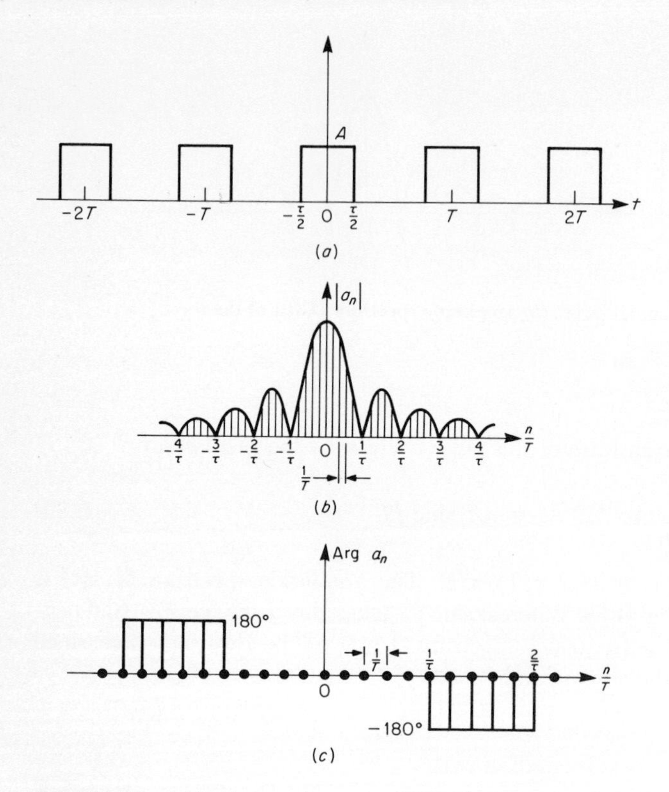

FIGURE B-2 *(a)* **Periodic pulse train.** *(b)* **Amplitude spectrum of the periodic pulse train.** *(c)* **Phase spectrum.**

Duality

If $g(t)$ and $G(f)$ are Fourier transform pairs [that is, if $g(t) \rightleftarrows G(f)$], then the function $G(t)$ will have as a Fourier transform the function $g(-f)$ [that is, $G(t) \rightleftarrows g(-f)$].

In Example B-1, it was found that the rectangular pulse

$$g(t) = \begin{cases} A & \text{for } -\dfrac{T}{2} < t < \dfrac{T}{2} \\ 0 & \text{for } |t| > \dfrac{T}{2} \end{cases}$$

had as a Fourier transform the sinc pulse

$$G(f) = AT \text{ sinc } fT$$

By this duality principle, the sinc pulse

$$g(t) = A \text{ sinc } t$$

has as a Fourier transform the rectangular pulse

$$G(f) = \begin{cases} A & -\frac{1}{2} < f < \frac{1}{2} \\ 0 & \text{for } |f| > \frac{1}{2} \end{cases}$$

Time Shift

If $g(t)$ and $G(f)$ are Fourier transform pairs, then

$$g(t - t_0) \leftrightarrows G(f)e^{-j2\pi f t_0}$$

This indicates that a time shift in the time domain represents a phase shift in the frequency domain.

Frequency Shift

If $g(t)$ and $G(f)$ are Fourier transform pairs, then

$$G(f - f_c) \leftrightarrows g(t)e^{j2\pi f_c t}$$

i.e., a frequency shift in the frequency domain represents multiplication by an exponential in the time domain. The relationship for frequency shifting could also be derived by applying the duality principle on the time-shift expression.

As an example, the Fourier transform of the RF pulse shown in Fig. B-3 is derived by using the transform of a rectangular pulse.

The time function representing the RF pulse is

$$g_1(t) = \begin{cases} g(t) \cos 2\pi f_c t & \text{for } -\dfrac{T}{2} < t < \dfrac{T}{2} \\ 0 & \text{for } |t| > \dfrac{T}{2} \end{cases}$$

where $g(t)$ is the rectangular pulse of Example B-1.

We can rewrite $g_1(t)$ as

$$g_1(t) = \tfrac{1}{2}g(t)\,(e^{j2\pi f_c t} + e^{-j2\pi f_c t})$$

Therefore

$$G_1(f) = \frac{AT}{2}\left[\frac{\sin \pi(f - f_c)T}{\pi(f - f_c)T} + \frac{\sin \pi(f + f_c)T}{\pi(f + f_c)T}\right]$$

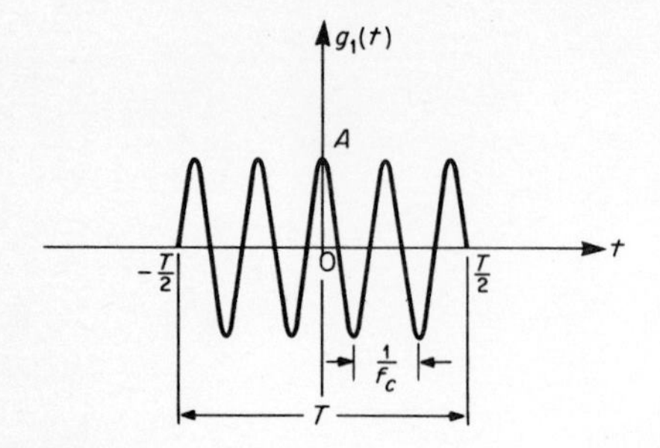

FIGURE B-3 RF pulse.

Differentiation in the Time Domain

If $g(t)$ and $G(f)$ are Fourier transform pairs, then

$$\frac{d}{dt}\, g(t) \leftrightarrows j2\pi f G(f)$$

that is, differentiation in the time domain corresponds to multiplication in the frequency domain.

Integration in the Time Domain

If $g(t)$ and $G(f)$ are Fourier transform pairs, then

$$\int_{-\infty}^{t} g(\tau)\, d\tau \leftrightarrows \frac{1}{j2\pi f}\, G(f) + \frac{G(0)}{2}\, \delta(f)$$

that is, integration in the time domain corresponds to a division in the frequency domain. The function $\delta(f)$ is the delta (Dirac) function in the frequency domain.

Conjugate Functions

If $g(t)$ and $G(f)$ are Fourier transform pairs, then

$$g^*(t) \leftrightarrows G^*(-f)$$

FOURIER TRANSFORMS OF SOME SPECIAL FUNCTIONS

Delta Function

The delta (Dirac) function $\delta(t)$ is defined by the relationships

$$\int_{-\infty}^{\infty} \delta(t - t_0)\, dt = 1$$

$$\delta(t - t_0) = 0 \qquad \text{for } t \neq t_0$$

The Fourier transform $\Delta(f)$ of the delta function $\delta(t)$ is given as

$$\Delta(f) = 1$$

Unit-Step Function

The unit-step function is defined as

$$u(t) = \begin{cases} 1 & \text{for } t > 0 \\ 0 & \text{for } t < 0 \end{cases}$$

The Fourier transform $U(f)$ of the unit-step function $u(t)$ is given by the relationship

$$U(f) = \frac{1}{j2\pi f} + \tfrac{1}{2}\,\delta(f)$$

where $\delta(f)$ is the delta function in the frequency domain.

The Signum Function

The signum function sgn (t) is defined as

$$\text{sgn } t = \begin{cases} 1 & \text{for } t > 0 \\ 0 & \text{for } t = 0 \\ -1 & \text{for } t < 0 \end{cases}$$

The Fourier transform of this function is given by the relationship

$$\text{SGN } f = \frac{1}{j\pi f}$$

By applying the duality principle, we can also derive

$$\frac{1}{\pi t} \leftrightarrows -j \text{ sgn } f$$

where sgn f is the signum function in the frequency domain.

Convolution

For two functions $g_1(t)$ and $g_2(t)$, the relationship

$$y(t) = \int_{-\infty}^{\infty} g_1(\tau)g_2(t - \tau)\, d\tau \qquad \text{(B-25)}$$

is defined as the convolution of $g_1(t)$ and $g_2(t)$, and it is usually expressed as

$$y(t) = g_1(t) \otimes g_2(t) \qquad \text{(B-26)}$$

If in Eq. (B-25) we express $g_2(t - \tau)$ in terms of its Fourier transform $G_2(f)$, we get

$$y(t) = \int_{-\infty}^{\infty} g_1(\tau) \int_{-\infty}^{\infty} G_2(f)e^{j2\pi f(t-\tau)}\, df\, d\tau$$

and, by exchanging the order of integration, we get

$$y(t) = \int_{-\infty}^{\infty} G_2(f)e^{j2\pi ft} \int_{-\infty}^{\infty} g_1(t)e^{-j2\pi ft}\, dt\, df$$

and, since the second integral gives the Fourier transform $G_1(f)$ of $g_1(t)$, we obtain

$$y(t) = \int_{-\infty}^{\infty} G_2(f)G_1(f)e^{j2\pi ft}\, df$$

which means that

$$g_1(t) \otimes g_2(t) \leftrightarrows G_1(f)G_2(f) \tag{B-27}$$

Equation (B-27) indicates that the Fourier transform of the convolution of two functions is the product of their Fourier transforms.

EXAMPLE B-3

The Fourier transform of the $\cos 2\pi f_c t$ function can be found easily by applying some of the above properties. Since

$$\cos 2\pi f_c t = \tfrac{1}{2}\,(e^{j2\pi f_c t} + e^{-j2\pi f_c t})$$

therefore

$$\cos 2\pi f_c t \leftrightarrows \tfrac{1}{2}\,\delta(f - f_c) + \delta(f + f_c)$$

EXAMPLE B-4

Consider the sinusoidal function $g(t) = A \cos(2\pi f_0 t + \phi)$. We can write

$$g(t) = \tfrac{1}{2}A\,(e^{j(2\pi f_0 t+\phi)} + e^{-j(2\pi f_0 t+\phi)})$$

but

$$Ae^{j(2\pi f_0 t+\phi)} = Ae^{j\phi}e^{j2\pi f_0 t} \leftrightarrows Ae^{j\phi}\,\delta(f - f_0)$$

and

$$Ae^{-j(2\pi f_0 t+\phi)} \leftrightarrows Ae^{-j\phi}\,\delta(f + f_0)$$

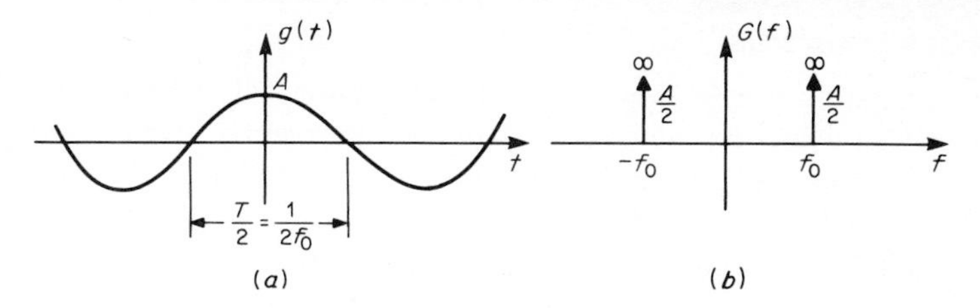

FIGURE B-4 (*a*) A cosine function with $\phi = 0$. (*b*) Frequency spectrum of a cosine function.

Therefore

$$A \cos (2\pi f_0 t + \phi) \leftrightarrows \tfrac{1}{2} A e^{j\phi}\, \delta(f - f_0) + \tfrac{1}{2} A e^{-j\phi}\, \delta(f + f_0)$$

and similarly,

$$A \sin (2\pi f_0 t + \phi) \leftrightarrows \frac{1}{2j} A e^{j\phi}\, \delta(f - f_0) - \frac{1}{2j} A e^{-j\phi}\, \delta(f + f_0)$$

For $\phi = 0$, we get

$$A \cos 2\pi f_0 t \leftrightarrows \tfrac{1}{2} A\, \delta(f - f_0) + \tfrac{1}{2} A\, \delta(f + f_0)$$

and the frequency spectrum of the cosine consists of two impulses in the frequency domain with area $A/2$ (see Fig. B-4).

ENERGY RELATIONSHIPS AND PARSEVAL'S THEOREM

If $g(t)$ and $G(f)$ are Fourier transform pairs, it can be shown that

$$\int_{-\infty}^{\infty} g(t) g^*(t)\, dt = \int_{-\infty}^{\infty} G(f) G^*(f)\, df \tag{B-28}$$

This relationship can be written as

$$\int_{-\infty}^{\infty} |g(t)|^2\, dt = \int_{-\infty}^{\infty} |G(f)|^2\, df \tag{B-29}$$

where $E = \int_{-\infty}^{\infty} |g(t)|^2\, dt$ represents the total energy of the function $g(t)$.

Equation (B-28) is known as Parseval's theorem, and it indicates that the total energy in the time domain equals the total energy in the frequency domain.

The function $W(f) = |G(f)|^2$ is called the "energy-density spectrum."

FAST FOURIER TRANSFORM

The fast Fourier transform (FFT) is a computer algorithm that computes the discrete Fourier transform extremely fast. The discrete Fourier transform could be obtained by sampling the continuous time function $g(t)$ at regular intervals. The resulting samples represent a function with discrete values. The Fourier transform of this discrete function will be a periodic function.

APPENDIX C

HILBERT TRANSFORMS

Given a function $g(t)$, the Hilbert transform of $g(t)$, namely $H[g(t)] = \hat{g}(t)$, is defined by the relationship

$$\hat{g}(t) = \frac{1}{\pi} \int_{-\infty}^{\infty} \frac{g(\tau)}{t - \tau} \, d\tau \tag{C-1}$$

From Eq. (C-1), it is obvious that the Hilbert transform of $g(t)$ is the convolution of $g(t)$ with the function $1/\pi t$, that is,

$$\hat{g}(t) = g(t) \otimes \frac{1}{\pi t} \tag{C-2}$$

However, a more useful interpretation of the Hilbert transform is the following: If $G(f)$ is the Fourier transform of $g(t)$, then the Fourier transform $\hat{G}(f)$ of $\hat{g}(t)$ will satisfy the relationship

$$\hat{G}(f) = \begin{cases} -jG(f) & \text{for } f > 0 \\ 0 & \text{for } f = 0 \\ +jG(f) & \text{for } f < 0 \end{cases} \tag{C-3}$$

or, in a more concise form,

$$\hat{G}(f) = -j \, (\text{sgn } f) \, G(f) \tag{C-4}$$

Equations (C-3) and (C-4) are easy to prove by considering Eq. (C-2) and the fact that the Fourier transform of $1/\pi t$ is $-j \text{ sgn } f$ (see Appendix B). Equation (C-4) indicates that deriving the Hilbert transform of a function $g(t)$ is equivalent to passing the function $g(t)$ through a two-port network with the transform characteristic shown in Fig. C-1.

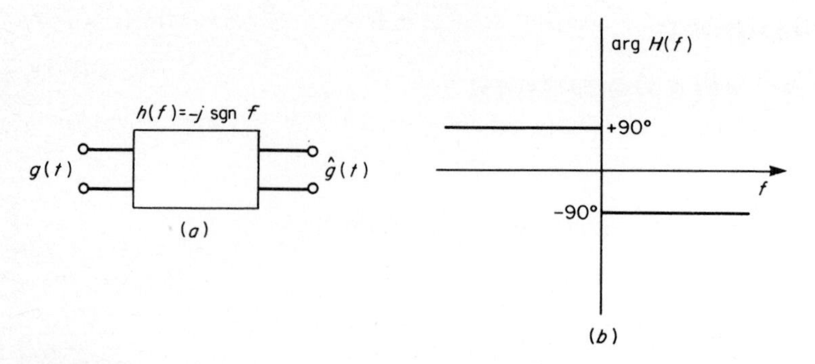

FIGURE C-1 **(a) Hilbert transform network. (b) Phase-transfer characteristic.**

When the phase angles of all the frequency components of a signal are shifted by 90°, the resulting function in the time domain is the Hilbert transform of the original function. The original function can be derived from its Hilbert transform by using the relationship

$$g(t) = -\frac{1}{\pi}\int_{-\infty}^{\infty}\frac{\hat{g}(\tau)}{t-\tau}\,d\tau \tag{C-5}$$

Equation (C-5) is easy to verify by multiplying $\hat{G}(f)$ by $-j \operatorname{sgn} f$. The result is $-G(f)$. But since $-j \operatorname{sgn} f$ is the Fourier transform of $1/\pi t$, the convolution of $1/\pi t$ and $\hat{g}(t)$ ought to give the inverse Fourier transform of $-G(f)$, namely, $-g(t)$.

EXAMPLE C-1 HILBERT TRANSFORM OF A CARRIER SIGNAL

Consider the signal

$$g(t) = m(t)e^{j2\pi f_c t} \tag{C-6}$$

Since $f_c > |f_m|$, where f_m is the maximum frequency for which the low-pass $M(f)$ is nonzero, by applying the frequency-shift relationship for the Fourier transform of $g(t)$ we get

$$G(f) = 0 \qquad \text{for } f < f_c - f_m \text{ or } f > f_c + f_m \tag{C-7a}$$

Therefore

$$G(f) = \begin{cases} M(f - f_c) & \text{for } f_c - f_m \le f \le f_c + f_m \\ 0 & \text{otherwise} \end{cases} \tag{C-7b}$$

By applying Eq. (C-3),

$$\hat{G}(f) = \begin{cases} -jM(f - f_c) & \text{for } f_c - f_m \le f \le f_c + f_m \\ 0 & \text{otherwise} \end{cases} \tag{C-8}$$

and by deriving the inverse Fourier transform of $\hat{G}(f)$, we obtain

$$\hat{g}(t) = -jm(t)e^{j2\pi f_c t} \tag{C-9}$$

This relationship is useful in obtaining certain Hilbert transform pairs.

If

$$g(t) = m(t)e^{j2\pi f_c t} = m(t)\cos 2\pi f_c t + jm(t)\sin 2\pi f_c t$$

then

$$\hat{g}(t) = -jm(t)e^{j2\pi f_c t} = m(t)\sin 2\pi f_c t - jm(t)\cos 2\pi f_c t$$

Consequently,

$$H[m(t)\cos 2\pi f_c t] = m(t)\sin 2\pi f_c t \tag{C-10}$$

$$H[m(t)\sin 2\pi f_c t] = -m(t)\cos 2\pi f_c t$$

and, for $m(t) = 1$,

$$H(e^{j2\pi f_c t}) = -je^{j2\pi f_c t}$$

$$H(\cos 2\pi f_c) = \sin 2\pi f_c t \tag{C-11}$$

$$H(\sin 2\pi f_c t) = -\cos 2\pi f_c t$$

EXAMPLE C-2 ORTHOGONALITY OF $g(t)$ AND $\hat{g}(t)$

For a real signal $g(t)$, we must prove that

$$I = \int_{-\infty}^{\infty} g(t)\hat{g}(t)\, dt = 0 \tag{C-12}$$

We have

$$I = \int_{-\infty}^{\infty} \hat{g}(t) \int_{-\infty}^{\infty} G(f)e^{j2\pi ft}\, df\, dt$$

$$= \int_{-\infty}^{\infty} G(f) \int_{-\infty}^{\infty} \hat{g}(t)e^{j2\pi ft}\, dt\, df$$

$$= \int_{-\infty}^{\infty} G(f)\hat{G}(-f)\, df = \int_{-\infty}^{\infty} G(f)\hat{G}^*(f)\, df$$

and, by applying Eq. (C-4), we obtain

$$I = \int_{-\infty}^{\infty} j \operatorname{sgn} f |G(f)|^2\, df$$

But, since $|G(f)|^2$ is positive and an even function of f, the integral I must be zero.

Preenvelope

The "analytic signal" $z(t)$ is defined as

$$z(t) = g(t) + j\hat{g}(t) \tag{C-13}$$

The function $z(t)$ is also called the preenvelope of $g(t)$. The Fourier transform $Z(f)$ will be given by the relationship

$$Z(f) = G(f) + j\hat{G}(f)$$

and, from Eq. (C-3),

$$Z(f) = \begin{cases} 2G(f) & \text{for } f > 0 \\ G(0) & \text{for } f = 0 \\ 0 & \text{for } f < 0 \end{cases} \tag{C-14}$$

The frequency spectrum of the preenvelope is zero along the negative-frequency semiaxis, and double that of the function itself along the semiaxis of positive frequencies.

APPENDIX D

FIELDS, VECTORS, AND MATRIX REPRESENTATIONS

FIELDS

A field F is a set of elements (scalars) for which two operations (i.e., addition and multiplication) and their inverses are defined. The addition and multiplication may not necessarily be the same as the addition and multiplication used with the ordinary numbers. In order for F to be a field, the following axioms must be true:

1. For any two elements c and d of the field, the element $e = c + d$ is an element of the field. In addition, $c + d = d + c$, which means that the commutative law under addition applies. This field is called abelian under addition.
2. For any two elements c and d of the field, the element $c \cdot d$ is also an element of the field. The field is also abelian under multiplication; i.e., the commutative law $c \cdot d = d \cdot c$ applies.
3. The associative law $a(bc) = (ab)c$ applies.
4. The distributive law $a(b + c) = ab + ac$, applies.
5. Every element has its inverse under addition and multiplication; that is, $a + (-a) = 0$ and $a \cdot a^{-1} = 1$.

Axioms 1 and 2, which demand that the elements $c + d$ and $c \cdot d$ also be elements of the field, are often described as satisfying closure under addition and multiplication.

The identity element 0 under addition is defined by the relationship $c + 0 = c$, and the identity element 1 under multiplication is defined by the relationship $1 \cdot c = c$. A field must have at least these two elements, that is, 0 and 1.

The addition and multiplication table of a field of two elements 0 and 1 is defined as follows:

+	0	1
0	0	1
1	1	0

·	0	1
0	0	0
1	0	1

(D-1)

The inverse element $-c$ of an element c under addition is defined by the relationship $c + (-c) = 0$.

The inverse element c^{-1} of an element c under multiplication is defined by the relationship $c \cdot c^{-1} = 1$.

The set of all real numbers forms a field, as does the set of all complex numbers.

A finite field consisting of a finite set of q elements that satisfies the previously mentioned condition is also called a Gulois field. A Gulois field having a finite set of q elements is designated as GF(q).

For example, a GF(5) modulo 5 will have the following multiplication and addition tables:

+	0	1	2	3	4
0	0	1	2	3	4
1	1	2	3	4	0
2	2	3	4	0	1
3	3	4	0	1	2
4	4	0	1	2	3

·	0	1	2	3	4
0	0	0	0	0	0
1	0	1	2	3	4
2	0	2	4	1	3
3	0	3	1	4	2
4	0	4	3	2	1

Subtraction is performed by adding the inverse of the element to be subtracted. For example,

$$2 - 3 = 2 + (2) = 4$$

We also have

$$\frac{2}{3} = 2 \cdot 3^{-1} = 2 \cdot 2 = 4$$

VECTORS

A set of ordered elements $(x_1, x_2, \ldots, x_n)$ could be considered to be the coordinates of a vector **x**. The **x** is said to be an element of a vector space S if all the vectors within the space satisfy certain specific conditions to be discussed later.

The ordered elements $(x_1, x_2, \ldots, x_n)$ may be the result of an experiment or observations obtained in the order $x_1, x_2, \ldots, x_n$.

The multiplication of a vector **v** $(v_1, v_2, \ldots, v_n)$ by a scalar c is defined to be $c\mathbf{v}$ $(cv_1, cv_2, \ldots, cv_n)$.

The inner or dot product of two vectors **v** and **w** is defined to be the scalar

$$\mathbf{v} \cdot \mathbf{w} = v_1w_1 + v_2w_2 + \cdots + v_nw_n$$

The sum of two vectors **v** and **w** is the ordered set of elements $(v_1 + w_1, v_2 + w_2, \ldots, v_n + w_n)$.

The vector **0** $(0, 0, \ldots, 0)$ is called the "null vector."

Two vectors **v** and **w** are said to be orthogonal if the inner product of these two vectors is zero, that is,

$$\mathbf{v} \cdot \mathbf{w} = v_1 w_1 + v_2 w_2 + \cdots + v_n w_n = 0 \tag{D-2}$$

VECTOR SPACES

Definitions

A set of vectors V forms a vector space if it satisfies the following axioms:

1. The product of a scalar c and a vector $\mathbf{v}$ $(v_1, v_2, \ldots, v_n)$, denoted by $c\mathbf{v}$ $(cv_1, cv_2, \ldots, cv_n)$, is a vector in the same vector space.
2. The addition of two vectors **v** and **w**, denoted by $\mathbf{v} + \mathbf{w}$ $(v_1 + w_1, v_2 + w_2, \ldots, v_n + w_n)$, is a vector in the same vector space. The addition is commutative; that is, $\mathbf{v} + \mathbf{w} = \mathbf{w} + \mathbf{v}$.
3. The distributive and associative laws apply; i.e., if c and d are scalars and **v** and **w** are vectors,

$$c(\mathbf{v} + \mathbf{w}) = c\mathbf{v} + c\mathbf{w}$$

$$(cd)\mathbf{w} = c(d\mathbf{w})$$

Axioms 1 and 2 satisfy closure under multiplication by a scalar and under addition of two vectors.

A set of vectors from a vector space V is a "subset" of the vector space if the set of vectors satisfies all the axioms of the vector space.

A vector space is called a "linear vector space" if the inner or dot product $\mathbf{v} \cdot \mathbf{w}$ is defined and the associative law $(\mathbf{vw})\mathbf{u} = \mathbf{v}(\mathbf{wu})$ applies when **v**, **w**, and **u** are vectors of this space. The fact that $\mathbf{v} \cdot \mathbf{w}$ is defined to be a vector in the same vector space satisfies closure under vector multiplication.

A set of vectors $\mathbf{v}_1, \mathbf{v}_2, \ldots, \mathbf{v}_n$ are linearly dependent if and only if a set of nonzero scalars $a_1, a_2, \ldots, a_n$ can be found to satisfy the relationship

$$a_1\mathbf{v}_1 + a_2\mathbf{v}_2 + \cdots + a_n\mathbf{v}_n = 0 \tag{D-3}$$

If such a set of scalars cannot be found, the set of vectors $\mathbf{v}_1, \mathbf{v}_2, \ldots, \mathbf{v}_n$ is linearly independent.

The maximum number of linearly independent vectors that can be found within a vector space constitutes a "basis" for the space, and the number of these vectors is called the "dimension" or "rank" of the space.

The totality of vectors obtained by the linear combination of a set of independent vectors forms a vector space, and the set of the independent vectors is a basis of the vector space.

Matrix Representation of Vector Spaces

A set of vectors, placed one under the other in rows, will form a matrix. In fact, a number n of vectors, each with k ordered elements, will form an $n \times k$ matrix.

Given the matrix of a vector space, the basis of the space can be found by employing the "sweep-out" method or by reducing the matrix to its echelon canonical form. One method is a slight variation of the other, and they both correspond to solving a set of linear equations by eliminating one variable at a time.

In reducing a matrix to its echelon canonical form, we operate on the rows of the matrix by multiplying the appropriate rows with a scalar and subtracting a row from another as necessary or interchanging two rows. The desired result is a matrix in which every nonzero row has a leading element 1, and all rows below have their leading element 1 to the right of the row above. All the other elements of the column in which a row has a leading element 1 should be zero. The nonzero rows of a matrix reduced to this form are linearly independent and constitute a basis. The number of the nonzero rows is the dimension of the matrix. As an example, consider the vector space represented by matrix 1 in the following set of matrices, which shows the reduction of matrix 1 to an echelon canonical form:

$$
\underset{\text{Matrix 1}}{\begin{vmatrix} 0&0&0&0&0\\ 0&1&0&1&0\\ 1&1&0&0&1\\ 0&0&1&0&1\\ 1&0&1&1&0\\ 0&1&1&1&1\\ 1&1&1&0&0\\ 1&0&0&1&1 \end{vmatrix}}
\underset{\text{Matrix 2}}{\begin{vmatrix} 1&0&0&1&1\\ 0&1&0&1&0\\ 1&1&0&0&1\\ 0&0&1&0&1\\ 1&0&1&1&0\\ 0&1&1&1&1\\ 1&1&1&0&0\\ 0&0&0&0&0 \end{vmatrix}}
\underset{\text{Matrix 3}}{\begin{vmatrix} 1&0&0&1&1\\ 0&1&0&1&0\\ 0&1&0&1&0\\ 0&0&1&0&1\\ 0&0&1&0&1\\ 0&1&1&1&1\\ 0&1&1&1&1\\ 0&0&0&0&0 \end{vmatrix}}
$$

$$
\underset{\text{Matrix 4}}{\begin{vmatrix} 1&0&0&1&1\\ 0&1&0&1&0\\ 0&0&1&0&1\\ 0&1&1&1&1\\ 0&0&0&0&0\\ 0&0&0&0&0\\ 0&0&0&0&0\\ 0&0&0&0&0 \end{vmatrix}}
\underset{\text{Matrix 5}}{\begin{vmatrix} 1&0&0&1&1\\ 0&1&0&1&0\\ 0&0&1&0&1\\ 0&0&1&0&1\\ 0&0&0&0&0\\ 0&0&0&0&0\\ 0&0&0&0&0\\ 0&0&0&0&0 \end{vmatrix}}
\underset{\text{Matrix 6}}{\begin{vmatrix} 1&0&0&1&1\\ 0&1&0&1&0\\ 0&0&1&0&1\\ 0&0&0&0&0\\ 0&0&0&0&0\\ 0&0&0&0&0\\ 0&0&0&0&0\\ 0&0&0&0&0 \end{vmatrix}}
\underset{\substack{\text{Matrix 7}\\ \text{(generator matrix)}}}{\begin{vmatrix} 1&0&0&1&1\\ 0&1&0&1&0\\ 0&0&1&0&1 \end{vmatrix}}
\qquad \text{(D-4)}
$$

In matrix 1, the vector (0 0 0 0 0) in the first row was exchanged with the vector (1 0 0 1 1) of the eighth row to give us matrix 2. In matrix 2, row 1 was added to rows 3, 5, and 7 in order to make the elements of the first column under the leading 1 of the first row each 0, as in matrix 3. In matrix 3, row 2 was added to row 3, row 4 to row 5, and row 6 to row 7 to give us (0 0 0 0 0) for rows 3, 5, and 7, which were then respectively interchanged with rows 5, 6, and 7 to give us matrix 4. In matrix 4, row 2 was added to row 4 to give us matrix 5. Finally, in matrix 5, by adding row 3 to row 4 we obtained matrix 6. This last matrix is the desired echelon canonical form with dimension 3, and the three vectors (1 0 0 1 1), (0 1 0 1 0), and (0 0 1 0 1) constitute a basis for the vector space represented in matrix 1. In matrix 6, we can eliminate the zero rows and the resulting matrix 7 is called the "generator matrix" G of the vector space of matrix 1.

If we have an $n \times n$ square matrix which in addition is a generator matrix, then this matrix will have 1s along the main diagonal and all the other elements will be 0. Such an $n \times n$ matrix is called an "identity matrix":

$$\begin{vmatrix} 1 & 0 & 0 & 0 & 0 \\ 0 & 1 & 0 & 0 & 0 \\ 0 & 0 & 1 & 0 & 0 \\ 0 & 0 & 0 & 1 & 0 \\ 0 & 0 & 0 & 0 & 1 \end{vmatrix} \tag{D-5}$$

Any $n \times n$ square matrix with n independent rows is called a "nonsingular matrix," and it can be reduced to an identity matrix.

The matrix that results from exchanging the rows of a matrix M with the columns is called "transpose matrix" M^T. Therefore an $n \times k$ matrix will have a $k \times n$ transpose matrix.

Two $n \times k$ matrices can be added by adding the corresponding row or column elements. If x_{ij} is the element in the ith row and jth column of the first matrix and y_{ij} is the corresponding element in the second matrix, $x_{ij} + y_{ij}$ will be the element in the ith row and jth column of the sum matrix.

An $n \times k$ matrix and an $k \times m$ matrix can be multiplied. This multiplication is feasible only if the number of the rows of the second matrix equals the number of the columns of the first matrix. The ith row and jth column element z_{ij} of the product matrix will be

$$z_{ij} = x_{i1}y_{1j} + x_{i2}y_{2j} + \cdots + x_{ik}y_{kj} \tag{D-6}$$

Of course, if the vector represented by the ith row of the first matrix is orthogonal to the vector represented by the jth column of the second matrix, the above term will be zero. If this is true for every pair of such vectors, the product matrix will have all 0 elements, which means that the two initial matrices are orthogonal. The vector space represented by the second matrix is called the "null space" of the first matrix, and vice versa.

Given a $k \times n$ generator matrix G, where $k \leq n$, the echelon canonical form can be found by operating on the k rows of the matrix. In addition, by exchanging rows, we can write matrix G in such a form that the first k columns taken together with the k rows resemble a $k \times k$ identity matrix I_k:

$$G = \left| \begin{array}{cccccccccc} 1 & 0 & 0 & \cdots & 0 & a_{11} & a_{12} & \cdots & a_{1,n-k} & k \\ 0 & 1 & 0 & \cdots & 0 & a_{21} & a_{22} & \cdots & a_{2,n-k} & \\ \cdots & \cdots & \cdots & \cdots & \cdots & \cdots & \cdots & \cdots & \cdots & \\ 0 & 0 & 0 & \cdots & 1 & a_{k1} & a_{k2} & \cdots & a_{k,n-k} & \end{array} \right| \tag{D-7}$$

This form of generator matrix G, called the "reduced echelon form," is very useful in representing $R = k/n$ linear block codes. The last $n - k$ columns resemble a $k \times (n - k)$ matrix P, where

$$P = \left| \begin{array}{cccc} a_{11} & a_{12} & \cdots & a_{1,n-k} \\ \cdots & \cdots & \cdots & \cdots \\ a_{k1} & a_{k2} & \cdots & a_{k,n-k} \end{array} \right| \tag{D-8}$$

The generator matrix G now can be written as

$$G = [I_k P] \tag{D-9}$$

If the vector space V is the row space of the generator matrix, this expression of G can be useful in deriving the null space H^T of the vector space V. In fact, if the matrix P is transposed to P^T, the matrix

$$H = [-P^T I_{n-k}] \tag{D-10}$$

where I_{n-k} is an $(n - k) \times (n - k)$ identity matrix, represents the transposed null space of V.

The proof is very simple. Find the product GH^T and show that every element is zero, so that

$$GH^T = [I_k P] \begin{bmatrix} -P \\ I_{n-k} \end{bmatrix} = 0 \tag{D-11}$$

In the case in which the only elements of the matrix P are 0s and 1s, $-P^T = P^T$, in accordance with the definition of the table

+	0	1
0	0	1
1	1	0

APPENDIX E

CHARACTERISTICS OF MAJOR SYSTEMS DEPLOYED FROM 1962 TO 1982

TABLE E-1 OPERATIONAL SYSTEMS DEPLOYED FROM 1962 TO 1982

U.S. and Allied military systems	International systems
Defense Satellite Communications System (DSCS)	Intelsat series
Fleetsatcom	Marecs (European)
United Kingdom Skynet	Marots (European)
NATO Series	Domestic and regional systems (non-U.S. and non-U.S.S.R.)
U.S. domestic systems	Anik (Canada)
American Satellite system	Palapa (Indonesia)
Americom-RCA system (Satcom)	Insat (India)
Western Union system (Westar)	Eutelsat (ECS-European)
AT&T system (Comstar)	Sacura (Japan)
GTE system	Yuri (Japan)
Satellite Business System (SBS)	
U.S.S.R systems	
Molniya	
Loutch	
Statsionar (Gorizont and Raduga)	
Volna	
Gals	

TABLE E-2 DEFENSE SATELLITE COMMUNICATIONS SYSTEMS (DSCS) FOR STRATEGIC COMMUNICATIONS

	Orbit	First launch	Launch vehicle	No. of spacecraft per launch	Weight of spacecraft in orbit	Stabili-zation	Frequency
DSCS-I*	Near-synchronous equatorial†	1966	Titan IIIC	8	100 lb (45 kg)	Spin	X band
DSCS-II	Geostationary	1971	Titan IIIC	2	1200 lb (545 kg)	Spin	X band
DSCS-III	Geostationary	1981	Titan IIIC or shuttle	1 DSCS-II and 1 DSCS-III per Titan	1600 lb (726 kg) dry 600 lb (272 kg) fuel	Three-axis	X band and UHF

*Initial Defense Communications Satellite Program (IDCSP)

†Spacecraft slowly drifting from west to east.

‡Electronically forms 1° circles on earth.

§Electronically forms 3° circles on earth.

Note: BOL, beginning of life; EOL, end of life; TWT, traveling-wave tube; EIRP, effective isotropic radiated power; MBA, multibeam antenna.

Power	Design life	Characteristics of communications system	
		Capacity	Beams
33 W (EOL)	3 yr	11 tactical voice channels or 1550 teletypewriter channels (equivalent to 5 commercial voice channels) One 20-MHz repeater	EIRP: 7 dBW (one beam)
520 W (BOL)	5 yr	100 Mb/s or 1300 duplex voice channels Two 20-W TWTs	EIRP: 28 dBW (1 global beam) 43 dBW (2 narrow beams) Beam width, 2.5° Steerable coverage, 10°
1100 W (BOL)	10 yr	Six channels: Two 40-W TWTs One 10-W TWT Bandwidth: 60 MHz (channels 1, 2, and 4) 85 MHz (channel 3)	EIRP for channels 1 and 2: 40 dBW (narrow MBA‡) 29 dBW (global MBA) 44 dBW (gimbaled dish§) EIRP for channels 3 and 4: 34 dBW (narrow MBA‡) 23 dBW (narrow MBA‡) 37.5 dBW (gimbaled dish,§ channel 4 only) 25 dBW (global horn) EIRP for channels 5 and 6: 25 dBW (global horn)

TABLE E-3 INTELSAT SERIES SPACE SEGMENT (Geostationary)

	First launch	Launch vehicle	Weight of spacecraft in orbit	Stabilization	Frequency	Power
Early Bird	1965	Thor-Delta	85 lb (39 kg)	Spin	C band	45 W (BOL) 33 W (after 3 yr)
II	1966	Improved Thor-Delta	189 lb (86 kg)	Spin	C band	100 W (BOL) 85 W (EOL)
III	1968	Improved Thor-Delta	335 lb (152 kg)	Spin	C band	178 W (BOL) 130 W (EOL)
IV	1971	Atlas-Centaur	1548 lb (702 kg)	Spin	C band	569 W (BOL) 460 W (EOL)
IVA	1975	Atlas-Centaur	1800 lb (800 kg)	Spin	C band	708 W (BOL) 600 W (EOL)
V	1981	Atlas-Centaur	2094 lb (950 kg)	Three-axis	C band/ K_u band	1566 W (BOL) 1300 W (EOL)

Note: TWT, traveling-wave tube; EIRP, effective isotropic radiated power; G/T, gain-to-temperature ratio.

Design life	Characteristics of communications system		
		Beams	
	Capacity	EIRP, dBW	G/T, dB/K
$1\frac{1}{2}$ yr	240 voice circuits or 1 television channel 25-MHz repeaters Two 6-W TWTs per repeater	100	
3 yr	240 voice circuits or 1 television channel 126-MHz repeaters Four 6-W TWTs per repeater	15.5	
5 yr	1200 voice circuits or 4 television channels 225-MHz repeaters Two 10-W TWTs per repeater	27	
7 yr	Twelve 36-MHz transponders	22.5 34.5	−18.6
7 yr	Twenty 36-MHz transponders	22 26 29	−11.8
	Bandwidth, MHz (no. of transponders)		
7 yr	C band: 77 (2) 72 (15) 36 (5) 41 (1) K_u band: 77 (2) 72 (2) 240 (2)	Global C band: 23.5 26.5 Hemizone C band: 29 26 Zone C band: 29 Spot K_u band: 44.1 44.4	 −18.6 −11.6 −8.6 −3.3

TABLE E-4 TYPICAL U.S. DOMESTIC SPACE SEGMENTS (Geostationary)

	First launch	Launch vehicle	Weight of spacecraft in orbit	Stabilization	Frequency	Power
Westar I	1974	Thor-Delta 2914	655 lb (300 kg)	Spin	C band	307 W 260 W (EOL)
Westar IV	1982	Thor-Delta 3910	1284 lb (584 kg)	Spin	C band	900 W (BOL)
Satcom I	1976	Thor-Delta 3914	1019 lb (460 kg)	Three-axis	C band	770 W (BOL) 550 W (EOL)
SBS-F1	1980	Thor-Delta 3910 or shuttle	1287 lb (584 kg)	Spin	K_u band	1000 W (BOL)
American Satellite	1985*	Thor-Delta 3910 or shuttle	1400 lb (635 kg)	?	C band/ K_u band	1000 W (BOL)

*Scheduled.

Note: BOL, beginning of life; EOL, end of life; EIRP, effective isotropic radiated power; G/T, gain-to-temperature ratio.

Design life	Bandwidth, MHz (no. of transponders)	Beams: EIRP, dBW	Beams: G/T, dB/K
7 yr	36 (12)	33	−6
10 yr	36 (24)	34	−5
8 yr	36 (24)	32 (Conus) 26 (Hawaii)	−5 (Conus) −10 (Hawaii)
7 yr	43 (10)	37 44	
10 yr	36 (12) 72 (6) 72 (6)	C band: 34 C band: 37 K_u band: 39 (Conus) 42 (spot beam)	

Characteristics of communications system

TABLE E-5 TYPICAL EARTH STATIONS FOR SATELLITE COMMUNICATIONS

Parabolic reflector	EIRP	*G/T*, dB/K
MILITARY TERMINALS (FOR DSCS)		
AN/FSC-78		
60 ft	127 dBm	39
UK SSC-2/Scot[*]		
3.5 ft	95.5 dBm	10.6
AMERICAN SATELLITE COMPANY (FOR DOMESTIC USE WITH WESTAR)		
Backbone terminal		
10 m	85 dBW	33
Satellite data exchange (SDX)		
10 m	77 dBW	31
5 m	62 dBW	22.5
Country	**Antenna gain, dB**	***G/T*, dB/K**
INTELSAT TERMINALS (WITH 30-m PARABOLIC REFLECTORS)		
United States	62.7	43.4
United Kingdom	62.5	41.9
Federal Republic of Germany	63.0	42.5
France	63.0	40.8
Italy	64.0	41.1

[*]British-developed shipboard terminal.

BIBLIOGRAPHY

Abramson, N.: "Packet Switching with Satellites," *AFIFS Conference Proceedings*, vol. 42, 1973 National Computer Conference, June 1973, pp. 695–702.

Acampora, A. S.: "Maximum Likelihood Decoding of Binary Convolutional Codes on Band-Limited Satellite Channels," *Conference Record, National Telecommunications Conference*, Dallas, 1976.

Anderson, J. B., and D. P. Taylor: "Trellis Phase Modulation Coding: Minimum Distance and Spectral Results," *EASCON 1977*, pp. 29-1A to 19-1G.

Ashton, S., and D. Silverman: "The American Satellite Communications System," AIAA 5th Communications Satellite Systems Conference, Paper 74-482, Los Angeles, April 22–24, 1974.

Assal, F.: "Approach to a Near-Optimum Transmitter-Receiver Filter Design for Data Transmission Pulse-Shaping Networks," *COMSAT Technical Review*, vol. 3, Fall 1973, pp. 310–322.

The ATS-F and -G Data Book, rev. ed., Goddard Space Flight Center, Greenbelt, Md., 1972.

Bachmann, E., K. G. Howe, and T. J. Petry: "Introduction of Digital Transmission Including TDMA/DSI as Viewed by a Communications Organization," *Proceedings of ICDSC-3*, Kyoto, November 1975, pp. 394–402.

Bakeman, D. C., K. W. Jenkins, J. J. Rodden, A. J. Iorillo, and T. B. Garber: "The Relative Merits of Three-Axis and Dual-Spin Stabilization Systems for Future Synchronous Communication Satellites," in N. E. Feldman and C. M. Kelley (eds.), *AIAA Progress in Astronautics and Aeronautics, vol. 26: Communication Satellites for the 70's—Systems*, MIT Press, Cambridge, Mass., 1971, chap. IX, pp. 605–653.

Bargellini, P., and M. Schaffner: "Switching in Orbit," AIAA/CASI 6th Communications Satellite Systems Conference, Montreal, April 1976.

Benedetto, S., and E. Biglieri: "Performance of *M*-ary PSK Systems in the Presence of Intersymbol Interference and Additive Noise," *Alta Frequenza*, vol. 41, 1977, pp. 225–239.

Benedetto, S., G. DeVincentiis, and A. Luvison: "Error Probability in the Presence of Intersymbol Interference and Additive Noise for Multilevel Digital Signals," *IEEE Transactions on Communications*, vol. COM-21, March 1973, pp. 181–190.

Bennet, W. R., H. E. Curtis, and S. O. Rice: "Interchannel Interference in FM and PM Systems under Noisy Loading Conditions," *Bell System Technical Journal*, vol. 34, May 1965, pp. 601–636.

Berlekamp, E. R.: "A Survey on Coding Theory for Algebraists and Combinatorialists," International Centre for Mechanical Science, Udine, Italy, 1970.

Biederman, L., and J. K. Omura: "Performance of Viterbi Demodulators for Nonlinear Satellite Channels with Intersymbol Interference," National Telemetry Conference, 1977.

Bond, F. E.: "Overview—DOD Satcom Systems," AIAA 5th Communications Satellite Systems Conference, Paper 74-456, Los Angeles, April 1974.

Boyes, J. L.: "A Navy Satellite Communications System," *Signal*, vol. 30, no. 6, March 1976.

Boyko, D. F., and P. J. McLane: "Error Rate Bounds for Digital PAM Transmission with Phase Jitter Intersymbol or Co-channel Interference," *IEEE Transactions on Communications*, vol. COM-25, May 1977, pp. 536–541.

Brown, J. P.: "Public Service Communication Satellite Program," *EASCON 1977*, Washington, D.C., September 26-27, 1977.

Campanella, S. J.: "Digital Speech Interpolation," *COMSAT Technical Review*, vol. 6, no. 1, Spring 1976.

Campanella, S. J., and H. G. Suyderhoud: "Performance of Digital Speech Interpolation Systems for Satellite Telecommunications," *National Telecommunications Conference—1975*, December 1975.

Campanella, S. J., H. G. Suyderhoud, and M. Wachs: "Frequency Modulation and Variable-Slope Delta Modulation in SCPC Satellite Transmission," *Proceedings of the IEEE*, vol. 65, no. 3, March 1977.

CCIR: "Comparison of Different Methods for the Digital Coding of Television Signals," CCIR Study Groups, period 1974–1978, Document CMTT/39-E.

CCIR: "Digital Transmission of Television Signals," CCIR 13th Plenary Assembly, 1974, vol. 12, Document 646, pp. 208–209.

CCIR: "The Specification of Performance for Transmission Circuits Which Employ Digital Methods," CCIR 13th Plenary Assembly, 1974, vol. 12, Document 645, pp. 206–208.

Celebiler, M. I., and G. M. Coupe: "Effects of Thermal Noise Filtering and Co-channel Interference on the Probability of Error in Binary Coherent PSK Systems," *IEEE Transactions on Communications*, vol. COM-26, no. 2, February 1978, pp. 257–267.

Celebiler, M., and G. Coupe: "Probability of Error for Coherent PSK in the Presence of Thermal Noise and Intersymbol, Interchannel and Co-channel Interferences," IEEE International Conference on Satellite Communication Systems Technology, London, April 1975.

Chen, C. C., and J. W. Burnett: "TDRS Multiple Access Channel Design," *National Telecommunications Conference*, 1977, pp. 19:2-1 to 19:2-7.

Chidester, L. G.: "Advanced Lightweight Solar Array Technology," *Proceedings of the AIAA 7th Communications Satellite Systems Conference*, San Diego, April 1978, pp. 55–60.

Chu, T. S.: "Rain-Induced Cross-Polarization at Centimeter and Millimeter Wavelengths," *Bell System Technical Journal*, vol. 53, 1974, p. 1557.

Clarke, A. C.: "Extra-terrestrial Relays," *Wireless World*, vol. 51, no. 10, October 1945, reprinted in *Progress in Astronautics and Aeronautics, vol. 19: Communication Satellite Systems Technology*, R. B. Marsten (ed.), 1966.

"Communications Systems Technology Assessment Study," NASA Lewis Research Center, Cleveland, June 1977.

Crandall, K. H.: "The 12 and 14 GHz Bands in Domestic Satellite Communications," *National Telecommunications Conference*, Paper 31D, November 1973.

Crane, R. K.: "Prediction of the Effects of Rain on Satellite Communications Systems," *Proceedings of the IEEE*, vol. 65, 1977, pp. 456–474.

Cuccia, C. L., and R. S. Davies: "Optimum Adjacent Channel Design Considerations for a Communications Satellite Carrying Both FDMA and TDMA," International Conference on Communications, 1976.

Cuccia, C. L., et al.: "Transponder and Antenna-Design Problems at Millimeter Wavelengths for 20–30 GHz Communications Satellites," *Proceedings of Symposium on Advanced Satellite Communications Systems*, Genoa, Italy, December 1977.

Curry, W. H.: "The Military Satellite Communications System Architecture," AIAA/CASI 6th Communications Satellite Systems Conference, Paper 76-268, April 1976.

Davisson, L. D.: "Universal Noiseless Coding," *IEEE Transactions on Information Theory*, vol. IT-19, November 1973, pp. 783–795.

Davisson, L. D., and L. B. Milstein: "On the Performance of Digital Communication Systems with Bandpass Limiters, Part I: One-Link Systems. Part II: Two-Link Systems," *IEEE Transactions on Communications*, vol. COM-20, October 1972, pp. 972–980.

Deal, J.: "TDMA Open-Loop Acquisition and Synchronization," *Proceedings of Joint Automatic Control Conference*, San Francisco, June 1977, pp. 1163–1169.

deBrouwer, J.: "Voice Coding for Digital Communication," *National Telecommunications Conference*, 1976.

Dill, G., J. Deal, and W. Maillet: "The Intelsat Prototype TDMA System," *1975 IEEE Intercon Record*, April 1975.

Dill, G., and A. Tomozawa: "Time-Domain Multiple-Access Frame and Burst Synchronization for Spot-Beam Satellites," *IEEE EASCON*, 1974.

Dorian, C., T. O. Calvit, and D. W. Lipke: "Marisat: Design and Operational Aspects," *Journal of the British Interplanetary Society*, vol. 29, no. 5, May 1976.

Duncan, J. W., S. J. Hamada, and P. G. Ingerson: "Dual Polarization Multiple-Beam Antenna for Frequency Reuse Satellites," AIAA/CASI 6th Communications Satellite Systems Conference, Paper 76-248, Montreal, April 1976.

Duncan, J. W., S. J. Hamada, and W. C. Wong: "Polarization Isolation Characteristics of a Dual-Beam Reflector Antenna," AIAA 4th Communications Satellite Systems Conference, Paper 72-531, 1972.

"The Evolution of Cost-Effective (Small Aperture) Earth Terminals for Communications," International Conference on Communications, Session 21, 1976.

Falconer, D. D., and J. Salz: "Optimal Reception of Digital Data over the Gaussian Channel with Unknown Delay and Phase Jitter," *IEEE Transactions on Information Theory*, vol. IT-23, no. 1, January 1977.

Fang, R., and O. Shimbo: "Unified Analysis of Digital Systems in Additive Noise and Interference," *IEEE Transactions on Communications*, vol. COM-21, October 1973, pp. 1075–1091.

Ferguson, M. E.: "Design of FM Single-Channel-per-Carrier Systems," International Conference on Communications, June 1975.

Forney, G. D., Jr.: "Maximum-Likelihood Sequence Estimation of Digital Sequences in the Presence of Intersymbol Interference," *IEEE Transactions on Information Theory*, vol. IT-18, no. 3, May 1972, pp. 363–378.

Forney, G. D., and E. K. Bower: "A High-Speed Sequential Decoder: Prototype Design and Test," *IEEE Transactions on Communication Technology*, vol. COM-19, October 1971, pp. 821–835.

Fredericson, S. A.: "Optimum Receiver Filters in Digital Quadrature Phase-Shift-Keyed Systems with a Nonlinear Repeater," *IEEE Transactions on Communications*, vol. COM-23, December 1975, pp. 1389–1400.

Fuenzalida, J. C., O. Shimbo, and W. L. Cook: "Time-Domain Analysis of Intermodulation Effects Caused by Nonlinear Amplifiers," *COMSAT Technical Review*, vol. 3, no. 1, Spring 1973, pp. 89–143.

Glave, F. E., and A. S. Rosenbaum: "An Upper Bound Analysis for Coherent Phase-Shift Keying with Cochannel, Adjacent-Channel, and Intersymbol Interference," *IEEE Transactions on Communications*, vol. COM-23, no. 6, June 1975, pp. 586–597.

Golding, L. S.: "Single Channel per Carrier Transmission for Satellite Communications," *National Telecommunications Conference*, December 1975.

Horwood, D., and R. Gagliardi: "Signal Design for Digital Multiple Access Communications," *IEEE Transactions on Communications*, vol. COM-23, March 1975, pp. 378–383.

Inagki, K., Y. Hirata, and A. Ogawa: "A New Technique for Data Transmission via TDMA Satellite Link," *National Telecommunications Conference*, 1977.

"International Conference on Satellite Communication Systems Technology," IEEE Conference Publication 126, April 1975.

"Introduction to Satellite Communications," CCP-105-5, HQ, U.S. Army Communications Command, April 1974.

Ippolito, L. J.: "Characterization of the CTS 12 and 14 GHz Communications Links—Preliminary Measurements and Evaluation," International Conference on Communications, June 1976.

Jayant, N. S.: "Adaptive Quantization with a One-Word Memory," *Bell System Technical Journal*, September 1973.

Jayant, N. S.: "Digital Coding of Speech Waveforms: PCM, DPCM and DM Quantizers," *Proceedings of the IEEE*, May 1974.

Johnston, W. A.: "ATS-6 Experimental Communications Satellite: A Report on Early Orbital Results," National Telecommunications Conference, December 1974.

Jones, J. J.: "Filter Distortion and Intersymbol Interference Effects on PSK Signals," *IEEE Transactions on Communications*, vol. COM-19, April 1971, pp. 120–132.

Kadar, Ivan (ed.): *AIAA Selected Reprint Series*, vol. XVIII: *Satellite Communications Systems*, January 1976.

Kantor, L. Ya., V. A. Polukhin, and N. V. Talyzin: "New Relay Stations of the Orbita-2 Satellite Communications System," *Telecommunications and Radio Engineering*, vol. 27, no. 5, May 1973.

Kasami, T., and S. Lin: "Coding for a Multiple-Access Channel," *IEEE Transactions on Information Theory*, vol. IT-22, March 1976, pp. 129–137.

Kohlenberg, A., and G. D. Forney, Jr.: "Convolutional Coding for Channels with Memory," *IEEE Transactions on Information Theory*, vol. IT-14, September 1968, pp. 618–620.

Korn, I.: "Probability of Error in Binary Communications with Casual Band-Limiting Filters, Part 1: Non Return-to-Zero Signal," *IEEE Transactions on Communication Technology*, August 1973, pp. 878–890.

LaVean, G. E., and E. J. Martin: "Communication Satellites: The Second Decade," *Astronautics and Aeronautics*, vol. 12, no. 4, April 1974.

Lee, W. U., and F. S. Hill, Jr.: "A Maximum-Likelihood Sequence Estimator with Decision-Feedback Equalization," *IEEE Transactions on Communications*, vol. COM-25, no. 9, September 1977, pp. 971–979.

Lesh, J. R.: "Signal to Noise Ratios in Coherent Soft Limiters," *IEEE Transactions on Communications Technology,* vol. COM-22, June 1974, pp. 803–811.

Limb, J. O., C. B. Rubinstein, and J. E. Thompson: "Digital Coding of Color Video Signals—A Review," *IEEE Transactions on Communications,* vol. COM-25, no. 11, November 1977.

Lindsey, W. C., et al.: "Investigation of Modulation/Coding Trade-Off for Military Satellite Communications," TR-7808-1276, LinCom Corporation, July 19, 1977.

Lipke, D., D. Swearingen, J. Parker, E. Steinbrecher, T. Calvit, and H. Dodel: "MARISAT—A Maritime Satellite Communications System," *COMSAT Technical Review,* vol. 7, no. 2, 1977.

Lombard, D., and G. Payet: "Digital Speech Interpolation in Satellite Systems," International Conference on Communications, June 1975.

Louie, M., and I. Rubin: "On Access Control Disciplines for a TDMA System with Multiple-Rate Real-Time Sources," *International Telemetering Conference,* Los Angeles, September 1976, pp. 326–336.

Lubell, P. D., and F. D. Rebhun: "Suppression of Co-Channel Interference with Adaptive Cancellation Devices at Communications Satellite Earth Stations," International Conference on Communications, 1977.

Magee, F. R., Jr., and J. G. Proakis: "Adaptive Maximum Likelihood Sequence Estimation for Digital Signaling in the Presence of Intersymbol Interference," *IEEE Transactions on Information Theory,* vol. IT-19, January 1973, pp. 120–124.

Melnick, M.: "Intelligibility Performance of Variable Slope Delta Modulation," *Proceedings ICC-73,* June 1973.

Mesiya, M. F., P. J. McLane, and L. L. Campbell: "Maximum Likelihood Sequence Estimation of Binary Sequences Transmitted over Bandlimited Nonlinear Channels," *IEEE Transactions on Communications,* vol. COM-25, no. 7, July 1977.

Mesiya, M. F., P. J. McLane, and L. L. Campbell: "Optimal Receiver Filters for BPSK Transmission Over a Bandlimited Nonlinear Channel," *National Telecommunications Conference,* 1976, pp. 13:5-1 to 13:5-5.

Messerschmitt, D. G.: "Speech Encoding for Digital Transmission," *EASCON 1976.*

Miller, J. H.: "Digital-to-Digital Conversion between an Adaptive Delta Modulation Format and Companded PCM," International Conference on Communications, 1976.

Morgan, W. L.: "Design Criteria for Communications Satellites," *EASCON '74 Conference Record,* October 1974.

Moulton, P. D.: "Current Status of Commerical Satellite Communications," *Signal,* April 1975.

Murphy, J. V.: "Binary Error Rate Caused by Intersymbol Interference and Noise," *IEEE Transactions on Communications,* vol. COM-21, September 1973, pp. 1039–1046.

"NASA Compendium of Satellite Communications Programs," Document X-751-73-178, Goddard Space Flight Center, Greenbelt, Md., August 1975.

Nash, S.: "Multiplexer Survey—Comments on Frequency and Time Division Multiplexing," *Telecommunications,* January 1970, pp. 36–39.

Nunes, P. R., and J. P. A. Albuquerque, "Statistical Delta Modulation," International Conference on Communications, 1976.

Nuttall, A. H.: "Error Probabilities for Equicorrelated M-ary Signals under Phase-Coherent and Phase-Incoherent Reception," *IRE Transactions on Information Theory,* vol. IT-18, July 1962, p. 305.

Odenwalder, J. P., A. J. Viterbi, I. M. Jacobs, and J. A. Heller: "Study of Information Transfer

Optimization for Communication Satellites," Final Report for NASA Contract NAS2-6810, Linkabit Corp., NASA CR 114561, N73-20889.

Omura, J. K.: "Optimal Receiver Design for Convolutional Codes and Channels with Memory via Control Theoretical Concepts," *Information Science*, vol. 3, July 1971, p. 243.

Palmer, L., and S. Lebowitz: "Adjacent Channel Interference Between Unfiltered and Filtered QPSK Signals," *IEEE International Conference on Communications Record*, Seattle, Washington, June 1973, pp. 31-25 to 31-29.

Poza, H. B.: "TDRSS Telecommunications Payload: An Overview," *National Telecommunications Conference*, 1977.

Prabhu, V. K.: "Bandwidth Occupancy in PSK Systems," *IEEE Transactions on Communications*, vol. COM-24, no. 4, April 1976, pp. 456–462.

Rhodes, S. A.: "Carrier Synchronization Techniques for Offset-QPSK Signals," *National Telecommunications Conference*, 1974, pp. 937–945.

Robinson, G., et al.: "PSK Signal Power Spectrum Spread Produced by Memoryless Nonlinear TWTs," *COMSAT Technical Review*, vol. 3, Fall 1973, pp. 227–256.

Robinson, J. J.: "Aeronautical Satellites—Progress Report on the Joint Aerosat Evaluation Programme," *Journal of the British Interplanetary Society*, vol. 29, no. 5, May 1976.

Roese, J. A., and W. K. Pratt: "Theoretical Performance Models for Interframe Transform and Hybrid Transform/DPCM Coders," *Proceedings SPIE*, vol. 87, August 1976, pp. 172–179.

Sabelhaus, A. B.: "Applications Technology Satellites F and G Communications Subsystem," *Proceedings of the IEEE*, vol. 59, no. 2, February 1971.

Satellite Communications Reference Data Handbook, Defense Communications Agency, Arlington, Va., September 1973.

Shimbo, D., and B. A. Pontano: "A General Theory for Intelligible Crosstalk between Frequency-Division Multiplexed Angle-Modulated Carriers," *IEEE Transactions on Communication*, vol. COM-24, no. 9, September 1976, pp. 999–1007.

Simmons, N.: "The United Kingdom Programme of Communications Satellites," *Progress in Astronautics and Aeronautics, vol. 32: Communications Satellite Systems*, P. L. Bargellini (ed.), 1974.

Sims, R. J., and R. P. Sherwin: "Communication Technology Trends in the DSCS," *National Telecommunications Conference 1974*, December 1974.

"Solar Cell Array Design Handbook," Jet Propulsion Laboratory, Pasadena, Report JPL SP 43-38, October 1976.

Sparks, R. H.: "Nickel-Cadmium Battery Technology Advancements for Geosynchronous Orbit Spacecraft," *Proceedings of the AIAA 7th Communications Satellite Systems Conference*, San Diego, April 1978, pp. 61–65.

Special Issue on ATS 6, *IEEE Transactions on Aerospace and Electronic Systems*, vol. 11, no. 6, November 1975.

Tankersley, B. C., and H. E. Bartlett: "Tracking and Data Relay Satellite Single Access Deployable Antenna," *National Telecommunications Conference*, 1977, pp. 19:4-1 to 19:4-6.

Taylor, D. P., and D. Cheung: "The Effect of Carrier Phase Error on the Performance of a Duobinary Shaped QPSK Signal," *IEEE Transactions on Communications*, vol. COM-25, July 1977, pp. 738–744.

Van Trees, H. L.: "Future Intelsat System (1986–1993) Planning," *Proceedings of the AIAA/CASI 6th Communications Satellite Systems Conference*, Paper 76-233, April 1976.

Viswanathan, R., E. Blackman, and J. Makhoul: "Variable Frame Rate Narrowband Speech Transmission Over Fixed Rate Noisy Channels," *EASCON 1977*.

Viterbi, A. J.: *Principles of Coherent Demodulation*, McGraw-Hill, New York, 1966, chaps. 7, 10.

Viterbi, A. J., and J. K. Omura: *Digital Communications and Coding*, McGraw-Hill, New York, 1979.

Wall, V. W.: "Military Communications Satellites," International Telecommunications Conference, October 1973.

Welti, G. R.: "PCM/FDMA Satellite Telephony with 4-Dimensionally Coded Quadrature Amplitude Modulation," *COMSAT Technical Review*, vol. 6, no. 2, Fall 1976, pp. 323–338.

Welti, G. R., and R. K. Kwan: "Comparison of Signal Processing Techniques for Satellite Telephony," *National Telecommunications Conference*, 1977, pp. 05:1-1 to 05:1-6.

Westcott, R. J.: "Investigation of Multiple F.M./F.D.M. Carriers through a Satellite T.W.T. Operating Near to Saturation," *Proceedings of the IEEE*, United Kingdom, vol. 114, no. 6, June 1976, pp. 726–750.

Wheelon, A. D.: "The Future Outlook for Communication Satellite Applications," World Telecommunications Forum, Geneva, 1975.

Witherspoon, J. T., and R. P. Sherwin: "Real-Time Adaptive Control for the Defense Satellite Communications System," *Proceedings of the AIAA/CASI 6th Communications Satellite Systems Conference*, Paper 76-272, April 1976.

Wolverton, C. T., T. Moeller, and L. M. Nirenberg: "Satellite Multiple Access Study," TOR-0078(3417)-1, The Aerospace Corporation, El Segundo, Calif., June 15, 1978.

Wozencraft, J. M., and I. M. Jacobs: *Principles of Communications Engineering*, Wiley, New York, 1965.

Wright, D. L., and J. W. B. Day: "The Communications Technology Satellite and the Associated Ground Terminals for Experiments," AIAA Conference on Communications Satellites for Health/Education Applications, Paper 75-904, July 1975.

Yeh, Y. S., and E. Y. Ho: "Improved Intersymbol Interference Error Bounds in Digital Systems," *Bell System Technical Journal*, vol. 50, October 1971, pp. 2585–2598.

INDEX

ABOUT THE AUTHOR

Emanuel Fthenakis is currently senior vice president of Fairchild Industries with responsibility for the space, communications, and electronics group. A graduate of Columbia University, he first worked at Bell Laboratories, then at General Electric's Space Division (as director of engineering), helping to pioneer the development of space technology. Between 1962 and 1969, he founded, organized, and directed the Philco-Ford Space Division, and he later accomplished the same tasks at the American Satellite Company. At both organizations, he was deeply involved in the development of communications satellite systems.

He presently teaches a graduate-level course in space and communications technology at the University of Maryland. In 1982, he was named "Man of the Year" by EASCON for his "imaginative pioneering work in digital communications via satellite." In the same year, he was appointed by President Reagan to the National Security Telecommunications Advisory Council.